Y0-EHW-964

Population Biology

THE EVOLUTION AND ECOLOGY
OF POPULATIONS

Population Biology

THE EVOLUTION AND ECOLOGY OF POPULATIONS

Philip W. Hedrick
UNIVERSITY OF KANSAS

Editorial offices: Jones and Bartlett Publishers, Inc., 30 Granada Court, Portola Valley, CA 94025.

Sales and customer service offices: Jones and Bartlett Publishers, Inc., 20 Park Plaza, Boston, MA 02116.

Library of Congress Cataloging in Publication Data

Hedrick, Philip W.
Population biology.

Bibliography
Includes index.
1. Population biology. I. Title.
QH352.H43 1984 574.5'248 84-7906
ISBN 0-86720-043-X

ISBN: 0-86720-043-X

Publisher: Arthur C. Bartlett
Production: Bookman Productions
Designer: Hal Lockwood
Copy editor: Suzanne Lipsett
Illustrator: Mary Ann Klotz
Composition: Polyglot Compositors

Printed in the United States of America

Printing number (last digit): 10 9 8 7 6 5 4 3 2 1

To Leigh and Scott

CONTENTS

Preface xi

Introduction 1

I. Models 3

II. Statistics 7

III. Probability 10

PART 1
POPULATION GENETICS 13

Chapter 1
Genetic Variation 15

I. Visible Variants 15

II. Chromosomal Variation 18

III. Electrophoretic Variation 20

Summary 25

Problems 25

Chapter 2
Allelic and Genotypic Frequencies 27

I. Observed Genotypic and Allelic Frequencies 28

II. The Hardy-Weinberg Principle 29

III. Multiple Alleles 34

IV. The χ^2 Test 36

V. Measures of Genetic Variation 38

VI. Genetic Counseling 40

Summary 43

Problems 44

Chapter 3
Selection 45

I. Recessive Lethal 48

II. Selection Against Recessives 52

III. Purifying versus Progressive Selection 54

IV. Heterozygote Advantage 55

V. Heterozygote Disadvantage 58

Summary 61

Problems 62

Chapter 4
Mutation 63

I. Allelic Frequency Change 65

II. Mutation–Selection Balance 67

III. Estimation of Mutation Rates 69

IV. The Incidence of Genetic Disease 71

Summary 73

Problems 73

Chapter 5
Gene Flow 75

I. The Continent–Island Model 75

II. General Model 78

III. Estimating Admixture 80

IV. Direct Observation of Gene Flow 81

V. Gene Flow and Selection 83

VI. Clines 85

Summary 86

Problems 86

Chapter 6
Genetic Drift 89

I. The Effect of Genetic Drift 91

II. Effective Population Size 96

III. Mutation in a Finite Population 98

Summary 100

Problems 100

Chapter 7
Nonrandom Mating 101

I. Complete Self-Fertilization 102

II. The Coefficient of Inbreeding 103

III. The Inbreeding Coefficient from Pedigrees 105

IV. Inbreeding and Selection 108

Summary 112

Problems 112

Chapter 8
Quantitative Traits 113

I. The Nature of Quantitative Traits 113

II. The Quantitative Genetics Model 119

III. Estimating Genetic Variance and Heritability 124

IV. Race and Intelligence 130

Summary 134

Problems 135

Conclusions to Part 1 137

PART 2
POPULATION ECOLOGY 139

Chapter 9
Factors Affecting Distribution and Abundance 141

I. Ecological Niche 142

II. Changes in Distribution 146

III. Factors Affecting Distribution 149
- a. Temperature 153
- b. Moisture 154
- c. Soil Nutrients and Type 155
- d. Competition 156
- e. Predation 157

Summary 158

Problems 159

Chapter 10
Population Density and Dispersion 161

I. Absolute Density 163
- a. The Quadrat Method 164
- b. The Capture–Recapture Method 166

II. Relative Density 168

III. Dispersion 169
- a. Index of Dispersion 169
- b. Nearest-Neighbor Distance 172

Summary 175

Problems 175

Chapter 11
Population Growth 177

I. Geometric Growth 178

II. Exponential Growth 181

III. Logistic Growth 182

IV. Modifications of the Growth Models 187
- a. Variation in Finite Growth Rate 189
- b. Minimum Critical Population Density 193

V. Human Population Growth 194

Summary 198

Problems 198

Chapter 12
Demography 199

I. Fecundity and Survival Schedules 199

II. Demographic Measures 202
- a. The Net Replacement Rate 202
- b. Generation Length 203

c. The Instantaneous Rate of Increase 203
d. Life Expectancy 204
e. Reproductive Value 210

Summary 223

Problems 223

Chapter 13
Interspecific Competition 225

I. Competitive Exclusion 228

II. Theoretical Models 229

III. Competition in Natural Populations 239

Summary 246

Problems 247

Chapter 14
Predator–Prey Interactions 249

I. Theoretical Models 251

II. Predation in Natural Populations 259

III. Extinction 263

Summary 270

Problems 271

Conclusions to Part 2 273

PART 3
ECOLOGICAL GENETICS AND EVOLUTIONARY ECOLOGY 275

Chapter 15
Ecological Models of Selection 277

I. The General Effect of Environmental Variation 277

II. Frequency-Dependent Selection 283

III. Density-Dependent Selection 289

Summary 291

Problems 292

Chapter 16
Adaptation and Speciation 293

I. Adaptation 293

II. Speciation 305
a. Allopatric Speciation 311
b. Sympatric Speciation 312
c. Parapatric Speciation 313

Problems 314

Chapter 17
Molecular Variation and Evolution 317

I. Theories Regarding the Maintenance of Molecular Variation 317

II. Evidence Relating to These Theories 319

III. DNA Variation 321

IV. The Molecular Clock 327

V. Phylogenetic Trees 330

Summary 336

Problems 336

Chapter 18
The Evolution of Social Behavior 337

I. Behavior as a Phenotypic Trait 340

II. Kin and Group Selection 345

Summary 351

Problems 351

Chapter 19
Demographic Genetics and Life-History Evolution 353

I. Demographic Genetics 353
a. Zygotic Selection 354
b. Sexual Selection 357
c. Gametic Selection 357
d. Fecundity Selection 358

II. Life-History Evolution 362

Summary 373

Problems 373

Chapter 20
Coevolution and the Evolution of Interspecific Interactions 375

I. Interspecific Competitive Ability and Character Displacement 376

II. Predator–Resource Evolution 379

III. Mutualism 386

Summary 390

Problems 392

Chapter 21
Applied Population Biology 393

I. The Conservation of Genetic Variation 393

II. The Development of Crops and Crop Varieties 396

III. The Control of Agricultural and Public-Health Pests 400

IV. The Harvesting of Natural Populations 405

Summary 410

Problems 410

Conclusions to Part 3 413

Glossary 415

Bibliography 421

Answers to Numerical Questions 433

Author Index 437

Subject Index 441

PREFACE

This book is designed for students who have been introduced to genetic and ecological principles in a general biology course. I have combined examples drawn from nature with conceptual material to illustrate the various topics, and have used basic theory expressed in introductory algebra (and some very elementary calculus) to convey them. The examples involve a wide variety of plants and animals, but, of necessity, most are organisms that can be easily observed or raised under controlled conditions.

The book is divided into three parts. Parts 1 and 2 develop the principles of population genetics and population ecology, respectively. These parts are intended to be independent of each other and their order may be reversed. The part on population genetics discusses the factors important in determining the *kind* of organisms in a population. That on population ecology discusses the factors important in determining the *number* of individuals in a population. Part 3 covers the joint effect of these genetic and ecological factors on both the kind and number of individuals in a population, recognizing that genetics and ecology often interact to influence the attributes of a population. The development of this perspective has been the most innovative and exciting aspect of the book to me. I hope that those interested in genetics, ecology, and evolutionary biology will find this viewpoint useful.

Following the text are an extensive glossary and a bibliography containing both general references and references cited in the text. Also included there are the answers to the numerical questions following each chapter.

The basis of this book is the population biology course that I have taught for a number of years at the University of Kansas. However, I was introduced to many of the principles of population biology through the stimulating presentations of Richard Lewontin and Richard Levins at the University of Chicago. The manuscript was greatly improved by the comments of the following people: Blaine Cole, Cedric Davern, Michael Gaines, James Hamrick, Daniel Hartl, Robert Holt, Subodh Jain, David Merrell, Allan Larson, Nancy Ross, Tom Schoener, Norman Slade, Robert Tamarin, and Robert Travis. My special thanks to Coletta Spencer for her expert typing, Mary Ann Klotz for the skillful artwork, and Hal Lockwood for his thoughtful and fastidious attention to the production of the book.

October 1983 *Philip W. Hedrick*

INTRODUCTION

A population is a group of individuals of one species. The factors that determine the kinds of individuals in a population—such as selection, gene flow, mutation, and genetic drift—are traditionally the basis of population genetics. The factors that determine the size of a population and its distribution—such as the physical conditions in the environment and numbers of competitors, predators, prey, parasites, and other interactive species—are traditionally the province of population ecology. In the 1960s, a number of researchers realized that the disciplines of population genetics and population ecology were limited because their respective ideas had not been integrated. For example, Birch (1960) states

> The ecological problems of populations have to do with the numbers of animals and what determines these numbers. The genetical problem of populations has to do with the kind or kinds of animals and what determines kind. These two disciplines meet when the questions are asked, how does the kind of animal (i.e., genotype) influence the numbers and how does the number of animals influence the kind, that is, the genetical composition of the population?

As a result, these investigators came to advocate a unified approach that included aspects of both population genetics and population ecology, *population biology*. The integration of ideas from these two disciplines has progressed at a slow pace, in part because both are complex in themselves and the combination of both areas is generally yet more complicated. However, our understanding of such topics as the maintenance of genetic variation, speciation, social behavior, life-history evolution, and coevolution has benefited by input from both population genetics and population ecology.

There are three basic approaches to investigating phenomena in population biology: the *empirical*, *experimental*, and *theoretical*. The traditional empirical approach in population ecology, for example, comprises the extensive observation of the numbers of individuals in a population, the variation in numbers over time and space, and the measurement of physical and biotic factors that may affect population numbers. In population genetics, similar large-scale empirical studies have been made of the amount of genetic variation and potential environmental factors that may affect genetic variation. These data may give associations between either population numbers or genetic variants and some environmental factors and suggest problems for further study. However, generally only experimental manipulation can provide support for hypotheses about the effect of particular factors on either population numbers or genetic variation. For example, removing a predator from an area may shed light on the importance of the predator in affecting the numbers of a prey species, or moving a sample of a population to a new environment may allow examination of the significance of an environment on genetic variants. Using information obtained either from empirical or experimental studies, one can construct a general theoretical model. Theoretical models provide a general framework in which to evaluate the impact of particular factors on population numbers or on genetic variation (the use and construction of theoretical models are discussed in the following section).

Figure I.1 shows the interconnections among these three approaches to population biology. Generally, empirical information is obtained first, and then experiments are carried out in an effort to disprove a particular

---

Example I.1. Populations of many small rodents, such as lemmings and voles, undergo fluctuations in population numbers. For example, voles may regularly change in density tenfold or more in what appears in some cases to be a three- to four-year-long oscillation. A number of explanations have been suggested as the basis of these density changes (see Table I.1). The first three listed here—disease, predation, and starvation—have been excluded as major causative factors by some field observations. The last hypothesis, stress, appears consistent with field observations and some experiments (Krebs et al., 1973). However, the actual mechanism by which stress results in changes in population density is not well understood. One interesting proposal is that the population changes genetically to favor more aggressive genotypes, which also have a lower fecundity, at high densities. As a result of this lowered fecundity, the population declines in density, making the less aggressive genotypes favored. These individuals have higher fecundity and consequently result in an increase in density and so forth. However, it is possible that, even if the behavioral changes exist, they may not be consequences of genetic change. As we will discuss later, it is difficult to demonstrate experimentally the genetic bases for alternative forms of behavior (see Chapter 18 for a discussion of behavior).

Table I.1
Some alternative hypotheses to explain temporal variation in small-rodent density and some related observations.

Hypothesis	*Observation*
Disease	No epidemics observed
Predation	No change in predator number
Starvation	No apparent malnutrition
Stress	Change in aggressive behaviour

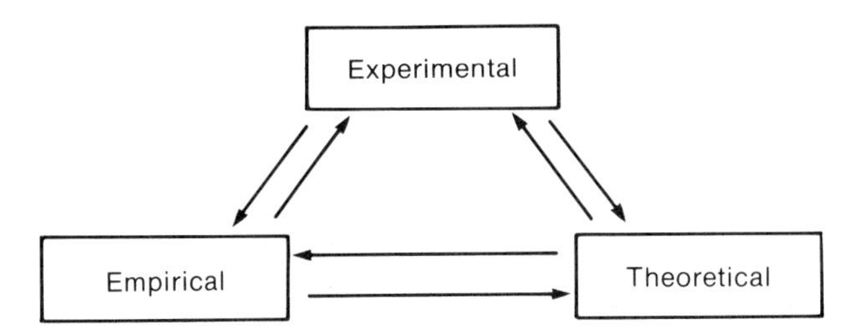

Figure I.1. The relationships among the empirical, experimental, and theoretical approaches to population biology.

hypothesis or to provide support for its validity. In other instances, a theoretical model is given to explain the genetic or ecological observations. The feedback among the three approaches then can allow the refinement and development of particular hypotheses.

For scientific knowledge to increase, alternative hypotheses to explain a given observation or situation must be developed, and then *critical experiments*—that is, experiments with alternative possible outcomes that will exclude one or more of the hypotheses—must be devised and properly executed (see Platt, 1964, for an excellent introduction to these ideas). This approach can result in the elimination, or *falsification*, of some hypotheses so that only one or several remain. The remaining hypotheses are then said to be *consistent* with the experimental and/or empirical results. Owing to the nature of the problem or the difficulty in designing critical experiments, it may not be possible to falsify all hypotheses but one. In fact, in many aspects of population biology, the causation of particular phenomena appears to be multifaceted, making experimentation difficult (see Example I.1).

I. MODELS

To illustrate concepts in population biology, we will use a series of models. *Models* are verbal, graphic, or mathematical representations of phenomena. Here we will generally use simple mathematical or graphical models. A model has the advantage of describing both a particular process and the organization of the parts of the process. A model should be an accurate enough description of a natural process to be consistent with actual observations.

A simple example of a mathematical model might be a household budget with a variety of its important aspects expressed in symbols. We can consider the amount of money on hand, the balance B, as a function of the income I minus the expenses E, or

$$B = I - E.$$

Although this expression may describe the overall situation, it does not provide the details necessary to understanding the process. For example, both

I and E may be composed of several quantities. Let us assume that E is primarily a function of food costs F, housing costs H, and medical costs M. Therefore, the model may be expanded to read

$$B = I - (F + H + M)$$

allowing a more complete description of the factors that contribute to the balance. Further extensions of the model would allow a nearly exact under-

---

Example I.2. Northern elephant seals are extremely large marine mammals that were hunted nearly to extinction in the late nineteenth century because of their blubber (see Figure I.2). In fact, they were thought extinct until approximately 20 were found on Isla Guadalupe off Mexico in 1890. Because northern elephant seals have been protected from hunting since this time, their population numbers have grown quickly, and they have resettled many areas from which they had been extirpated.

The generation length in northern elephant seals is about 8 years, and the net replacement rate is approximately 2. If we let t be the actual year in this case, so that the number in 1890 is $N_{1890} = 20$, then

$$\begin{aligned} N_{1898} &= RN_{1890} \\ &= 2(20) \\ &= 40. \end{aligned}$$

Therefore, it is predicted that there were 40 individuals in 1898. The same process can be carried out again so that the number in 1906 is predicted to be

$$\begin{aligned} N_{1906} &= RN_{1898} \\ &= 2(40) \\ &= 80. \end{aligned}$$

If this process is continued, then the predicted population number in 1978 is 40,960.

The actual population numbers were estimated to be about 125 in 1911 and approximately 60,000 in 1977 (Le Boeuf and Kaza, 1981). The predicted values and these estimated values are plotted in Figure I.3 (notice that the vertical scale is not linear). It appears that the predictions of this model are reasonably consistent with the observations. Of course, the number of observations is very small, and the model does not incorporate many of the complexities that are probably important in determining population growth (see Chapter 11 for a more complete discussion of population growth models). However, despite our omission of these complexities, we have constructed a satisfactory model of the seals' population growth.

---

standing of the factors that determine the amount of money on hand, although the degree of detail necessary in a model in some respects depends upon the researcher's particular interest.

Models have several advantages that are important in population biology. First, they allow a *basic understanding* and a focus upon the factors important in a particular process. This is especially helpful in teaching the general concepts of a new topic and in designing experiments to examine the

Figure 1.2. Two male northern elephant seals contesting a breeding territory. The smaller animals in the background are females (Le Boeuf and Koza, 1981).

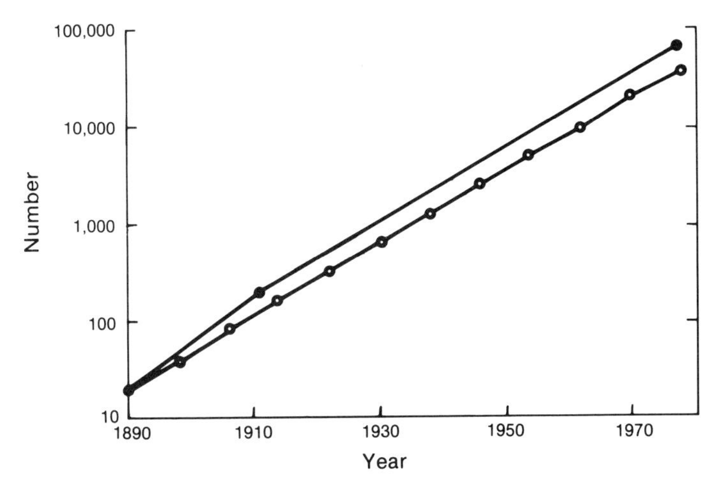

Figure I.3. The observed (closed circles) and predicted (open circles) numbers of the northern elephant seal over time.

importance of various factors. Second, a model can provide an *exact description* of a situation or process. This descriptive use of models may be used to explain in detail, for example, the factors involved in a pest outbreak, making a detailed scenario available for similar future situations, and thereby ultimately reducing the damage from the pest.

Last, models can be used for the *prediction of future events*. For example, given a model and some basic information, one might predict the population numbers at a particular time. Predictability is also important in determining the appropriate model. For instance, several models might predict the number of individuals in a population at some future time, but an experiment that monitors the change in numbers over time could exclude one or more of these models where their predictions are inconsistent with the observations. It should be noted that, in general, verbal models (and some mathematical models) are not precise enough to make predictions that would allow their exclusion. Furthermore, verbal models cannot predict counterintuitive results, while mathematical models may.

As an example, let us consider a simple model to predict population numbers and then apply it to a population of elephant seals. First, let N_t and N_{t+1} be the number of individuals in generation t and generation $t + 1$, respectively. Let R be the ratio of the numbers in generation $t + 1$ over the number in generation t, or

$$R = \frac{N_{t+1}}{N_t}.$$

For example, if this ratio, usually called the net replacement rate or the average number of offspring per parent (see Chapter 11), is 2, then the population number would double each generation. This expression can be rearranged to predict the number in a future generation; that is, the number in generation $t + 1$ is

$$N_{t+1} = RN_t.$$

In other words, the number in the next generation is a product of the numbers in the previous generation and the net replacement rate (see Example I.2).

It is often useful to examine the change in numbers (or other values) in a given period of time. For example, if we let

$$\Delta N = N_{t+1} - N_t$$

then by substitution

$$\begin{aligned}\Delta N &= RN_t - N_t \\ &= N_t(R - 1).\end{aligned}$$

This is called a *difference equation*, and with it we can calculate the expected change in numbers at different times. We will use difference equations to describe the changes in allelic frequencies due to various factors such as selection and gene flow.

On the other hand, if population growth is continuous (where the time intervals become very small), the change in population numbers is best described by a *differential equation* such as

$$\frac{dN}{dt} = rN$$

where r is another measure of population growth called the instantaneous rate of growth (see Chapter 11). We will use differential equations to examine changes in population numbers due to various factors such as competition and predation.

II. STATISTICS

In evaluating various attributes of populations, such as the amount of genetic variation and the numbers of individuals, it is necessary to use some descriptive statistics. Here we will briefly discuss how to calculate the *mean*, a measure of the average of a group of values, and the *variance*, a measure of the variability around that central value. Because populations are often large, generally a sample is obtained from the population and its characteristics are evaluated.

Let us assume we are considering the heights of a sample of n different plants. If we symbolize the height of plant i as x_i, then we may calculate the mean plant height as

$$\bar{x} = \frac{1}{n}\sum_{i=1}^{n} x_i$$

where Σ (Greek letter sigma) indicates the summation of all n plant heights in the sample. For example, if there are five plants in the sample and their heights in centimeters are $x_1 = 28$, $x_2 = 30$, $x_3 = 27$, $x_4 = 32$, and $x_5 = 33$, then

$$\begin{aligned}\bar{x} &= \tfrac{1}{5}(x_1 + x_2 + x_3 + x_4 + x_5)\\ &= \tfrac{1}{5}(28 + 30 + 27 + 32 + 33) = 30.\end{aligned}$$

In other words, the mean plant height in this sample of individuals is 30.

Where a sample of individuals has a given mean, different amounts of dispersion may exist around that mean. For example, there may be no dispersion if all plants are exactly the same height, or there may be an extreme amount of dispersion if there are equal numbers of small and large individuals. The *variance* is a measure of the dispersion around the mean and is calculated as

$$V_x = \frac{1}{n-1}\sum_{i=1}^{n} (x_i - \bar{x})^2.$$

In other words, the variance of x is the average of the sum of the squared deviations of individual values from the mean. (Here, $n - 1$ is used instead of n because variance estimated in this manner is unbiased; see a statistics book for a discussion.) The estimate of the variance for the data above is

$$\begin{aligned} V_x &= \tfrac{1}{4}[(x_1 - \bar{x})^2 + (x_2 - \bar{x})^2 + (x_3 - \bar{x})^2 + (x_4 - \bar{x})^2 + (x_5 - \bar{x})^2] \\ &= \tfrac{1}{4}[(28 - 30)^2 + (30 - 30)^2 + (27 - 30)^2 + (32 - 30)^2 + (33 - 30)^2] \\ &= 6.5. \end{aligned}$$

Example I.3. For statistical analysis, it is often assumed that the distribution of values around a mean is close to a normal distribution. In fact, samples from natural populations are often close to a normal distribution. To illustrate this, the distribution of height in a company of 175 soldiers is given in Figure I.5 (from Stern, 1973). Notice that only a few individuals are either very short or very tall, and that most are in the intermediate height range (see Table I.2). The mean height for this sample is 67.31 inches, which can be obtained by summing the product of values in the first two columns of Table I.2 and dividing by 175, or

$$\begin{aligned} \bar{x} &= \tfrac{1}{175}[58(1) + 61(1) + 62(5) + \cdots + 72(4) + 73(4) + 74(1)] \\ &= 67.31. \end{aligned}$$

Figure I.5. The distribution of height in a company of soldiers (from Stern, 1973).

Often another measure of the dispersion around the mean, the standard deviation, is used. The *standard deviation* is the square root of the variance and is, therefore, a measurement on the same scale as the mean. As a result, the standard deviation may be used for distributions that are bell-shaped or normal, to state the proportion of the values within a given part of the distribution. Figure I.4 gives a theoretical normal distribution, with the mean indicated by the tickmark in the center. The total area under the distribution is unity. The darkly shaded area within one standard deviation on either side of the mean encompasses 68 percent of the total area, and the total shaded area

---

The variance may be obtained in an analogous manner using the second and third columns in Table I.2 so that

$$V_x = \tfrac{1}{174}[86.7(1) + 39.8(1) + \cdots + 32.4(4) + 44.8(1)]$$
$$= 7.29.$$

The standard deviation, the square root of the variance, is then 2.70. If we use the assumptions of the normal distribution, then approximately 68 percent of the individuals in the company would be between 64.61 and 70.01 inches in height and 95 percent would be between 61.91 and 72.71 inches in height. In fact, seven individuals are less than 62 or greater than 72 inches in height, 4 percent of the company, very close to the 5 percent expected.

Table I.2
The distribution of height for a company of 175 soldiers (from Stern, 1973).

Height (inches)	*Number*	$\sum(x_i - \bar{x})^2$
58	1	86.7
61	1	39.8
62	5	28.2
63	7	18.6
64	7	11.0
65	22	5.3
66	25	1.7
67	26	0.1
68	27	0.5
69	17	2.9
70	11	7.2
71	17	13.6
72	4	22.0
73	4	32.4
74	1	44.8
	$\bar{x} = 67.31$	$V_x = 7.29$

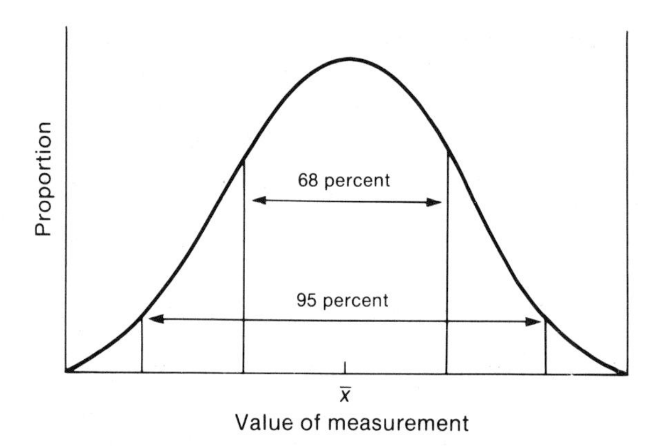

Figure I.4. The normal distribution indicating the mean $\bar{x}$ and areas encompassing 68 percent and 95 percent of the distribution.

within two standard deviations on either side of the mean encompasses 95 percent. The data on human height in Example I.3 closely fit this theoretical normal distribution.

III. PROBABILITY

In designing experiments to test various aspects of population biology, it is important to understand the elements of probability. As a simple example, assume that there are only two possible outcomes of an event, such as the flipping of a coin, which may result in either a head or a tail. Therefore, the sum of the probability of the first outcome, $Pr(1)$ or a head, and that of the second outcome, $Pr(2)$ or a tail, is unity, or

$$Pr(1) + Pr(2) = 1.$$

Using a fair coin in which the probability of a head is $Pr(1) = 0.5$ and the probability of a tail is $Pr(2) = 0.5$, then $0.5 + 0.5 = 1$. Of course, in another situation, such as the probability of getting or not getting an infection, the probability of the two outcomes may not be equal, for example, $Pr(1)$ could be 0.2 and $Pr(2)$ would then be 0.8.

When a number of trials occur—for example, a number of coins are tossed—then the *expected number* of outcomes of a particular type is equal to the probability of the outcome times the number of trials n. For outcome 1, it is

$$E(1) = Pr(1)n$$

where $E(1)$ is the expected number of outcomes of type 1. For example, if a coin is tossed 20 times, then the expected number of heads, outcome 1, is

$$E(1) = 0.5(20) = 10.$$

In fact, in a real situation the *observed number* of a particular outcome often does not equal the expected number. For example, the observed number of heads in the above coin toss might be 8 instead of the expected 10. In Chapter 2, we will discuss an approach comparing expected and observed results for some genetic data.

Another aspect of probability that is important in population biology is the probability of particular outcomes in several consecutive events. For example, we may need to know the probability of outcome 1 occurring in two different events. If we assume that the two events are independent of each other—that is, the outcome of event 1 is unrelated to the outcome of event 2 and vice versa—then the joint probability of outcome 1 occurring both times is the product of the probability of it occurring each time, or

$$Pr(1, 1) = Pr(1)Pr(1).$$

Example I.4. Many plants cannot withstand freezing temperatures, and, in fact, the limit of the distribution in some species is determined by the incidence of cold weather. For example, on the northern edge of the Sonoran Desert in Arizona (see Chapter 9), the probability of temperatures below freezing each year, $Pr(1)$, is approximately 0.1 (a value obtained from data over many decades). Observations at a weather station over a ten-year period showed that freezing temperatures occurred in three of those years. The expected number of years with freezing temperatures $E(1)$ in this case was

$$E(1) = Pr(1)n = 0.1(10) = 1.$$

The observed number of three years is substantially higher than this expected value.

In addition, in those years with freezing temperatures, the freezing temperatures occurred at a number of different weather stations. Assuming that the freezes were independent events at different stations, the joint probability of this occurring for three different weather stations would be

$$\begin{aligned} Pr(1, 1, 1) &= Pr(1)Pr(1)Pr(1) \\ &= (0.1)(0.1)(0.1) = 0.001. \end{aligned}$$

Because this joint probability is so low, the weather at different sites in this area appears not to be independent. In other words, the weather patterns may influence large areas, so that freezing temperatures may simultaneously affect a number of areas.

This rule can be extended to any number of independent events. For example, the joint probability of obtaining a head from a fair coin four times in a row (or from a simultaneous toss of four fair coins) is

$$\begin{aligned} Pr(1,1,1,1) &= Pr(1)\,Pr(1)\,Pr(1)\,Pr(1) \\ &= (0.5)(0.5)(0.5)(0.5) = 0.0625. \end{aligned}$$

As suggested above, in many situations the individual probabilities would not equal 0.5 (see Example I.4).

PART 1

Population Genetics

The study of the genetics of populations is important for three reasons. First, the basic mechanism for evolutionary change involves phenomena described by population genetics. For example, populations of insect pests that have become resistant to insecticides generally have evolved from susceptible populations that initially contained only a small proportion of genetically resistant individuals. Second, application of the principles of population and quantitative genetics is important in the success of programs to improve breeds of plants and animals for agricultural use. Last, a knowledge of population genetics is important in understanding the frequency of genetic diseases in humans and the differences and similarities among human racial groups.

A number of factors affect the rate of evolutionary change or the frequency of particular traits in a population. The major factors are selection, mutation, migration, the chance effects of genetic drift, and the mating system. These factors may have particular effects—for example, mutation that always increases the amount of genetic variation, or genetic drift that always reduces the average amount of genetic variation. Other factors, such as selection or gene flow, may either increase or decrease the amount of genetic variation, depending upon the situation. Combinations of two or more of these factors may generate virtually any amount or any pattern of genetic variation.

In Chapter 1, we begin discussing population genetics by considering the amount and types of genetic variation in a population. Next, we focus in individual chapters on the effects on genetic variation of selection, mutation, migration, genetic drift, and the mating system. In some cases, we also consider the joint effects of two factors, such as selection and mutation, on the amount of genetic variation. Last, in Chapter 8, we briefly discuss quantitative genetic models.

CHAPTER 1

Genetic Variation

The Lord in His wisdom made the fly
And then forgot to tell us why.

OGDEN NASH

Humans have long been aware of individual variations among themselves and that such variants tend to "run in families." Differences between strains or breeds of domesticated plants and animals also were assumed to be inherited long before the mechanisms of genetics were discovered. For example, the differences between breeds of dogs, cats, and horses were recognized centuries ago, but only in the 1930s was there a substantial effort to document the amount of genetic variation in natural populations—that is, populations not manipulated by humans for breeding purposes.

The early studies of genetic variation in natural populations concentrated on easily detectable variants, such as color variants in butterflies, chromosomal inversions in fruit flies, or red blood loci variants in humans. Although these are important variants, they did not allow an estimate of the total amount of genetic variation for all the genes of the organisms studied. Only in the last 15 years have biochemical techniques yielded an approach that allows an estimate of overall genetic variation in a population.

This chapter covers the amounts and types of variation present in natural populations. Often the term *gene pool* is used to describe the total variety and amount of genetic variation in a population or species. Although much remains to be learned about specific types of genetic variation, a great deal of information exists about different types of genetically determined discrete phenotypes, such as color variants and blood group types in many species. In Chapter 8, we will consider the extent of variation for traits simultaneously affected by a number of genes generally resulting in continuous phenotypic variation. A particular emphasis in many studies is the comparison of genetic variation in different populations and at different times—in other words, spatial and temporal differences and similarities in genetic variation.

I. VISIBLE VARIANTS

The first genetic variants known were those that affected color, shape, pattern, and other morphological aspects. Visible variants are known in virtually every kind of organism—for example, flower color variants in many plants, shell

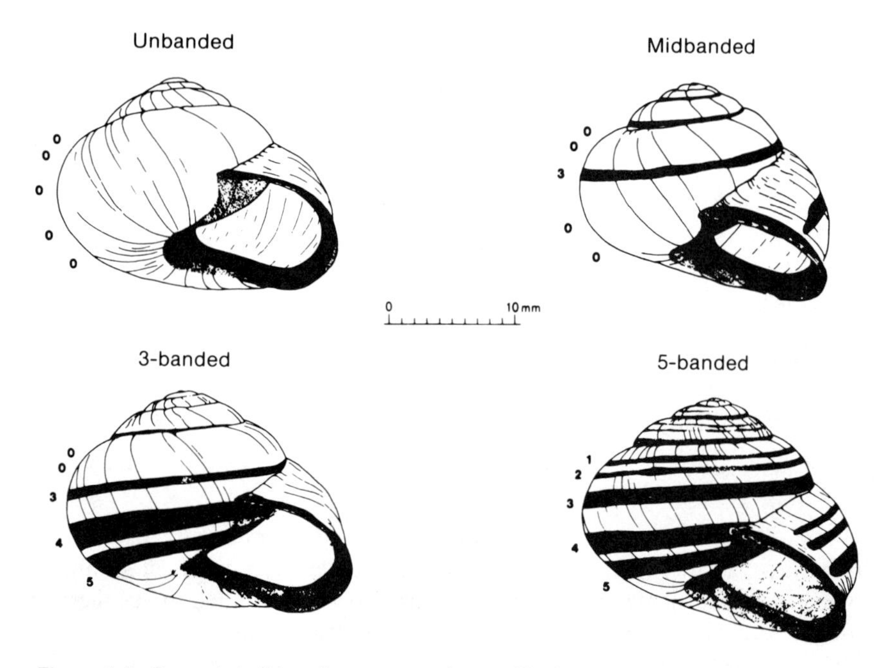

Figure 1.1. Several shell-banding patterns observed in the snail *Cepaea nemoralis* (from Jones et al., 1977).

Table 1.1
The frequencies of color forms in a fossil population and three living populations of an African land snail (after Owen, 1966).

Color form	Fossil	Living a	Living b	Living c
Streaked	0.610	0.546	0.406	0.337
Broken-streaked	0.052	0.089	0.042	0.098
Pallid 1	0.039	0.240	0.070	0.019
Pallid 2	0.283	0.119	0.432	0.373
Pallid 3	0.016	0.006	0.050	0.173

color and pattern variants in snails, wing shape variation in butterflies, and melanic forms in rodents. Here we concentrate on several examples in which the mode of inheritance is known and information exists on the temporal and/or spatial frequencies of these variants.

One of the classic examples of morphological variation in natural populations is in the European land snail, *Cepaea nemoralis*. Several genes are involved in determining the color and banding pattern of their shells (see Figure 1.1). The snails vary from yellow, unbanded types to brown, banded types. Different populations vary dramatically in the frequency of these types, with, for example, yellow high in frequency in one population and low in another population. It appears that these forms confer protective camouflage from the predation of birds although other factors also appear important (see Jones et al., 1977, for a review).

An African land snail also exhibits variations in shell color and patterns and, as in *Cepaea*, adjacent populations of this species may vary considerably in the frequency of different forms involving variation in streaking and pigmentation (Owen, 1966). Fossil snails, 8,000 to 10,000 years old, that appear to be the same species as that present today have been found in areas of Western Uganda that were covered by volcanic ash. These fossil snails have the same shell color and patterns as living populations in the same area. In fact, the frequencies of these types are quite similar to the fossil frequencies for some of the populations sampled (see Table 1.1). The three populations given in Table 1.1 are all fewer than 30 km from the fossil site, indicating that the variant forms may be quite ancient and constant in frequencies.

The variant forms of the peppered moth, *Biston betularia*, have been extensively investigated. This moth is generally a mottled light color, but in a number of areas of England one can observe dark, melanic forms (see Figure 1.2). The frequencies of melanic, darkly pigmented forms of *B. betularia* have been observed in England for more than a century. The melanics were

Figure 1.2. Photographs of two morphs, the melanic and typical, of the peppered moth nesting (a) on a lichen-covered tree trunk and (b) on a dark tree trunk (from Kettlewell, 1973).

quite rare but have increased so much in frequency that in some contemporary urban populations they may constitute nearly 100 percent of the population. The melanic coloration confers protective camouflage from bird predation in polluted areas (Chapter 3). The frequency differences of melanics in urban areas and nearby rural areas are quite striking. For example, nearly 100 percent melanics are found in the industrial city of Liverpool compared with fewer than 10 percent melanics 50 to 60 kilometers away in rural Wales.

II. CHROMOSOMAL VARIATION

In a number of species, variation in the structure of chromosomes has been observed among different individuals. Recently, techniques developed to examine morphological details of chromosomes have uncovered more chromosomal variants, and it is likely that many small variations in chromosome structure, such as deletions, duplications, and inversions, are still to be discovered. In humans, several inversions exist in substantial frequency, and the Y chromosome varies in size among different populations. Some plant species have high frequencies of translocations—that is, exchanges of large chromosomal regions among nonhomologous chromosomes.

One of the most thoroughly studied examples of chromosomal variation is the inversion variants of the third chromosome of the fruit fly, *Drosophila pseudoobscura*, a widespread species in the mountains of western North America. In *Drosophila* and other diptera, large chromosomes (the homologues are paired) in the salivary glands allow the recognition of banding patterns on these chromosomes. For example, there are 23 known gene arrangements on the third chromosome of *D. pseudoobscura*, each differing from the others by inversions of one or more chromosomal segments. Heterozygotes between two inversion types (heterokaryotypes) can be recognized by the loop patterns, as shown in Figure 1.3. The loop pattern is

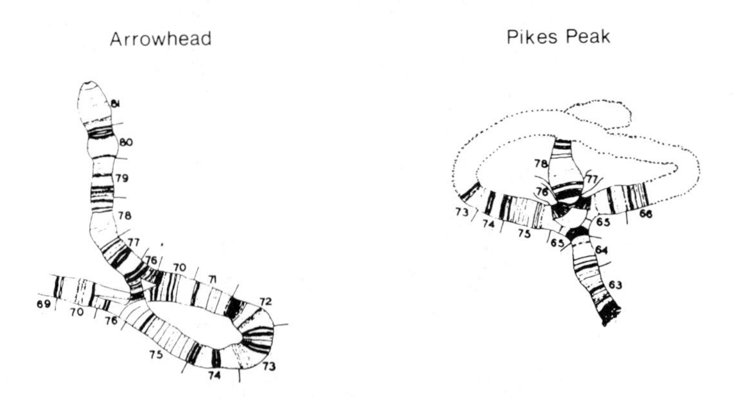

Figure 1.3. Two of the common third-chromosome arrangements in *D. pseudoobscura* as heterokaryotypes with the standard gene arrangement (after Spiess, 1977).

caused by the strong affinity between homologous chromosomal regions even though the order may differ on the two chromosomes in a heterokaryotype.

Many of these inversions are quite frequent within populations of *D. pseudoobscura*. Part of a summary of three decades of surveys of inversion frequencies in *D. pseudoobscura* is given in Figure 1.4 (Anderson et al., 1975). This figure gives the frequencies of the most common arrangements in the Panamint Mountains in California and the Capitan Area of New Mexico, two widely separated sites, over a number of years. Because at least several generations of *D. pseudoobscura* populations occur per year at these sites, these data illustrate frequencies over a 50- to 100-generation period.

In examining these data, notice that there are striking differences between locations, with ST (Standard) the most frequent inversion type in California and AR (Arrowhead) and PP (Pikes Peak) in New Mexico. Also, different temporal patterns are found at the various sample sites, with an increase in TL (Treeline) in California in recent samples, a fluctuation of AR and PP in New Mexico, and relative constancy of other arrangements at both sites.

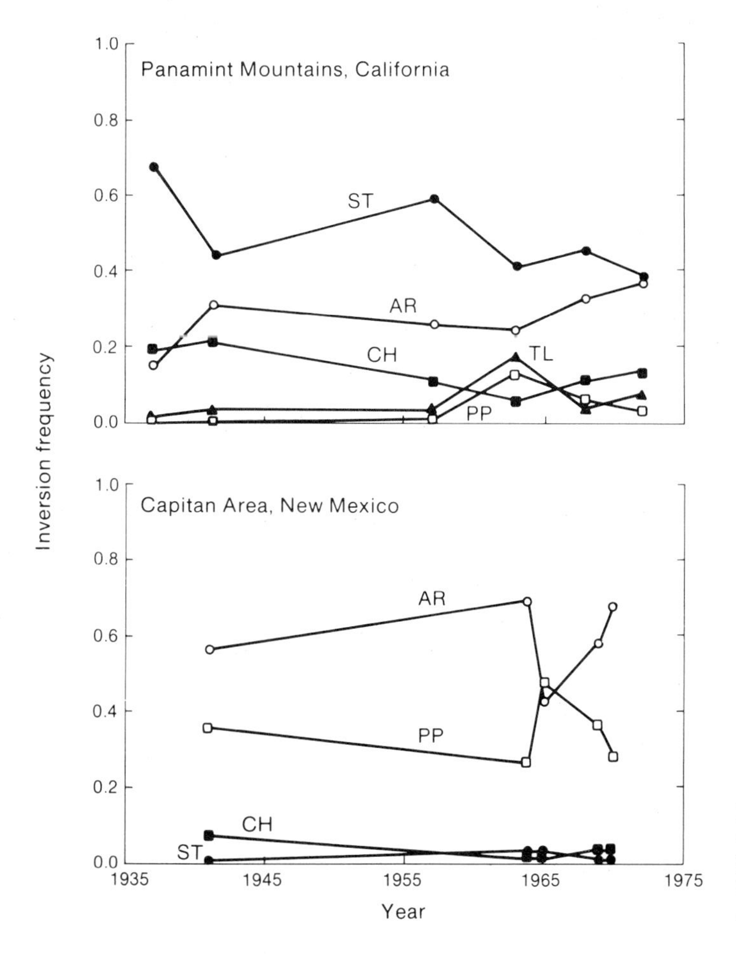

Figure 1.4. The frequencies of inversions on the third chromosome in *D. pseudoobscura* in two localities over three decades (data from Anderson et al., 1975).

Furthermore, there is a relatively stable presence of infrequent arrangements in all populations. For example, in addition to the inversions shown, TL, EP (Estes Park), and OL (Olympic) are at levels of 2 percent or fewer four out of five sampling times in New Mexico. The complete explanation for these patterns is complicated and not fully understood, but a major factor affecting these frequencies is that the viability of heterokaryotypes is higher than that of homokaryotypes.

III. ELECTROPHORETIC VARIATION

Variation in visible or chromosomal variants does not indicate the total amount of variation in a population. To fully measure the amount of genetic variation in a population, we would need to know the DNA sequence of every individual. It is possible to sequence specific segments of DNA, and some researchers are applying such techniques to population studies. Many recent population genetic surveys have focused on the variation in proteins in order to estimate the variation in the DNA sequence that determines the amino acid sequence of these proteins. For example, a difference in the DNA sequence of two alleles may result in a change in the amino acid sequence of the proteins produced from these alleles. See Chapter 17 for a discussion of the possible factors maintaining this molecular variation and its evolutionary implications.

Variant forms of an enzyme are generally detected by *electrophoresis*, a technique that allows the separation of different proteins extracted from blood, tissues, or whole organisms. The process is carried out by imposing an electric field across a supporting medium into which the protein has been placed. The proteins are allowed to migrate for a specific amount of time and then are stained with various protein-specific chemicals, resulting in bands on the gel that permit the relative mobility of a specific protein to be determined (see Figure 1.5). Relative mobility is generally a function of the size, charge, and shape of the molecule. If two proteins that are the products of different alleles have different amino acid sequences, then they often have different mobilities, since the differences in sequence result in a change in size and/or charge of the molecule. However, amino acid differences that do not result in significant charge or size differences may be undetectable as electrophoretic differences.

In surveys of electrophoretic variation, proteins from a group of individuals (often 12 or 20) are usually run simultaneously on a single gel. Figure 1.6 is an example of such a gel showing variation in two enzymes (both are leucine aminopeptidases) coded by different loci in the brown snail. Because these enzymes have only one polypeptide subunit, homozygotes show a single stained region, or band, and heterozygotes are double-banded. The upper enzyme is variable for two electrophoretic bands and the lower enzyme for three. Because migration of the proteins goes from the bottom to the top of the gel pictured, the phenotypes (presumed genotypes) are as given in Figure 1.6, where *F*, *M*, and *S* indicate alleles of fast, intermediate, and slow migration, respectively. For example, the individual on the far left is

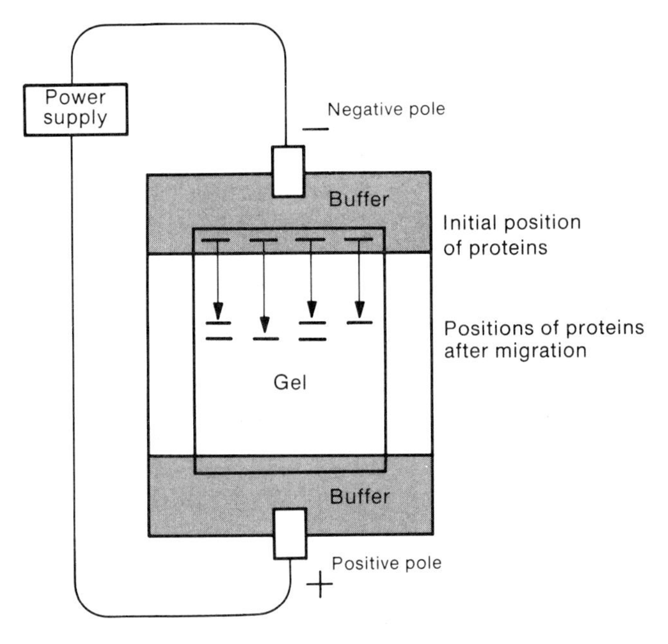

Figure 1.5. A diagram of a gel electrophoresis apparatus. The buffers are used to conduct electricity and ensure a given pH.

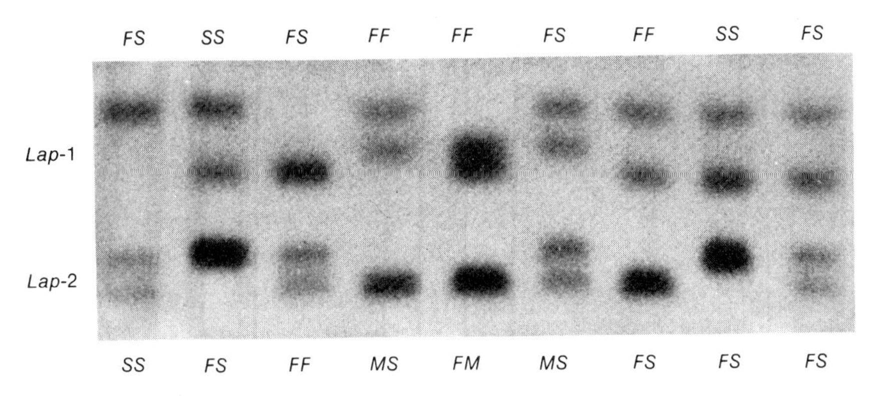

Figure 1.6. Variation in two enzymes in the brown snail. The upper system is variable for two alleles (*F* and *S*) and the lower system is variable for three alleles (*S*, *M*, and *F*). The genotypes are indicated above and below the gel for *Lap*-1 and *Lap*-2, respectively, for the nine different individuals analyzed (from Selander, 1976).

heterozygous *FS* for the top enzyme and homozygous *SS* for the bottom enzyme. (It is generally assumed that each band represents a different allele at a locus. But because electrophoretic types may be heterogeneous and may actually be a class of alleles and not a single allele, the term *electromorph* is often used to describe a particular band.)

Electrophoretic techniques used to investigate population genetics problems have invigorated the field of population genetics, because they permit an objective assessment of the amount of genetic variation. These techniques allow the study, involving relatively little equipment and expertise, of genetic variation in virtually any species. As a result, a number of studies have concentrated on quantifying the amount of electrophoretic variation in various biologically interesting species. It appears that some electromorphs are often widespread in a species and in closely related species. On the other hand, species or populations may differ in the presence or frequency of various electromorphs. In Chapter 17, we will discuss the factors that have been suggested to be important in affecting genetic variation determined by electrophoresis and other molecular techniques.

An excellent example of spatial variation for an electrophoretic variant is in the brown snail, *Helix aspera*, which was collected in a two city-block area of Bryan, Texas (Selander and Kaufman, 1975). The brown snail was introduced to Bryan in the early 1930s and is now found widely in gardens and yards. The frequencies of the various electromorphs for a *malic dehydrogenase* (*Mdh*-1) locus are indicated in the pie diagrams in Figure 1.7. Even over such

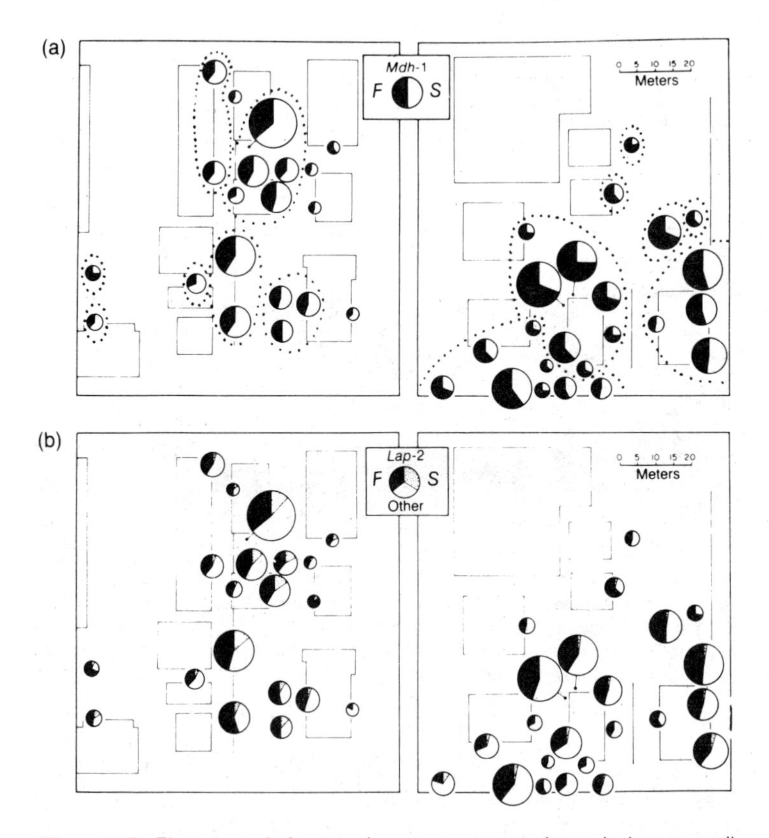

Figure 1.7. Electromorph frequencies at an enzyme locus in brown snail colonies. Circle size is proportionate to colony size, and the proportions within the circles indicate electromorph frequency (from Selander and Kaufman, 1975).

a small area, there was extensive variation. For example, the frequency of F varied from greater than 0.75 to approximately 0.30 in different colonies.

An example of both spatial and temporal variation was obtained in a thirty-month study of the prairie vole, *Microtus ochrogaster* (Gaines et al., 1978). The prairie vole is a small rodent that undergoes large fluctuations in population density in approximately three- to four-year cycles. The average generation length is estimated to be 6 to 8 weeks, so that during the study period, there were approximately 20 generations. Data for an electromorph at the *leucine aminopeptidase* (*Lap*) locus for four sites near Lawrence, Kansas, are given in Figure 1.8. Sites A, B, and D were within 1.2 km of each other and site C is 3.6 km from the main study area. There was considerable variation in the electromorph frequency over time and between sites, varying from zero early in the sampling period to nearly 0.5 at other times.

One of the major results from electrophoretic studies is the discovery of extensive variation in a number of genes in many different organisms. Let us briefly introduce two measures used to quantify the amount of variation in a population (these are discussed further in Chapter 2). These measures are the proportion of all loci that are polymorphic (have genetically variant forms), P, and the proportion of heterozygotes over all loci in the particular population, $\bar{H}$. Polymorphism refers to the occurrence of different (genetic) forms in the same population, such as the A, B, O, and AB blood group phenotypes for the ABO red blood cell locus in humans. If there are 50 such blood group loci known and 10 have different forms, then the proportion of polymorphic loci is $10/50 = 0.2$.

Likewise, the proportion of heterozygotes is the proportion of diploid genotypes that are composed of two different alleles. To illustrate, let us use another polymorphic human blood group locus, the MN locus. At this locus there are three possible genotypes: two homozygotes, MM and NN, and one

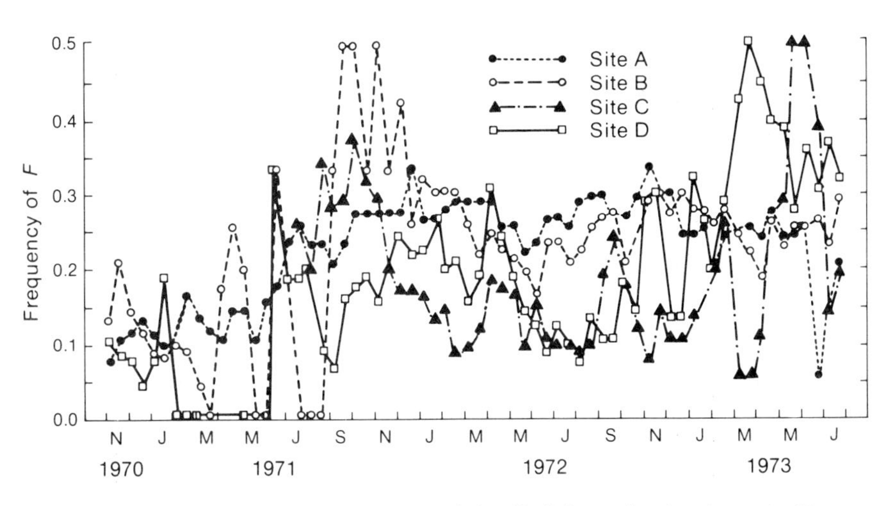

Figure 1.8. The changes in the frequency of the *F* allele at the *Lap* locus in *Microtus ochrogaster* over time on four live-trapped grids (from Gaines et al., 1978).

heterozygote, MN. Let us assume that, in a population of 200 individuals, 90 are MN heterozygotes. Therefore, the proportion of heterozygotes for this locus in this population, H, is $90/200 = 0.45$. If the heterozygosity is known for a number of loci, then these values can be averaged over loci to give a mean value of heterozygosity, $\bar{H}$.

An example to illustrate the extent of variation is given in Table 1.2 for

Table 1.2
The heterozygosity for 10 of the polymorphic loci of the 71 allozyme loci in humans (Harris and Hopkinson, 1972).

Locus	*Heterozygosity* (H)
51 monomorphic loci	0.00
Peptidase C	0.02
Pancreatic amylase	0.09
Adenosine deaminase	0.11
Acetyl cholinesterase	0.23
Mitochondrial malic enzyme	0.30
Peptidase A	0.37
Pepsinogen	0.47
Alcohol dehydrogenase-3	0.48
RBC acid phosphatase	0.52
Placental alkaline phosphatase	0.53

Table 1.3
A summary of the average proportion of polymorphic loci (P) and the average heterozygosity ($\bar{H}$) for a variety of organisms (from Nevo, 1978; and Hamrick, 1979).

(*a*) *Organisms*	*Number of loci*	P	$\bar{H}$
Humans	71	0.28	0.067
Northern elephant seal	24	0.0	0.0
Horseshoe crab	25	0.25	0.057
Elephant	32	0.29	0.089
Drosophila pseudoobscura	24	0.42	0.12
Barley	28	0.30	0.003
Tree frog	27	0.41	0.074
(*b*) *Taxonomic groups*	*Number of species*	P	$\bar{H}$
Plants	—	0.305	0.104
Insects (excluding *Drosophila*)	23	0.329	0.074
Drosophila	43	0.431	0.140
Amphibians	13	0.269	0.079
Reptiles	17	0.219	0.047
Birds	7	0.150	0.047
Mammals	46	0.147	0.036
Average		0.271	0.078

71 enzyme loci surveyed in humans. Of these loci, 51 did not have variable forms (were monomorphic) and 20 were polymorphic, making $P = 20/71 = 0.282$. The heterozygosity values ranged from 0.02 to 0.53 for the polymorphic loci, so that the average heterozygosity over all loci (both monomorphic and polymorphic), $\bar{H}$, was 0.067.

The extent of variation for loci studied electrophoretically has been examined in many species. Table 1.3 gives the proportion of polymorphic loci and the heterozygosity for (a) several different organisms and (b) a summary of data from both animals and plants. Values for P and $\bar{H}$ in these individual species range from 0.0 for northern elephant seals to high P and H values of 0.42 and 0.12, respectively, in *Drosophila pseudoobscura* (some other individual species actually appear to have more variation than *D. pseudoobscura*). Even in species with low heterozygosity, there is still potentially a great deal of variation if these loci can be thought of as representative of the 5,000 or more protein-coding loci that probably exist in these species. In data for different taxonomic groups, the proportion of polymorphic loci, P, and the average heterozygosity, $\bar{H}$, range from a low of 0.147 and 0.036 in mammals to a high of 0.431 and 0.140 in *Drosophila*. Overall, the most variation appears to exist in invertebrates and plants and the least variation in vertebrates, although there are exceptions to this trend.

SUMMARY

In most species, there exists extensive genetic variation as indicated by polymorphism at enzyme loci. In addition, in many species genetic variation exists at genes that determine color, shape, and other morphological characteristics. All patterns of genetic variation in time and space, including apparently stable frequencies over long periods and dramatic short-term changes, have been observed.

PROBLEMS

1. Compare the amount of variation in visible, chromosomal, and electrophoretic variants over different populations and over time.
2. Which type of variation do you think is most important in evolution—visible, chromosomal, or electrophoretic?
3. Why do you think different organisms may exhibit different amounts of electrophoretic variation?
4. The frequencies of different shell patterns of *Cepaea* vary in different populations. What factors do you think might be responsible? Design one or more experiments to evaluate the importance of these factors.
5. The frequencies of inversions in *Drosophila* vary over time. What factors do you think might be responsible? Design one or more experiments to evaluate the importance of these factors.

CHAPTER 2

Allelic and Genotypic Frequencies

It is clear that descriptions of the genetic variation in populations are the fundamental observations on which evolutionary genetics depends. Such observations must be both dynamically and empirically sufficient if they are to provide the basis for evolutionary explanations and predictions.

RICHARD LEWONTIN

How much genetic variation exists in a given population? Do populations or species differ in the extent of genetic variation? To answer these questions, we must be able first to describe genetic variation as discussed in Chapter 1 and, second, to quantify the amount of genetic variation in a standardized way. Therefore, in this chapter we will introduce some techniques for quantifying genetic variation that we will use later in explaining the influence of various factors on genetic variation.

However, to begin we must define the genetic connotation of the term *population*. As a simple, ideal definition, we can say that a population is a group of interbreeding individuals of the same species that exist together in time and space. It is usually assumed that a population is geographically well defined, although this is often not true. In any case, this ideal and generally theoretical definition is useful in developing the concepts of population genetics.

Let us go a step further in specifying this ideal population. Assume that a population of sexually mature individuals exists together and that there are two sexes. If the probability of each possible mating between females and males is equal, then the population is called a *random-mating population*. Stated another way, if individuals are specified as to genotype or phenotype, then a random-mating population is a group in which the probability of a mating between individuals of particular genotypes or phenotypes is equal to the product of their individual frequencies in the population. (Sometimes a random-mating population is referred to as a *panmictic population*.) In some real-life situations, such as in dense populations of insects or outcrossing plants, this ideal may be nearly correct. To develop the basic concepts of population genetics, we will consider the ideal of a large, random-mating population.

I. OBSERVED GENOTYPIC AND ALLELIC FREQUENCIES

The genotypic or allelic frequencies at a particular locus for a population are generally calculated after the individuals of the different genotypes have been counted. Assume that there are N individuals in the population, or that a sample of N individuals is categorized, and that there are N_{11} of genotype A_1A_1, N_{12} of genotype A_1A_2, and N_{22} of genotype A_2A_2 ($N_{11} + N_{12} + N_{22} = N$). If we assume that this is a sample from a larger population, then the estimated frequencies of the three genotypes A_1A_1, A_1A_2, and A_2A_2 in the population are

$$P = \frac{N_{11}}{N}$$

$$H = \frac{N_{12}}{N}$$

$$Q = \frac{N_{22}}{N}$$

respectively (see Table 2.1). The sum of the frequencies of the three genotypes is unity, that is, $P + H + Q = 1$.

Because all the alleles in the two homozygotes, A_1A_1 and A_2A_2, are A_1 and A_2, respectively, and half the alleles in the heterozygote are A_1 and half are A_2, the frequencies of alleles A_1 and A_2, p and q, respectively, in terms of the genotypic frequencies are then

$$p = P + \tfrac{1}{2}H$$

$$q = Q + \tfrac{1}{2}H$$

respectively. The estimated frequencies of A_1 and A_2 can also be calculated from the sample as

$$p = \frac{N_{11} + \frac{1}{2}N_{12}}{N}$$

Table 2.1
The three genotypes at a locus with two alleles, the numbers in a sample, and the frequencies of the genotypes.

Genotype	A_1A_1	A_1A_2	A_2A_2	Total
Number	N_{11}	N_{12}	N_{22}	N
Observed frequency	$P = \frac{N_{11}}{N}$	$H = \frac{N_{12}}{N}$	$Q = \frac{N_{22}}{N}$	1

$$q = \frac{\frac{1}{2}N_{12} + N_{22}}{N}$$

which is the best way to estimate allelic frequencies. Because there are only two alleles in this case, $p + q = 1$.

II. THE HARDY-WEINBERG PRINCIPLE

Although earlier studies focused on changes in populations as the result of natural or artificial selection, the beginning of population genetics was in the first decade of this century. In 1908, shortly after the rediscovery of Mendel's work, an English mathematician, G. H. Hardy, and a German physician, W. Weinberg, formulated what has become known as the Hardy-Weinberg principle. This principle describes a relationship that is of basic importance to population genetics because it allows the description of the genetic content in diploid populations to be made in terms of allelic frequencies. The Hardy-Weinberg principle strictly holds only in the absence of selection, mutation, migration, and genetic drift, and in the presence of random mating. In fact, though, Hardy-Weinberg proportions occur in many instances where these conditions are not met.

The *Hardy-Weinberg principle* states that single-locus genotypic frequencies after one generation of random mating can be represented as a function of the allelic frequencies. This principle allows a great simplification in the description of the genetic content of a population by reducing the number of frequencies that must be considered. For example, the Hardy-Weinberg principle allows us to describe a population by considering only the frequencies of the two alleles at a biallelic locus rather than the three genotypes. Furthermore, in the absence of factors that change allelic frequency (selection, mutation, migration, and genetic drift) and the continued presence of random mating, the Hardy-Weinberg proportions of the genotypes remain constant over time.

To understand the Hardy-Weinberg principle, assume that a population is segregating for two alleles, A_1 and A_2, at locus A in frequencies of p and q $(p + q = 1)$, respectively. Assume that male and female gametes unite at random to form zygotes, as shown graphically in Figure 2.1. The unit length across the top of the square is divided into proportions p and q representing the frequencies of the female gametes A_1 and A_2, respectively. Likewise the side of the square is divided into proportions representing the frequencies of the male gametes. Therefore, the areas within the square represent the proportions of different progeny zygotes—for example, the upper left square has an area of p^2, the frequency of genotype A_1A_1, and the lower right square has an area of q^2, the frequency of A_2A_2. Overall, the three progeny zygotes (A_1A_1, A_1A_2, A_2A_2) are formed in proportions $(p^2, 2pq, q^2)$.

In most organisms, random union of gametes is quite unlikely, because it is the parental genotypes that pair and then produce gametes that unite. Therefore, let us consider the situation in which reproductive individuals

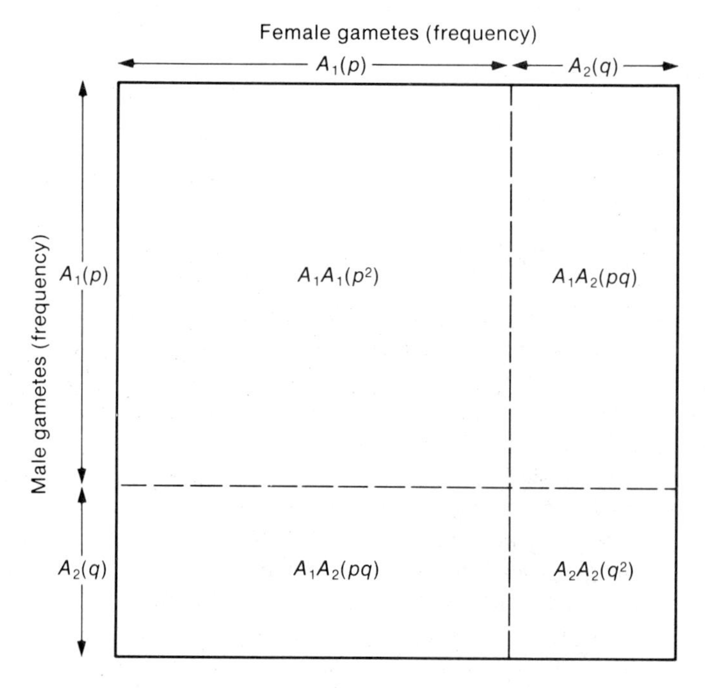

Figure 2.1. The Hardy-Weinberg proportions as generated from the random union of gametes using a unit square.

Table 2.2
The frequency of different mating types for two alleles at a locus when there is random mating.

Female genotypes (frequencies)	*Male genotypes (frequencies)*		
	$A_1A_1(P)$	$A_1A_2(H)$	$A_2A_2(Q)$
$A_1A_1(P)$	P^2	PH	PQ
$A_1A_2(H)$	PH	H^2	HQ
$A_2A_2(Q)$	PQ	HQ	Q^2

randomly pair to mate. If we consider a diploid organism, three possible genotypes, A_1A_1, A_1A_2, and A_2A_2, are present in the population in frequencies P, H, and Q, respectively (remember that $P + H + Q = 1$). Let us assume that there is random mating in the population, making the nine possible combinations of matings between the male and female genotypes shown in Table 2.2. The frequency of a particular mating type, given random mating, is equal to the product of the frequencies of the genotypes that constitute that mating type—for example, the frequency of mating type $A_1A_1 \times A_1A_1$ is P^2.

Only six mating types are to be distinguished here because reciprocal matings—for example, $A_1A_1 \times A_1A_2$ and $A_1A_2 \times A_1A_1$, where the first genotype is the female—have the same genetic consequences.

The mating types, their frequencies, and the expected frequencies of their offspring genotypes are given in Table 2.3. For example, the mating $A_1A_1 \times A_1A_1$ produces only A_1A_1 progeny, the mating $A_1A_1 \times A_1A_2$ produces $\frac{1}{2}A_1A_1$ and $\frac{1}{2}A_1A_2$, and so on. If the frequency of A_1A_1 progeny contributed by each mating type is summed up (adding down the first progeny column), then it is found that $P^2 + PH + \frac{1}{4}H^2 = (P + \frac{1}{2}H)^2$ of the progeny are A_1A_1. From the expression for the frequency of A_1 above, we know that $p = P + \frac{1}{2}H$ and, therefore, p^2 of the progeny are A_1A_1. Similarly, the frequency of A_1A_2 and A_2A_2 progeny are $2(P + \frac{1}{2}H)(Q + \frac{1}{2}H) = 2pq$ and $(Q + \frac{1}{2}H)^2 = q^2$, respectively. The crucial point here is that, given any set of initial genotypic frequencies (P, H, Q) after one generation of random mating, the genotypic frequencies are in the proportions (p^2, $2pq$, q^2). For example, given the initial genotypic frequencies (0.2, 0.4, 0.4) in which $p = 0.4$ and $q = 0.6$, after one generation the genotypic frequencies become $[(0.4)^2, 2(0.4)(0.6), (0.6)^2] = (0.16, 0.48, 0.36)$. Furthermore, the genotypic frequencies will stay in these exact proportions generation after generation, given continued random mating and the absence of factors that change allelic frequencies.

Of course, different populations may have different allelic frequencies. To illustrate the effect of different allelic frequencies on genotypic frequencies, Figure 2.2 gives the frequencies of the three different genotypes, assuming Hardy-Weinberg proportions for the total range of allelic frequencies. Notice that the heterozygote is the most common genotype for intermediate allelic frequencies, while one of the homozygotes is the most common when the allelic frequency is not intermediate. The maximum frequency of the heterozygote occurs when $q = 0.5$ ($p = 0.5$), and in this case half of the individuals are heterozygotes (see Example 2.1 for an illustration of the calculation of Hardy-Weinberg proportions).

Table 2.3
A demonstration of the Hardy-Weinberg principle assuming random mating in the parents and normal segregation to produce the progeny.

		Progeny		
Mating type	*Frequency*	A_1A_1	A_1A_2	A_2A_2
$A_1A_1 \times A_1A_1$	P^2	P^2	—	—
$A_1A_1 \times A_1A_2$	$2PH$	PH	PH	—
$A_1A_1 \times A_2A_2$	$2PQ$	—	$2PQ$	—
$A_1A_2 \times A_1A_2$	H^2	$\frac{1}{4}H^2$	$\frac{1}{2}H^2$	$\frac{1}{4}H^2$
$A_1A_2 \times A_2A_2$	$2HQ$	—	HQ	HQ
$A_2A_2 \times A_2A_2$	Q^2	—	—	Q^2
Total	1.0	$(P + \frac{1}{2}H)^2 = p^2$	$2(P + \frac{1}{2}H)(Q + \frac{1}{2}H) = 2pq$	$(Q + \frac{1}{2}H)^2 = q^2$

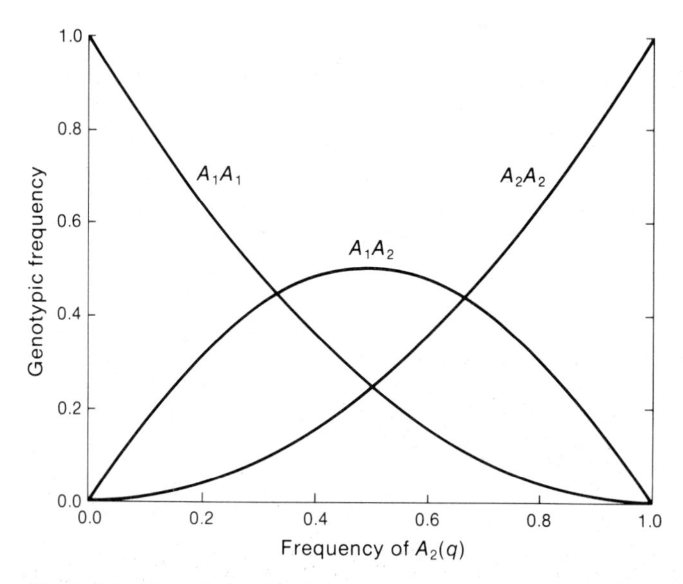

Figure 2.2. The relationship between the allelic frequency and the three genotypic frequencies for a population in Hardy-Weinberg proportions.

In the allelic frequency estimates given above, all alleles were assumed to be expressed in heterozygotes. In many cases—for example, metabolic disorders in humans and color polymorphisms in many organisms—the heterozygote is indistinguishable from one of the homozygotes. In these cases, the population is usually assumed to be in Hardy-Weinberg proportions so that the allelic frequencies can be estimated. For example, with two alleles and dominance, it is assumed that the proportion of A_2A_2 genotypes in the population is

$$\frac{N_{22}}{N} = q^2.$$

Therefore, an estimate of the frequency of A_2 is

$$q = \left(\frac{N_{22}}{N}\right)^{1/2}.$$

In metabolic diseases in humans, it is often important to know the proportion of heterozygotes (carriers) in the population (later in the chapter we will discuss an application to genetic counseling). An estimate of the proportion of carriers is

$$H = 2pq$$

where q is the estimate of allelic frequency above and $p = 1 - q$. Nongeneticists are often surprised at how many individuals are carriers of recessive disease alleles. For example, assuming that only 1 in 10,000 individuals has a particular rare recessive disease (is an A_2A_2 homozygote) such as albinism,

---

Example 2.1. The Hardy-Weinberg principle can be illustrated by the *MN* red blood cell locus in humans. For this locus, there are three genotypes, *MM*, *MN*, and *NN*, each distinguishable from the others by antigen–antibody reactions. In Table 2.4, the number of *MN* types for a sample of 1,000 English blood donors is given (Cleghorn, 1960). From these data, the estimated frequency of allele *M* is

$$p = \frac{298 + \frac{1}{2}(489)}{1000} = 0.542$$

and that of allele *N* is

$$q = 1 - p = 0.458.$$

The Hardy-Weinberg expected frequencies (p^2, $2pq$, and q^2) and the expected genotypic numbers can then be calculated. The expected genotypic numbers are the product of the expected frequencies and the number in the sample. For example, the expected number of *M* individuals is

$$p^2N = (0.542)^2 1000 = 294.3.$$

In this case, the observed numbers and the expected numbers are extremely close to each other (see Table 2.4). Later we will discuss a technique to test statistically whether a population is in Hardy-Weinberg proportions.

Table 2.4
The observed numbers and numbers expected from Hardy-Weinberg proportions for the *MN* blood group locus in an English population (from Cleghorn, 1960).

Phenotype	*Genotype*	*Observed number*	*Expected number*
M	*MM*	298	294.3
MN	*MN*	489	496.4
N	*NN*	213	209.3
Total		1,000	1,000.0

---

then $q = 0.01$ and $H = 2(0.01)(0.99) = 0.0198$. In other words, nearly 2 percent of such a population are carriers for the albino allele (see Example 2.2 for another application).

III. MULTIPLE ALLELES

As for the two-allele case, the genotypic and allelic frequencies can be estimated from a population for a locus with multiple alleles. This extension is particularly important because many loci, particularly loci identified by electrophoresis in recent years, have been found to be segregating for more than two alleles. As a direct extension of the two-allele model, the frequency of an allele when there are multiple alleles is equal to the sum of the frequency of its homozygote plus half the frequency of each heterozygote that contains the allele. For example, assume that a sample of N individuals is categorized for the genotypes from three alleles (see Table 2.6). There are six possible

---

Example 2.2. In the leopard frog, *Rana pipiens*, there exists a variant color form in some natural populations. The normal phenotype (pipiens) is spotted, while the variant phenotype (burnsi) is unspotted (see Figure 2.3). The inheritance of these forms is under the control of two alleles, with the burnsi form dominant over the wild-type pipiens. Merrell and Rodell (1968) collected nearly 2,000 frogs (both living and dead) from

Figure 2.3. The wild-type pipiens (left) and burnsi (right) morphs of the leopard frog, *Rana pipiens* (from Merrell, 1965).

genotypes, three homozygotes (A_1A_1, A_2A_2, and A_3A_3), and three heterozygotes (A_1A_2, A_1A_3, and A_2A_3). The frequency of allele A_1 is calculated in the same manner as for two alleles except that A_1 alleles are in two different heterozygotes, A_1A_2 and A_1A_3. Therefore, the frequency of A_1 is

$$p_1 = \frac{N_{11} + \frac{1}{2}N_{12} + \frac{1}{2}N_{13}}{N}.$$

Likewise, the frequencies of alleles A_2 and A_3 are

$$p_2 = \frac{N_{22} + \frac{1}{2}N_{12} + \frac{1}{2}N_{23}}{N}$$

and

$$p_3 = \frac{N_{33} + \frac{1}{2}N_{13} + \frac{1}{2}N_{23}}{N}$$

Minnesota lakes in the early spring of several years. A composite of their data is given in Table 2.5, and the allelic frequencies among the living and dead frogs are estimated. The allelic frequency of the *pipiens* allele is quite high—$q = (834/887)^{1/2} = 0.970$ and $q = (1050/1089)^{1/2} = 0.982$ among the living and dead frogs, respectively. The similarity in allelic frequencies between living and dead frogs suggests that individuals with these morphs have similar overwintering survival.

The frequencies of heterozygotes for the two groups can be estimated if we assume Hardy-Weinberg proportions. Assuming that the frequency of the *burnsi* allele ($p = 1 - q$) is 0.028 and 0.018, the estimated proportion of heterozygotes in the living and dead frogs is 0.054 and 0.035, respectively.

Table 2.5
The observed numbers of burnsi and pipiens and their allelic frequency estimates in living and dead leopard frogs, *Rana pipiens*, from Minnesota (after Merrell and Rodell, 1968).

	Phenotype	*Genotype*	*Observed number*	*Allelic frequency estimate*
Living	burnsi	*BB*, *B*+	53	0.028
(N = 887)	pipiens	++	834	0.970
Dead	burnsi	*BB*, *B*+	39	0.018
(N = 1,089)	pipiens	++	1,050	0.982

Table 2.6
The six genotypes at a locus with three alleles, the numbers in a sample, and the Hardy-Weinberg proportions of the genotypes.

Genotype	A_1A_1	A_1A_2	A_1A_3	A_2A_2	A_2A_3	A_3A_3	Total
Number	N_{11}	N_{12}	N_{13}	N_{22}	N_{23}	N_{23}	N
Hardy-Weinberg	p_1^2	$2p_1p_2$	$2p_1p_3$	p_2^2	$2p_2p_3$	p_3^2	1

respectively. We can demonstrate how the Hardy-Weinberg principle applies to multiple alleles by assuming either random mating or random union of gametes. The Hardy-Weinberg proportions are given for three alleles in the bottom row of Table 2.6 (see Example 2.3 for an application).

IV. THE χ^2 TEST

In Example 2.1, the observed genotypic numbers are quite close to the expected genotypic numbers calculated from Hardy-Weinberg. In some cases, such as in plants that have self-fertilization (see Chapter 7), the observed and expected genotypic frequencies may differ greatly. Therefore, a statistical test is necessary that will allow us to decide if the fit is sufficiently close to know that the population is in the expected Hardy-Weinberg proportions. The common test employed for this purpose is the χ^2 (chi-square) test. To determine whether the observed numbers are consistent with Hardy-Weinberg predictions, a χ^2 value can be calculated as follows:

$$\chi^2 = \sum_{i=1}^{k} \frac{(O - E)^2}{E}$$

where O and E are the observed and expected *numbers* of a particular genotype and k is the number of genotypic classes. In other words, a chi-square value is the squared difference of the observed and expected numbers divided by the expected numbers and summed over all genotypic classes. From the calculated value of χ^2 and with the knowledge of another value, the degrees of freedom, the probability that the observed numbers would deviate from the expected numbers as much or more by chance can be obtained from a χ^2 table (see Table 2.8).

The *degrees of freedom* used to determine the significance of a χ^2 value are equal to the number of genotypic classes, minus 1, and then minus the number of allelic frequencies estimated from the data. The first degree of freedom is always lost, since not all the genotypic classes are independent—that is, the numbers in all classes must sum to the total sample size, N, and given the numbers in all classes but one, the number in the last class is known.

Using the Hardy-Weinberg expected frequencies for the two-allele case where all genotypes are distinguishable, the χ^2 expression becomes

$$\chi^2 = \frac{(N_{11} - p^2N)^2}{p^2N} + \frac{(N_{12} - 2pqN)^2}{2pqN} + \frac{(N_{22} - q^2N)^2}{q^2N}$$

where the three terms represent the squared, standardized deviations from the

---

Example 2.3. Hebert (1974) examined the frequencies of electrophoretic types at several variable enzyme loci in *Daphnia magna*, a protozoan, from ponds near Cambridge, England. Some sample data from the *malate dehydrogenase* (*Mdh*) locus are given in Table 2.7, where *S*, *M*, and *F* are the three different alleles observed. From the numbers of different electrophoretic types and using the expression above (assuming alleles *S*, *M*, and *F* have allelic frequencies p_1, p_2, and p_3, respectively), the estimates of allelic frequencies are

$$p_1 = \frac{3 + \frac{1}{2}(8) + \frac{1}{2}(19)}{114} = 0.145$$

$$p_2 = \frac{15 + \frac{1}{2}(8) + \frac{1}{2}(37)}{114} = 0.329$$

$$p_3 = \frac{32 + \frac{1}{2}(19) + \frac{1}{2}(37)}{114} = 0.526.$$

From these estimates, the expected numbers for the six different genotypes can be calculated using the Hardy-Weinberg principle. For example, the expected number of *SM* individuals is

$$2p_1p_2N = 2(0.145)(0.329)114 = 10.9.$$

Overall, the observed genotypic numbers are very close to those expected, assuming Hardy-Weinberg proportions.

Table 2.7
The observed and expected numbers from Hardy-Weinberg proportions for the *Mdh* locus in a *Daphnia* population (from Hebert, 1974).

Phenotype	*Genotype*	*Observed number*	*Expected number*
S	*SS*	3	2.4
SM	*SM*	8	10.9
SF	*SF*	19	17.5
M	*MM*	15	12.3
MF	*MF*	37	39.5
F	*FF*	32	31.5
Total		114	114.1

Table 2.8
The critical values of the χ^2 above which the value is said to be statistically significant or highly significant. In other words, the probability of a χ^2 value this large occurring by chance is only 0.05 or 0.01, respectively.

Degrees of freedom	*Significant (0.05)*	*Highly significant (0.01)*
1	3.84	6.64
2	5.99	9.21
3	7.82	11.34
4	9.49	13.28
5	11.07	15.09

expected numbers for genotypes A_1A_1, A_1A_2, and A_2A_2, respectively. As before, N_{11} is the observed number of A_1A_1, and p^2N is the expected number of A_1A_1 (see Example 2.4).

V. MEASURES OF GENETIC VARIATION

In order to document the amount of genetic variation in a standarized way, we need measures that allow us to quantify genetic variation. Several approaches can be used to describe the genetic variation at a single locus or a number of similar loci. The most widespread measure of genetic variation in a population is the amount of *heterozygosity*. As suggested in Chapter 1, the observed frequencies of heterozygotes in a population may be used. In many cases, the observed heterozygosity and the expected heterozygosity assuming Hardy-Weinberg proportions are quite similar. Even in situations in which the genotypes are not in Hardy-Weinberg proportions (as with inbreeding; see Example 2.5 and Chapter 7, the Hardy-Weinberg heterozygosity is useful for comparisons with other species. When data are available for several loci in a population, an average heterozygosity over loci can be calculated. For example, for four loci this would be

$$\bar{H} = \tfrac{1}{4}(H_A + H_B + H_C + H_D)$$

where H_A is the heterozygosity at locus A.

Another approach mentioned earlier to quantify the amount of genetic variation is the proportion of loci that are polymorphic. A genetic *polymorphism* can be defined as the occurrence in the same population of two or more alleles at one locus, each with appreciable frequency. In a population survey what is considered an appreciable frequency must be defined. A practical approach to defining polymorphism is to decide arbitarily on a limit for the frequency of the most common allele—that is, polymorphic loci are those for which the frequency of the most common allele is less than 0.99 or less than 0.95. Both of these arbitrary cutoff points have been used, but 0.99 is used most frequently if the sample size is fairly large (approximately 100 individuals or more). To estimate the proportion of polymorphic loci (P) for a population where a number of loci have been examined, first the number of polymorphic loci must be counted, and then the proportion these represent of all the loci

---

Example 2.4. Let us use the data from previous examples to calculate χ^2 values. First, a χ^2 value can be calculated from the blood group data on *MN* data given in Table 2.4. For these data, the χ^2 value is

$$\frac{(298 - 294.3)^2}{294.3} + \frac{(489 - 496.4)^2}{496.4} + \frac{(213 - 209.3)^2}{209.3} = 0.22.$$

In this case, there is one degree of freedom because there are three genotypic classes, one degree of freedom is always lost, and another degree of freedom was lost owing to the estimation of the frequency of A_1 (A_2 does not need to be estimated because $p + q = 1$). The χ^2 value is much smaller than the critical value 3.84, making the numbers consistent with Hardy-Weinberg expectations.

To illustrate the problems encountered when there are low expected numbers in some classes, consider the data on the *Mdh* locus from *Daphnia* given in Table 2.7. The expected numbers in the *S* class are only 2.4 and, because of a statistical rule of thumb that the expected numbers in any class should be greater than 5, this class should be combined with another class. Let us arbitrarily combine the *S* and *SM* classes in this case so that the observed and expected numbers in this combined class are 11 and 13.3, respectively. The χ^2 value is then

$$\frac{(11 - 13.3)^2}{13.3} + \frac{(19 - 17.5)^2}{17.5} + \frac{(15 - 12.3)^2}{12.3} + \frac{(37 - 39.5)^2}{39.5} + \frac{(32 - 31.5)^2}{31.5} = 1.23$$

In determining the degrees of freedom, first note that there are only five genotypic classes, because *S* and *SM* have been combined. One degree of freedom is automatically lost and two more are lost because the frequency of the *S* and *M* alleles were estimated (*F* did not need to be separately estimated because $p_1 + p_2 + p_3 = 1$). Therefore, there are two degrees of freedom remaining, and the χ^2 value is much smaller than the critical value 5.99, making these data consistent with Hardy-Weinberg expectations.

A χ^2 test can not be calculated on the leopard frog example because there are no degrees of freedom. There were only two genotypic classes, and after estimating one allelic frequency, zero degrees of freedom are left.

---

examined must be calculated. In other words, the proportion of polymorphic loci is

$$P = \frac{x}{m}$$

where x is the number of polymorphic loci in a sample of m loci (see Example 2.5).

VI. GENETIC COUNSELING

The importance of genetic disease has increased substantially in the last several decades as the incidence of infectious disease has declined. For example, the major cause of infant mortality in many countries used to be various infectious diseases, such as smallpox and diptheria, but now genetic disease is one of the leading factors. In addition, it is estimated that up to

---

Example 2.5. The breeding systems of many plants result in genotypes that are not in Hardy-Weinberg proportions. For example, Table 2.9 shows the genotypic frequencies for two loci in *Oenothera*, the evening primrose (from Levy and Levin, 1975). The first locus (A) has three alleles in the population but only three of the six possible genotypes. The observed heterozygosity is 0.210 + 0.685 = 0.895. Using the estimated frequencies of the alleles, the expected Hardy-Weinberg heterozygosity is

$$\begin{aligned} 2p_1p_2 + 2p_1p_3 + 2p_2p_3 &= 2(0.105)(0.553) + 2(0.105)(0.342) \\ &\quad + 2(0.553)(0.342) \\ &= 0.566. \end{aligned}$$

In other words, the observed heterozygosity is much higher than the expected heterozygosity. This seems to occur because of a special breeding system that results in the production of many heterozygotes.

The second locus is monomorphic for the B_1 allele and as a result has a heterozygosity of 0.0. Therefore, the average observed heterozygosity for these loci is

$$\bar{H} = \tfrac{1}{2}(H_A + H_B) = \tfrac{1}{2}(0.895 + 0.0) = 0.498$$

and the average expected heterozygosity is

$$\bar{H} = \tfrac{1}{2}(0.566 + 0.0) = 0.283.$$

---

10 percent of newborns suffer from a genetic defect of some sort (Trimble and Doughty, 1974). Some of these diseases follow a single-gene pattern of inheritance, others are polygenic and have significant environmental components, and still others are the result of aberrant chromosomal numbers or types. Although particular single-gene diseases are generally rare, approximately 3,000 different anomalies are known (McKusick, 1983).

Insight from population genetics is useful in *genetic counseling*, whereby the risks for genetically defective infants are determined. Often after a child is born with a genetic disease the parents will seek advice concerning the probability of a second child having the same disease. Such genetic-counseling advice is offered at a number of medical centers, and the calculation of the probability of recurrence for single-gene traits is straightforward. Once the disease is properly diagnosed and its mode of inheritance known, then the *recurrence risk*, the probability that the condition will recur in a given family, can be calculated.

A large proportion of known genetic diseases are inherited as autosomal recessives. Many of these are inborn errors of metabolism—that is, enzymatic

The proportion of polymorphic loci is

$$P = \tfrac{1}{2} = 0.5$$

because only one (the A locus) of the two loci is polymorphic.

Table 2.9
The frequency of the genotypes at two loci in an *Oenothera* population and the calculated allelic frequencies and heterozygosities (from Levy and Levin, 1975).

Locus A		*Locus B*	
Genotype	*Frequency*	*Genotype*	*Frequency*
A_1A_1	0.0	B_1B_1	1.0
A_1A_2	0.210		
A_1A_3	0.0		
A_2A_2	0.105		
A_2A_3	0.685		
A_3A_3	0.0		
Allele	*Frequency*	*Allele*	*Frequency*
A_1	0.105	B_1	1.0
A_2	0.553		
A_3	0.342		

defects that result in a buildup of a particular chemical or the absence of an important biochemical product. Most of these diseases are quite rare, with their incidence generally less than 1 in 10,000 individuals. However, because there are a large number of such diseases, their cumulative incidence is substantial.

The expected probability of the occurrence of a genetic disease in a child is a function of the disease state (affected or unaffected) of the parents (or other relatives) and whether there already have been affected offspring. For example, Table 2.10 gives the risk of affected children from parents with different disease states when the frequency of heterozygotes, H, in the population is known. As stated above, the incidence of most recessive diseases (the frequency of homozygotes) is low, so that the frequency of the disease allele is also low, generally 0.01 or less. If it is assumed that approximate Hardy-Weinberg proportions exist in the population, then the proportion of heterozygotes is generally 0.02 or less. Therefore, whenever H is part of a risk in Table 2.10, the risk is generally less than 1 percent, lower than the general risk of genetic disease in newborns. Only in a few cases—as for sickle-cell anemia in Africans or populations of African ancestry, where H may be 0.10, or for cystic fibrosis in Caucasian populations, where H may be 0.05—is the proportion of heterozygotes for recessive disease alleles higher.

Obtaining the risk values in Table 2.10 is straightforward. For example, the probability of a first child of unaffected parents having a recessive disease is the probability that both are heterozygotes, H^2, times the probability of a recessive offspring from two heterozygous parents, $\frac{1}{4}$. A more complicated situation occurs when the parents are unaffected but one has an affected sib. To understand this situation, let us examine the pedigree for such a family (Figure 2.4). Here circles and squares symbolize females and males, respectively, and the oldest generation is at the top (see Chapter 7 for more discussion of pedigrees). The shaded individual is affected with the disease and the diamond shape indicates an individual of either sex. In this case, the probability that the parent with the affected sib, individual a, is heterozygous is

Table 2.10
The probability of affected offspring for an autosomal recessive, given information about the parents, with H the frequency of heterozygotes in the population.

Parents	*First child affected*	*Second child affected, given first child affected*
Both unaffected	$\frac{1}{4}H^2$	$\frac{1}{4}$
One affected	$\frac{1}{2}H$	$\frac{1}{2}$
Both unaffected, one has affected sib	$\frac{1}{6}H$	$\frac{1}{4}$
Both affected	1	1

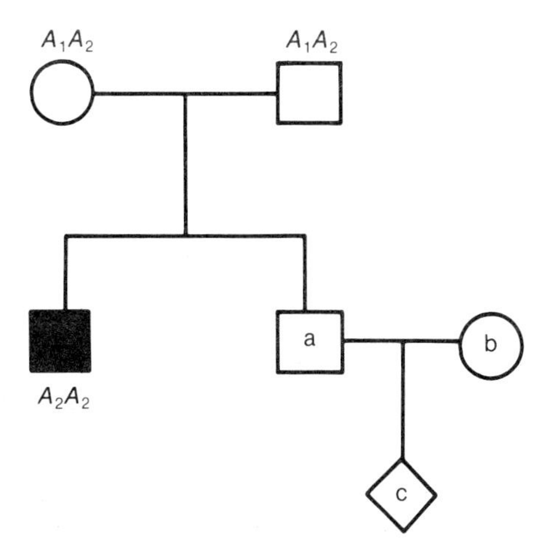

Figure 2.4. A pedigree in which the sib (shaded) of a potential parent, individual a, is affected with a recessive disease.

$\frac{2}{3}$, because this is the proportion of heterozygotes among unaffected sibs. The probability that the other parent, b, is heterozygous is H, and the segregation probability is $\frac{1}{4}$, making the overall probability that individual c will be a recessive offspring $(\frac{2}{3})(\frac{1}{4})H = H/6$.

Another situation in which genetic counseling can be useful occurs when parents strongly wish to have children and some likelihood exists that their children will have a genetic disease. For example, if it is known that both parents are carriers of a recessive disease allele (this may be known because heterozygotes can be detected for sickle-cell anemia, Tay-Sachs, and several other recessive diseases), then the probability of sibship of size N with no affected progeny is $(\frac{3}{4})^N$. Obviously, with a small family of one or two children, the chances for a family with no affected offspring is greater than 0.5 (0.75, and 0.5625, respectively), a risk that some people may find acceptable.

SUMMARY

Genotypic frequencies can be used to estimate allelic frequencies in a population. The Hardy-Weinberg principle allows the description of the genetic variation in terms of allelic frequencies. The chi-square test can be used to determine if the observed genotypic frequencies differ from the expected Hardy-Weinberg proportions. The amount of genetic variation in a population may be quantified using either the extent of heterozygosity or the proportion of polymorphic loci. An application of the use of genotypic frequencies to genetic counseling is discussed.

PROBLEMS

1. Calculate the frequency of alleles A_1 and A_2 for the following populations:

Population	N_{11}	N_{12}	N_{22}
a	20	40	40
b	30	8	12
c	72	122	91

   What are the Hardy-Weinberg frequencies and expected genotypic numbers for the three populations? Calculate the χ^2 values for these populations. Are these populations statistically different from Hardy-Weinberg proportions?

2. Assume that the frequencies of genotypes A_1A_1, A_1A_2, and A_2A_2 are 0.1, 0.4, and 0.5, respectively. Use these values to illustrate that random mating will produce Hardy-Weinberg proportions in one generation, using Table 2.3.

3. If $q = 0.3$ and there are Hardy-Weinberg proportions, what is the frequency of the most common genotype?

4. If A_2 is recessive and 4 out of 400 individuals are A_2A_2 (the rest are the dominant phenotype), what is the estimated frequency of A_2? What proportion of the population is expected to be heterozygotes?

5. Calculate the frequency of alleles A_1, A_2, and A_3 in the following populations:

Population	N_{11}	N_{12}	N_{13}	N_{22}	N_{23}	N_{33}
a	10	20	30	10	20	10
b	40	—	—	40	—	120

   What are the expected heterozygosities for these two populations? Compare them to the observed heterozygosities. Calculate the χ^2 values for these populations. Are these populations statistically different from Hardy-Weinberg proportions?

CHAPTER 3

Selection

The importance of the great principle of Selection mainly lies in the power of selecting scarcely appreciable differences . . . which can be accumulated until the result is made manifest to the eyes of every beholder.

CHARLES DARWIN

Selection is the most familiar factor affecting genetic variation. In Chapter 1, we mentioned that bird predation appears to be important in determining the frequency of coloration patterns of some snails and moths (see also Example 3.2). In addition, the recessive genetic diseases mentioned at the end of Chapter 2 result in strong selection against homozygotes. One of these diseases, sickle-cell anemia, is particularly interesting from a population genetics perspective owing to its high frequency today in populations of African ancestry. This unusually high incidence is the result of past selection in Africa favoring the sickle-cell heterozygote. Detailed testing has shown that sickle-cell heterozygotes have a higher survival in areas where malaria is common. However, because there is no advantage to the sickle-cell heterozygotes in temperate climates and there is a disadvantage to sickle-cell homozygotes who have anemia, the frequency of the sickle-cell allele can be expected to decline in the black populations in the United States and other countries without endemic malaria.

Selection can occur in several different ways, but here we will assume for simplicity that selection results from differential survival of genotypes. For example, 80 percent of genotype A_1A_1 may survive to adulthood while only 60 percent of A_2A_2 may survive to adulthood, making the survival rate of A_1A_1 higher than that of A_2A_2. Intuitively, one would then expect in a population composed of these two genotypes that the frequency of A_1A_1 would increase over time and that the frequency of A_2A_2 would decrease.

In this chapter, we will develop some models that allow us to predict how selection affects genetic variation. There are two basic consequences of selection on the amount of genetic variation. First, selection favoring a particular allele may lead to the reduction of genetic variation and, consequently, homozygosity for the favored allele, as for melanic alleles in moths in polluted areas. Second, selection may result in the maintenance of two or more alleles in a population, as for the banding polymorphism in snails or for sickle-cell anemia. To illustrate these consequences in a general way,

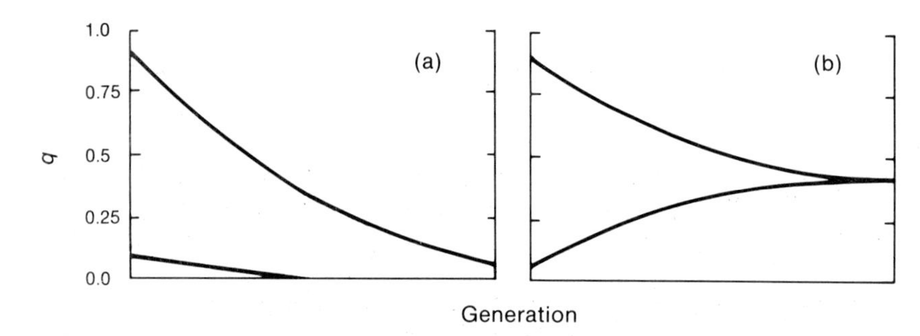

Figure 3.1. The frequency of A_2 over time when there is (a) selection favoring A_1 (against A_2) and (b) selection for maintenance of both alleles.

Figure 3.1 shows the expected frequency of A_2 over time when there is (a) selection against A_2 (favoring A_1) and (b) selection for the maintenance of both alleles, respectively. When there is selection against A_2, the frequency decreases from either an initial frequency of 0.9 or 0.1 so that the population becomes monomorphic for A_1 by the end of the time period. On the other hand, when selection favors the maintenance of both alleles, then from either an initial frequency of 0.9 or 0.1, the frequency of A_2 approaches the polymorphic state, which in this case is $p = 0.6$ and $q = 0.4$.

In order to understand the effect of selection on genetic variation, we must consider the relative fitnesses of the different genotypes. The *relative fitness* can be defined as the relative ability of different genotypes to pass on their alleles to future generations. For example, if, as above, the relative fitness of two genotypes is based solely on their viabilities, these values can be used to calculate relative fitness values. Generally, the highest relative fitness value is assumed to be 1.0, and the others are standardized against this value. Therefore, in the example above the relative fitness of genotype A_1A_1 is 1.0 and that of genotype A_2A_2 is $0.6/0.8 = 0.75$.

Owing to the complexities of natural selection, studies of selection generally make simplifying assumptions. This approach has been invaluable in establishing the feasibility and, hence, the potential significance of natural selection in evolution. Many of what are thought to be important biological complexities have since been incorporated into selection models to make them more realistic (for example, Chapter 15 discusses selection that is related to ecological factors). This will be our approach in examining selection—that is, to begin with simple models and examine the properties of these systems. In this discussion, we will concentrate on the dynamics of genetic change due to selection—that is, the rate of change in allelic frequencies—and the statics of genetic variation—that is, those situations in which there is no change or maintenance of genetic variation as the result of a selection mechanism.

Let us begin by considering the general situation in which the relative fitnesses of three possible genotypes A_1A_1, A_1A_2, and A_2A_2 are designated as

Table 3.1
The frequency of genotypes before and after selection, assuming Hardy-Weinberg proportions before selection.

Genotype	A_1A_1	A_1A_2	A_2A_2	Total
Relative fitness	w_{11}	w_{12}	w_{22}	—
Frequency before selection	p^2	$2pq$	q^2	1.0
Weighted contribution	p^2w_{11}	$2pqw_{12}$	q^2w_{22}	$\bar{w}$
Frequency after selection	$\frac{p^2w_{11}}{\bar{w}}$	$\frac{2pqw_{12}}{\bar{w}}$	$\frac{q^2w_{22}}{\bar{w}}$	1.0

w_{11}, w_{12}, and w_{22}, respectively. Because here we will consider only selection resulting from different viabilities of the genotypes, in this context the relative fitness of a genotype is its relative probability of surviving from a newly formed zygote until reproduction. We can give the genotypic frequencies before selection in terms of allelic frequencies, assuming Hardy-Weinberg proportions (Table 3.1). The weighted contribution of the three genotypes to the next generation is the product of the frequency before selection of that genotype and their relative fitnesses. The *mean fitness* of the population, $\bar{w}$, is the sum of the weighted contributions of the different genotypes, or

$$\bar{w} = p^2w_{11} + 2pqw_{12} + q^2w_{22}.$$

Because the relative fitness values are generally 1.0 or less, the mean fitness is also generally less than one.

If the relative contributions are standardized by the mean fitness value, then the frequencies of the genotypes after selection can be obtained as shown in the bottom line of Table 3.1. The frequency of A_2 after selection (q') is equal to half the frequency of the heterozygote, A_1A_2 (because half the genes in A_1A_2 are A_2 alleles), plus the frequency of the homozygote A_2A_2, or

$$\begin{aligned} q' &= \tfrac{1}{2}\left(\frac{2pqw_{12}}{\bar{w}}\right) + \frac{q^2w_{22}}{\bar{w}} \\ &= \frac{pqw_{12} + q^2w_{22}}{\bar{w}}. \end{aligned}$$

Therefore, the frequency of A_2 after selection is a function of the frequency before selection and the relative fitnesses of the genotypes.

The amount of allelic frequency change in one generation is defined as

$$\Delta q = q' - q.$$

By substituting q' from above, then

$$\Delta q = \frac{pqw_{12} + q^2w_{22}}{\bar{w}} - q$$

$$= \frac{pqw_{12} + q^2w_{22} - q\bar{w}}{\bar{w}}.$$

Substituting the value of $\bar{w}$ from above into the numerator of this expression (and remembering that $q = 1 - p$), then this expression can be rewritten as

$$\Delta q = \frac{pqw_{12} + q^2w_{22} - q(p^2w_{11} + 2pqw_{12} + q^2w_{22})}{\bar{w}}$$

$$= \frac{q(pw_{12} + qw_{22} - p^2w_{11} - 2pqw_{12} - q^2w_{22})}{\bar{w}}$$

$$= \frac{q(pqw_{22} - pqw_{12} - p^2w_{11} + p^2w_{12})}{\bar{w}}$$

$$= \frac{pq[q(w_{22} - w_{12}) - p(w_{11} - w_{12})]}{\bar{w}}.$$

We will modify this general expression later by substituting in various fitness values that represent different types of selection.

I. RECESSIVE LETHAL

Some genes that cause diseases in humans, such as Tay-Sachs and cystic fibrosis, seem to affect only the phenotype of the recessive homozygote. In individuals who are recessive homozygotes, such diseases generally result in prereproductive death. The relative fitnesses for a recessive lethal can be represented by the fitness values in row 1 of Table 3.2. The fitness of one homozygote, A_1A_1, and the heterozygote are the same, but the other homozygote, A_2A_2, is lethal and, thus, has a relative fitness of zero.

Table 3.2
The fitness values for the different fitness relationships examined.

	Genotype		
	A_1A_1	A_1A_2	A_2A_2
General fitnesses	w_{11}	w_{12}	w_{22}
(1) Recessive lethal	1	1	0
(2) Recessive allele	1	1	$1 - s$
(3) Heterozygote advantage	$1 - s_1$	1	$1 - s_2$
(4) Heterozygote disadvantage	$1 + s_1$	1	$1 + s_2$

The allelic frequency after selection for these fitness values can be determined by using the general equation above and substituting in these fitness values. Therefore, the allelic frequency after selection is

$$q' = \frac{pq(1) + q^2(0)}{\bar{w}}$$

$$= \frac{pq}{\bar{w}}.$$

Using the expression for the mean fitness and these fitness values, then

$$\bar{w} = p^2(1) + 2pq(1) + q^2(0)$$

$$= p^2 + 2pq$$

$$= p(1 + q).$$

The allelic frequency after selection then becomes

$$q' = \frac{pq}{p(1 + q)}$$

$$= \frac{q}{1 + q}.$$

The relationship given in the expression above is a recursive relationship—that is, the allelic frequency in generation $t + 1$ is a function of the frequency in generation t, or

$$q_{t+1} = \frac{q_t}{1 + q_t}$$

where q_t is the frequency of A_2 in generation t. Therefore, the allelic frequency in the first generation is a function of the frequency in the zero generation, and is

$$q_1 = \frac{q_0}{1 + q_0}.$$

Likewise, the frequency in the second generation is a function of that in the first generation, or

$$q_2 = \frac{q_1}{1 + q_1}.$$

Substituting the value of q_1 from above in this expression, the frequency of A_2 in the second generation becomes

Table 3.3
The number of generations (t) needed to reduce the allelic frequency from an initial value of q_0 to q_t for a recessive lethal.

q_0	q_t	t
	0.25	2
0.5	0.1	8
	0.01	98
	0.05	10
0.1	0.01	90
	0.001	990
	0.005	100
0.01	0.001	900
	0.0001	9900

$$q_2 = \frac{q_0}{1 + 2q_0}.$$

This relationship can be generalized to give the frequency in generation t as a function of the frequency in generation 0 as

$$q_t = \frac{q_0}{1 + tq_0}.$$

Furthermore, this expression can be solved for t as

$$t = \frac{1}{q_t} - \frac{1}{q_0}.$$

These last two expressions allow us to understand the rate of change of allelic frequency over several generations for recessive lethals. For example, we may calculate how many generations it will take for the allelic frequency to change by a specific amount. Table 3.3 gives the values calculated from this expression for several examples. When the frequency of the lethal is high, it is reduced very quickly—for example, from 0.5 to 0.1 in eight generations. However, when the lethal frequency initially is low, it is reduced slowly; thus, a change of frequency from 0.01 to 0.005, a halving of the frequency, takes 100 generations.

To examine experimentally the predictions of such a model, one needs an organism that has a number of experimental attributes. For example, the organism should be diploid, have two sexes, have a fast generation time, and produce large population numbers in a small space. As a result, many organisms, such as humans, most plants, and bacteria, are unsuitable for examining these selection models. Perhaps the most appropriate, and obviously the favorite organism, is the fruit fly, *Drosophila melanogaster*, found on rotting fruit in many kitchens. The experimental examples in this chapter will often utilize *D. melanogaster* as a "model" organism for examining the effects of selection (see Example 3.1).

The slow change for a lethal allele at low allelic frequencies occurs because most of the lethal alleles are "hidden" in heterozygotes and not subject to selection. To illustrate, we can calculate the proportion of A_2 alleles in the heterozygotes relative to that in homozygotes. When we remember that only half the alleles in the heterozygote are A_2, this ratio becomes

$$\frac{pq}{q^2} = \frac{p}{q}.$$

Example 3.1. The expressions above predict that when the frequency of a lethal is high, the allelic frequency is reduced very quickly. An experiment designed to compare this prediction to that in a laboratory experiment is given in Figure 3.2 for the lethal *Glued* allele in *D. melanogaster* (Clegg et al., 1976). Besides being lethal as a homozygote, the *Glued* mutant reduces eye size and affects eye appearance in heterozygotes. The four replicate populations given in Figure 3.2 were initiated with all heterozygotes, so that $q_0 = 0.5$, and then were followed for several generations. The observed decline in heterozygosity correlated quite well with the expected change in heterozygosity given by theoretical predictions.

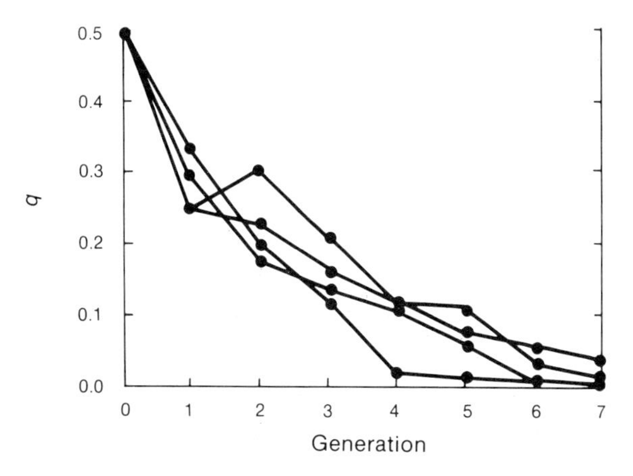

Figure 3.2. The allelic frequency of a lethal over generations. Four replicate populations are shown (after Clegg et al., 1976).

When the frequency of A_2 is low, this ratio becomes quite large. For example, if $q = 0.01$ ($p = 0.99$), the ratio is 99, meaning that for every A_2 allele in a homozygote and exposed to selection, there are 99 A_2 alleles masked by dominance in heterozygotes and not subject to selection.

This fact is particularly important in consideration of proposals to reduce the incidence of rare recessive diseases (the study and practice of such efforts are known as *eugenics*) by prohibiting or persuading individuals with such a disease not to reproduce. It is obvious from the discussion above that therapeutic abortions, sterilization, or other measures designed to eliminate recessives or substantially lower the fitness of recessives will be ineffective in reducing the frequency of detrimental recessives, since most of the alleles are in heterozygotes and not in individuals with the disease.

II. SELECTION AGAINST RECESSIVES

In many instances, selection against a homozygote is incomplete, and the relative fitness of the homozygote is thus only partially reduced compared to the other genotypes. For many human genetic diseases, such as albinism, the recessive homozygote can survive and produce progeny, although the probability of this occurring generally is greatly reduced compared to that of an unaffected individual's reproduction. In *Drosophila*, mice, corn, and other organisms that have been investigated genetically, there are many examples of recessive morphological mutants that somewhat reduce fitness.

To represent the relative fitnesses for the type of selection where a recessive has a reduced fitness, we can use the values in row 2 of Table 3.2. In other words, the relative fitnesses of A_1A_1 and A_1A_2 are the same, and the fitness of A_2A_2 is smaller by an amount s, which is the *selective disadvantage* (or *selection coefficient*) of the homozygote. The selective disadvantage has a maximum value of unity, in which case the allele is a recessive lethal, and a minimum value of zero where the three genotypes have the same relative fitness. If these fitness values are substituted in the general equation given above, then the frequency of A_2 after selection becomes

$$\begin{aligned} q' &= \frac{pq(1) + q^2(1 - s)}{p^2(1) + 2pq(1) + q^2(1 - s)} \\ &= \frac{q(p + q) - sq^2}{p^2 + 2pq + q^2 - sq^2} \\ &= \frac{q(1 - sq)}{1 - sq^2}. \end{aligned}$$

The change in allelic frequency becomes

$$\begin{aligned}\Delta q &= q' - q \\ &= \frac{q(1 - sq)}{1 - sq^2} - q \\ &= \frac{q(1 - sq) - q + sq^3}{1 - sq^2} \\ &= -\frac{spq^2}{1 - sq^2}.\end{aligned}$$

Notice that the change in allelic frequency is negative, indicating that selection is always reducing the frequency of A_2.

As an example, let us assume that the selective disadvantage of A_2A_2 is 0.2 and calculate the change in the frequency of A_2 over several generations. If the initial frequency is 0.6, then

$$\begin{aligned}q' &= \frac{0.6[1 - 0.2(0.6)]}{1 - (0.2)(0.6)^2} \\ &= 0.569.\end{aligned}$$

The same formula can be used to calculate the frequency in the next generation, assuming that $q = 0.569$ before selection, so that

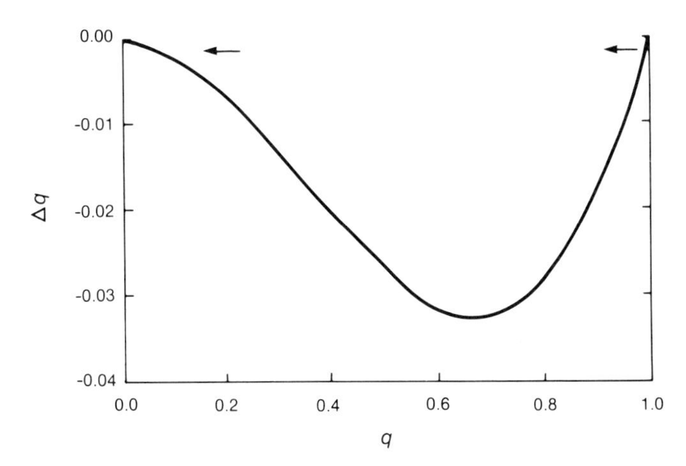

Figure 3.3 The change in the frequency of A_2 in one generation for different initial frequencies when there is selection against the recessive genotype A_2A_2.

$$q' = \frac{0.569[1 - 0.2(0.569)]}{1 - (0.2)(0.569)^2}$$
$$= 0.539.$$

This process can be repeated over subsequent generations to calculate the allelic frequency. In this case after 20 generations, the frequency of A_2 has been reduced from 0.6 to approximately 0.2.

The change in allelic frequency is a function of the allelic frequency and the selective disadvantage. Figure 3.3 illustrates how Δq change for different initial frequencies when the selective disadvantage is 0.2. The change in frequency is 0.0 when either $p = 0$ or $q = 0$ because the population is monomorphic for either the A_2 or A_1 alleles. The change in frequency is greatest for an intermediate allelic frequency and becomes quite low when q approaches 0.0 because most of the A_2 alleles are in heterozygotes and not subject to selection, just as for a recessive lethal.

III. PURIFYING VERSUS PROGRESSIVE SELECTION

We have imagined that selection was against homozygotes containing allele A_2, thereby reducing its frequency, as with many human diseases. This phenomenon can be thought of as *purifying selection*, selection that reduces the frequency of detrimental alleles in a population. Such alleles are usually already uncommon, and purifying selection further reduces their frequency.

One can, however, view directional genetic change from a somewhat different perspective and imagine that a population is continually adapting to its environment and that new, better adapted alleles are continually being introduced at low frequencies and increasing to high frequencies, thereby

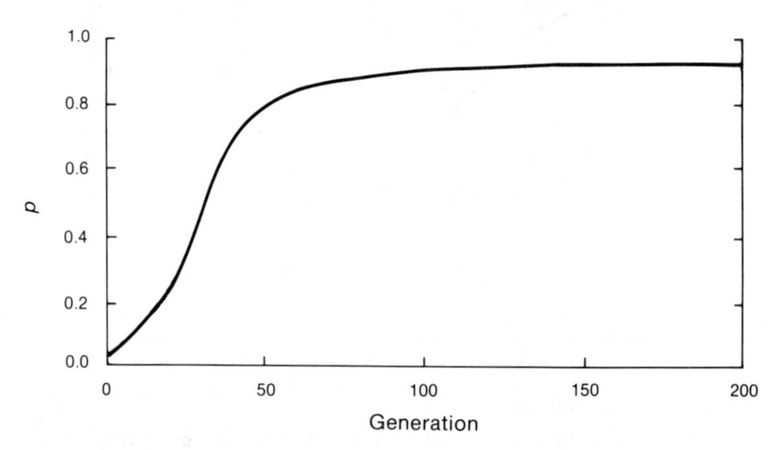

Figure 3.4. The frequency of A_1 over 200 generations when $s = 0.1$ and the frequency of A_1 is initially 0.01.

replacing old alleles. This process, called *progressive selection*, is selection for alleles that are advantageous in a given environment, as with melanic forms in peppered moths. The dynamics of genetic change are the same for purifying and progressive selection, but the general emphasis is different. In focusing on progressive selection one considers the increase of an allele, say A_1, from a low initial frequency, assuming it was introduced by mutation or migration, to a high frequency.

As an illustration of progressive selection, let us examine the allelic frequency over time for an allele with a low initial frequency. An example is given in Figure 3.4, where the initial frequency of A_1 is 0.01 and $s = 0.1$. The initial rise in allelic frequency given is quite fast, and the frequency is greater than 0.75 by generation 50 and then slowly approaches 1.0. The slow change in the latter generations occurs because most of the A_2 alleles are in heterozygotes (see Example 3.2).

IV. HETEROZYGOTE ADVANTAGE

All of the previous selection models lead to the eventual fixation of one allele and, as a result, reduce the genetic variation in the population. On the other hand, when the heterozygote has a higher fitness than both homozygotes, then two alleles can be maintained in the population. The classic example for this type of selection is sickle-cell anemia, as mentioned earlier; another example is resistance to the rodenticide warfarin in rats, discussed in Chapter 21.

To investigate the heterozygous advantage model, we will use the fitness array in row 3 of Table 3.2, where the heterozygote, A_1A_2, has the maximum relative fitness, 1, and the fitnesses of the homozygotes are $1 - s_1$ and $1 - s_2$, where s_1 and s_2 are the selective disadvantages of the homozygotes A_1A_1 and A_2A_2, respectively. Putting these fitness values in the general expression and simplifying, the change in allelic frequency due to selection is

$$\Delta q = \frac{pq(s_1 p - s_2 q)}{\bar{w}}.$$

In order to maintain both alleles in the population, Δq must be zero for some value of q between 0 and 1. The point where this occurs is called the *equilibrium* frequency of allele A_2—that is, the allelic frequency for which the change in frequency is zero. In this expression there are what are termed trivial equilibria, which occur when Δq is zero because either $p = 0$ or $q = 0$. No change occurs in allelic frequency in these cases, of course, because only one allele is present in the population.

There is another value of q for which this expression is equal to zero, and it occurs when the term in parentheses is zero or

$$s_1 p - s_2 q = 0.$$

By solving this equation for q, one can show that an equilibrium occurs when

$$q_e = \frac{s_1}{s_1 + s_2}$$

Example 3.2. Melanism in the peppered moth, *Biston betularia*, is one of the most widely studied examples of a genetic polymorphism. The major selective agent is thought to be differential predation by birds, which favors melanics in polluted areas and the nonmelanic or typical form in unpolluted areas. This hypothesis has been tested in an experiment by Clarke and Sheppard (1966), in which dead moths were exposed to predation either on dark or pale backgrounds (Table 3.4). On the dark background, melanics have a higher relative survival, while on the pale background the typicals have a higher relative survival. The results of the experiment are entirely consistent with the view that protective camouflage is important in the survival of the two forms of the moth and has led to an increase in the frequency of melanics in areas where much of the background is darkened because of pollution.

Let us examine the viabilities in the experiment in which the moths were placed on dark backgrounds. The survival of melanics is $58/70 = 0.83$ and that of typicals is $39/70 = 0.56$. If these are standardized into survival relative to that of melanics (because they have the highest survival), then the relative survival of melanics is $0.83/0.83 = 1.0$ and that of the typicals is $0.56/0.83 = 0.67$. Because the melanic is dominant, the relative fitnesses become 1.0, 1.0, and 0.67 for A_1A_1, A_1A_2, and A_2A_2, respectively, where A_1 and A_2 are the melanic and typical alleles. Assuming that the relative fitness of A_2A_2 individuals is $w_{22} = 1 - s$, then the selective disadvantage of A_2A_2 individuals is

$$\begin{aligned} s &= 1 - w_{22} \\ &= 1 - 0.67 = 0.33. \end{aligned}$$

Table 3.4
The "survival" of dead melanic or typical moths placed on either dark or pale backgrounds (from Clarke and Sheppard, 1966).

	Melanic		*Typical*	
	Exposed	*Survived*	*Exposed*	*Survived*
Dark background	70	58	70	39
Pale background	40	24	40	32

---

where q_e signifies the equilibrium frequency of allele A_2. In other words, for this allelic frequency there is no change in allelic frequency and both alleles are maintained in the population. Notice that the equilibrium is a function only of the selection coefficients for the two homozygotes.

We can use this example to illustrate numerically the expression we have used for allelic frequency change (Table 3.5). Assume that $q = 0.8$ ($p = 0.2$) and that we would like to know how much selection under the conditions of a dark background would change allelic frequencies. Table 3.5 shows step by step how the change in genotypic frequencies can be calculated. The allelic frequency after selection is

$$q' = \tfrac{1}{2}(0.406) + 0.544 = 0.747$$

and

$$\Delta q = 0.747 - 0.8 = -0.053.$$

Using the expression for allelic frequency change above

$$\Delta q = -\frac{0.33(0.2)(0.8)^2}{1 - (0.33)(0.8)^2}$$
$$= -0.054$$

the same as the step-by-step answer (except for rounding error). The same process can be repeated for any number of generations, and the pattern of increase is that of progressive selection as shown in Figure 3.4.

Table 3.5
The frequency of genotypes before and after selection, assuming $s = 0.33$ and $q = 0.3$.

Genotype	A_1A_1	A_1A_2	A_2A_2	Total
Phenotype	melanic	melanic	typical	
Relative fitness	1	1	0.67	
Frequency before selection	0.04	0.32	0.64	1.0
Weighted contribution	0.04	0.32	0.429	0.789
Frequency after selection	0.051	0.406	0.544	1.0

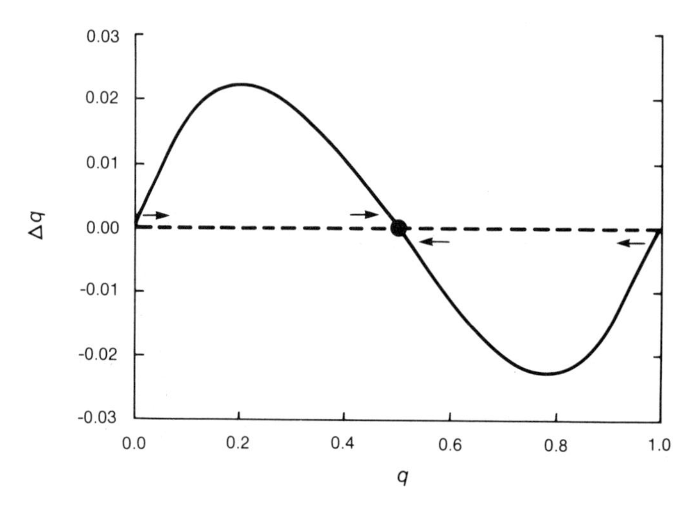

Figure 3.5. The change in the frequency of A_2 in one generation for different initial frequencies when there is a heterozygote advantage.

The properties of the equilibrium can be seen in another way by rewriting Δq using the equilibrium frequency so that

$$\Delta q = -\frac{pq(s_1 + s_2)(q - q_e)}{\bar{w}}.$$

From this formulation one can see that when q is greater than q_e, then Δq is negative, so that the allelic frequency will decrease towards the equilibrium. Likewise, when q is less than q_e, Δq is positive and the allelic frequency will increase towards the equilibrium. Owing to these properties, the equilibrium is called a *stable equilibrium*, because after a perturbation away from the equilibrium, the allelic frequency will change back toward the equilibrium frequency.

The change in allelic frequency is a function of the allelic frequency and the selection coefficients. An example is given in Figure 3.5, where the fitnesses of the three genotypes are 0.8, 1.0, and 0.8, so that $s_1 = s_2 = 0.2$. As a result of these fitness values, the equilibrium frequency is $q_e = 0.2/(0.2 + 0.2) = 0.5$. If q is smaller than 0.5, the change in allelic frequency is positive, with a maximum value just above $q = 0.2$. When value q is greater than 0.5, the change in allelic frequency is negative, with a maximum value just below 0.8 (see Example 3.3).

V. HETEROZYGOTE DISADVANTAGE

In some cases, the heterozygote may have a fitness smaller than that of either homozygote. Such a fitness array may occur in hybrids between species,

Example 3.3. A demonstration of the effect of heterozygote advantage was given by Prout (1971) using two mutants on the very small fourth chromosome in *Drosophila melanogaster*. (There is virtually no recombination among genes on this chromosome, so different types of chromosomes may be considered different alleles.) Using the recessive eye mutant, *eyeless* (*ey*) and the recessive bristle mutant, *shaven* (*sv*), Prout constructed individuals that were one of the recessive phenotypes when homozygous and the wild-type phenotype when heterozygous. Estimates of the fitness values of these genotypes indicated a strong heterozygous advantage (see Table 3.6). The predicted equilibrium frequency for one allele was 0.6 (broken line in Figure 3.6). In a laboratory population, Prout started two populations above the equilibrium and two below the equilibrium. As shown in Figure 3.6, after the perturbations above and below this value, the frequency of the allele quickly returned to near the predicted equilibrium value. This result supports the general validity of this model to maintain genetic variation.

Table 3.6
The genotypes and phenotypes and the relative fitness values in the experiment of Prout (1971).

Genotype	*ey+/ey+*	*ey+/+sv*	*+sv/+sv*
Phenotype	*eyeless*	*wild type*	*shaven*
Fitness	0.60	1.0	0.40

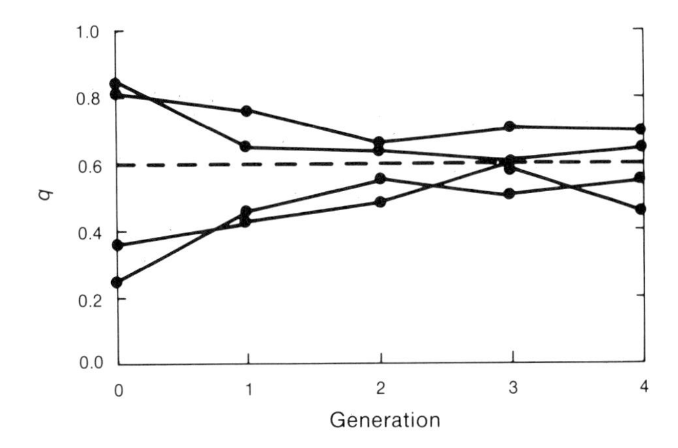

Figure 3.6. The observed change in the frequency of a mutant in *Drosophila melanogaster* in several replicates where there is a heterozygote advantage (after Prout, 1971).

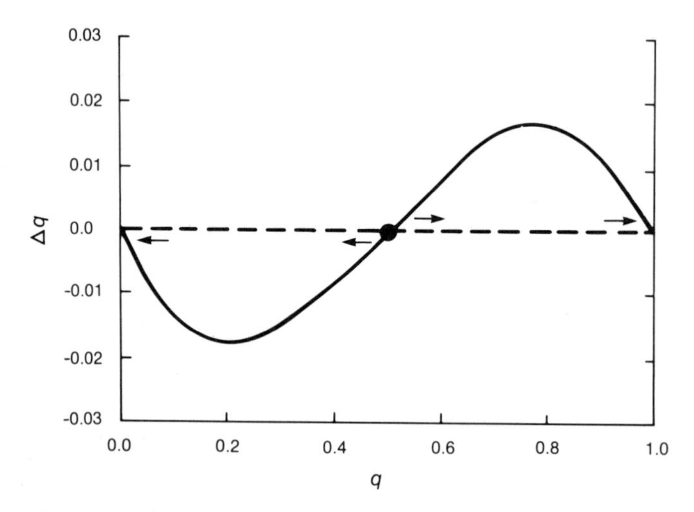

Figure 3.7. The change in allelic frequency in one generation when there is a heterozygote disadvantage.

although many genes are probably involved in such a case, and a single-gene model may not be always appropriate. The heterozygote disadvantage model may also be important in genetic control of pest populations, as we will briefly discuss in Chapter 21.

To examine the heterozygote disadvantage model, let us use the fitness array in row 4 of Table 3.2, where the heterozygote has a fitness of 1.0 and the homozygotes A_1A_1 and A_2A_2 have fitnesses of $1 + s_1$ and $1 + s_2$, respectively. With this array of fitness values, the change in allelic frequency is

$$\Delta q = \frac{pq(s_2 q - s_1 p)}{\bar{w}}.$$

This equation, like that for a heterozygous advantage, can be shown to have an equilibrium at

$$q_e = \frac{s_1}{s_1 + s_2}.$$

However, in this case Δq is positive above the equilibrium frequency and negative below it. As a result, this is an *unstable equilibrium* and, with a slight perturbation away from it, the allelic frequency continues to move away.

Again, the change in allelic frequency is a function of the allelic frequency and the selection coefficients. An example is given in Figure 3.7, where selection results in an unstable equilibrium at 0.5. When the frequency of A_2 is greater than 0.5, Δq is positive, and when it is smaller than 0.5, Δq is negative. Therefore, all populations with this fitness array will eventually become fixed for A_1 if the initial frequency is smaller than 0.5 or for A_2 if the initial frequency is greater than 0.5 (see also Example 3.4).

SUMMARY

Selection can lead to a change in allelic frequencies or maintenance at a given value. Equations are given that can predict allelic frequency changes for different fitness arrays, such as lethality, selection against recessives, and selection for or against heterozygotes.

Example 3.4. Foster et al. (1972) examined the change in allelic frequency when there is heterozygous disadvantage. Their example is the basis of a proposed insect control program using chromosomal variants and utilizing *D. melanogaster* as a model organism (see a discussion in Chapter 21). Because individuals heterozygous for chromosomal variants often produce unbalanced gametes and thereby have lower fertility, their fitness is much smaller than for homozygotes. Figure 3.8 gives an example where the unstable equilibrium is at 0.5. In a matter of a few generations, all the replicates were fixed either for one chromosomal type or the other. As expected, the chromosomal type for which individual replicates became fixed depended upon whether the initial frequency (actually the frequency in generation 1) was above or below the unstable equilibrium.

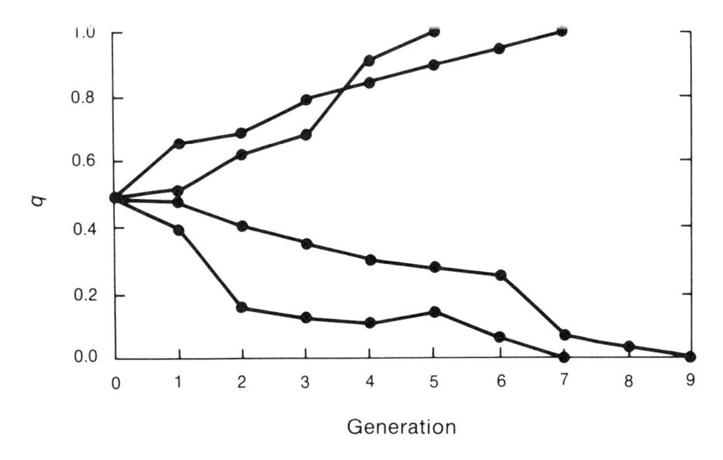

Figure 3.8. The observed change in the frequency of a chromosomal variant in *Drosophila melanogaster* in several replicates where there is a heterozygote disadvantage (after Foster et al., 1972).

PROBLEMS

1. Assuming that the initial frequency of a lethal is 0.02, how many generations does it take to reduce the frequency to 0.01? To 0.005?
2. For a recessive with $s = 0.4$, what is Δq when $q = 0.0$, 0.25, 0.5, 0.75, and 1.0? Graph these results on a Δq curve. For a recessive with $s = 0.1$, what is the allelic frequency after one and two generations when the initial frequency is 0.2?
3. Calculate the relative fitness values from the "survival" of melanics and typicals on pale backgrounds (Table 3.4). What is the frequency of the melanic allele after one generation, given that the initial frequency is 0.6 (use the general expression for Δq)?
4. If the relative fitnesses of genotypes A_1A_1, A_1A_2, and A_2A_2 are 0.8, 1.0, and 0.2, respectively, what is the expected equilibrium frequency of A_2? Calculate Δq for one q value above the equilibrium and one q value below the equilibrium (not $p = 0$ or $q = 0$).
5. If the relative fitnesses of genotypes A_1A_1, A_1A_2, and A_2A_2 are 1.5, 1.0, and 2.5, respectively, what is the expected equilibrium frequency of A_2? Calculate Δq for one q value above the equilibrium and one q value below the equilibrium (not $p = 0$ or $q = 0$).

CHAPTER 4

Mutation

> I think it would be better to ... constantly maintain (what I certainly believe to be the fact) that variations of every kind are always occurring in every part of every species, and therefore that favorable variations are always ready when wanted.
>
> ALFRED WALLACE in a letter to CHARLES DARWIN

Mutation is a particularly important process in population genetics and evolution because it is the original source of genetic variation in a population. The process of mutation is multilevel and may involve only one DNA base, several bases, the major part of a chromosome, whole chromosomes, or sets of chromosomes. The immediate cause of a mutation may be, for example, a mistake in DNA replication, a physical breakage of the chromosome by a mutagenic agent, an insertion of a transposable element (a piece of DNA that can move from place to place in the genome), or a failure in disjunction of meiosis. The discussion here will center on *spontaneous mutations*, those that appear without apparent explanation.

It should be kept in mind that mutagens can greatly increase the mutation rate in a given population. Such mutagenic agents as ultraviolet light, background radiation, and chemical pollutants can have a significant effect on the mutation rate and in turn can influence the basic amount of genetic variation present in a population. Mutations that are known to be caused by a particular mutagenic agent are termed *induced mutations*. In many cases, the rate of induced mutations appears to increase linearly with the dose of the mutagen—thus, for example, a doubling of the mutagenic dose approximately doubles the mutation rate.

Mutation at a single gene may involve different amounts of change in the DNA and the subsequent protein sequence. For example, a mutation may be simply a change in a single base pair in the DNA molecule; such a change accounts for the difference between the sickle cell and the normal allele. On the other hand, a mutation can affect a long stretch of DNA, since an insertion or deletion of a base changes all the subsequent codons in a transcript and thereby may result in only a small part of a protein being synthesized. The insertion of a transposable element, recently found to be the basis of many mutants in *D. melanogaster*, can also completely disrupt protein synthesis.

Some single-base changes in the DNA sequence may not result in a change in an amino acid; these are referred to as silent mutations. For example, many of the triplets or codons that code for the same amino acid differ only in the third nucleotide position of the codon. When the first two nucleotides are UC, then the resultant amino acid is serine for all four possible nucleotides in the third position (UCA, UCC, UCG, and UCU). Some other redundancies occur in the genetic code—for example, the amino acids leucine and arginine are both coded for by six different codon sequences.

It is difficult to determine the fitness of newly produced mutants. In general, visible mutations induced by radiation or chemical mutagens cause a reduction in fitness, and spontaneous mutations that change the phenotype also usually result in a reduction in fitness. A hypothetical distribution of the fitness of new mutants is given in Figure 4.1. The distribution is basically bimodal, with some mutations being so detrimental that they result in lethality or near lethality (shaded region on left). Another group of mutations, presumably many single-base substitutions, result in only a small change in fitness (the shaded region on the right). This category can be thought of as neutral or nearly neutral mutations. A third, probably quite small but important category includes the advantageous mutations that have an increased relative fitness. The relative sizes of these categories are open to speculation, but Mukai et al. (1972) estimated that approximately 10 to 20 times as many mutations have a slight effect on viability (are nearly neutral) in *D. melanogaster* as result in lethality.

In order to evaluate the impact of mutation on genetic variation, let us assume that at each locus there are two principal types of alleles, wild-type and detrimental alleles. Furthermore, assume that mutation is reversible and can occur from the wild-type to the detrimental category or from the detrimental

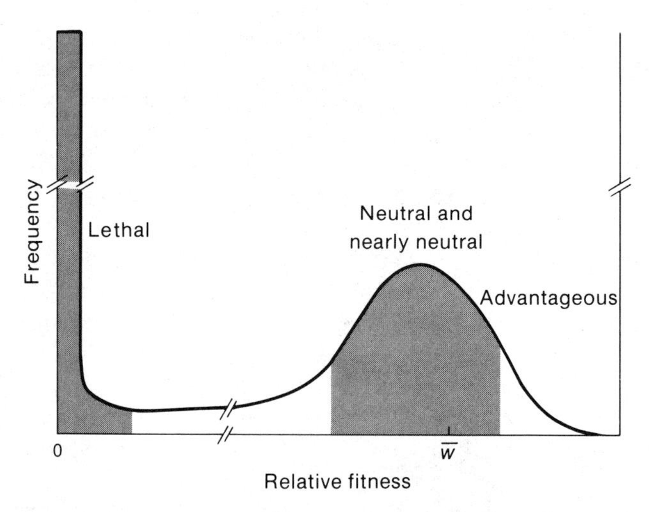

Figure 4.1. A hypothetical distribution of the effects on fitness of new mutants.

to the wild-type category. These types of mutations are conventionally called forward and backward mutations, respectively. Generally, it is assumed that forward mutations occur more frequently because they are mutations that result in a malfunction of the gene. Such a malfunction in a normal allele presumably can occur in a multitude of ways, while the repair of such a problem, a backward mutation, probably occurs much less frequently, since only a limited number of mutations can compensate for the original mutation other than a specific reversion (see Example 4.1). However, mutations caused by the insertion of a transposable element may easily revert by the removal of the element.

I. ALLELIC FREQUENCY CHANGE

In order to examine the effect of mutation on genetic variation in a population, let us assume that the rate of mutation from wild-type alleles (A_1) to detrimental alleles (A_2) is u, the forward mutation rate, and the rate of mutation from detrimental to wild-type alleles, the backward mutation rate, is v. Then the change in the frequency of A_2 due to mutation alone is

$$\Delta q = up - vq.$$

In other words, the proportion of alleles that can mutate from A_1 to A_2 is p with a rate u, and q is the proportion that can mutate from A_2 to A_1 with a rate v. If we rewrite this expression as

$$\Delta q - u - q(u + v)$$

it is obvious that the change in frequency is linearly related to the allelic frequency. Notice also that the maximum positive value for Δq is u when $q = 0$, and that the maximum negative value is v when $q = 1$. Because u and v are generally quite small (see the section on mutation-rate estimation below), the expected change due to mutation is also quite small. Figure 4.2 illustrates this relationship when the forward and backward mutation rates are the same, 10^{-5}, and when the backward mutation rate is 10^{-6} and the forward mutation rate is still 10^{-5}. For example, if $u = 10^{-5}$, $v = 10^{-6}$, and $q = 0.4$, then,

$$\begin{aligned}\Delta q &= (0.00001)(0.6) - (0.000001)(0.4) \\ &= 0.000056\end{aligned}$$

quite a small change in allelic frequency.

It is obvious from Figure 4.2 that there can be a stable equilibrium as a result of a balance between forward and backward mutation rates. The equilibrium frequency can be obtained by setting the above expression equal

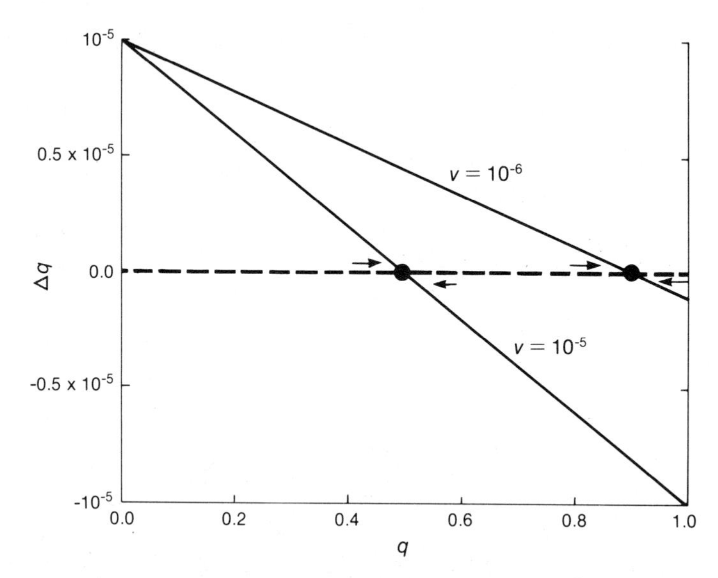

Figure 4.2. The change in allelic frequency in one generation as a function of q when the backward mutation rate is 10^{-5} or 10^{-6}.

to zero ($0 = up - vq$) and solving for q, so that the equilibrium frequency is

$$q_e = \frac{u}{u + v}.$$

Notice that this equilibrium is a function only of the mutation rates. If the forward mutation rate is greater than the backward mutation rate ($u > v$), then the frequency of detrimental alleles would be expected to be greater than that of wild-type alleles. For example, if $u = 10^{-5}$ and $v = 10^{-6}$, then

$$\begin{aligned} q_e &= \frac{0.00001}{0.00001 + 0.000001} \\ &= 0.909. \end{aligned}$$

This, of course, does not generally occur, because purifying selection reduces the frequency of detrimental alleles and counteracts the effect of forward mutation, a topic discussed in the next section.

How effective a force is mutation in changing the allelic frequency in a population over a number of generations? If we assume that v is small compared with u, so that $\Delta q \cong up$, then the frequency of A_1 after one generation with mutation is

$$\begin{aligned} p_1 &= p_0 + \Delta p \\ &= p_0 - up_0 \\ &= (1 - u)p_0 \end{aligned}$$

where p_0 is the initial frequency of A_1 and $\Delta p = -\Delta q$. This relationship may be generalized (using the same technique as we used for lethals in Chapter 3), so that the frequency in generation t is

$$p_t = (1 - u)^t p_0.$$

From this one can calculate, for example, that it takes nearly 70,000 generations to halve the frequency of the wild-type allele ($p_t = \frac{1}{2}p_0$) when $u = 10^{-5}$. In other words, mutation by itself is extremely slow in changing allelic frequencies, even when the forward mutation rate is fairly large and there is no backward mutation.

II. MUTATION–SELECTION BALANCE

Selection is the major force that keeps detrimental alleles from increasing in frequency. For example, individuals with cystic fibrosis usually do not survive to adulthood and very seldom reproduce, and *Drosophila* males with eye mutants are at a great disadvantage in mating. As a result of this purifying selection, the frequency of detrimental alleles is kept quite low. In the following discussion, which illustrates the joint effects of selection and mutation, we will assume that there are two alleles and that one of those, the mutant A_2, causes a reduction in viability.

If the detrimental allele is recessive, then the change in allelic frequency due to selection is

$$\Delta q_s = -\frac{spq^2}{1 - sq^2}$$

as given in Chapter 3. The increase in allelic frequency due to mutation is approximately

$$\Delta q_{mu} \cong up$$

if we assume the backward mutation rate v is small compared with the forward mutation rate u. Because the two forces have opposite effects on allelic frequency, they balance each other so that at some point

$$\Delta q_{mu} + \Delta q_s = 0$$

or

$$up - \frac{spq^2}{1 - sq^2} = 0.$$

If q^2 is small because the frequency of the detrimental allele is low, then the denominator of the expression giving the change due to selection can be assumed to be approximately unity, so that

$$up - spq^2 = 0.$$

The expression then can be solved for the equilibrium frequency of the recessive genotype as

$$q_e^2 = \frac{u}{s}$$

and the equilibrium allelic frequency as

$$q_e = \left(\frac{u}{s}\right)^{1/2}.$$

Of course, if the allele is a recessive lethal, then $s = 1.0$, and the equilibrium genotypic and allelic frequencies are the mutation rate, u, and $(u)^{1/2}$, respectively. The equilibrium allelic frequency is obviously increased with a higher mutation rate and decreased with a higher selective disadvantage.

The equilibrium can be understood by examining Δq_{mu} and Δq_s at low frequencies of A_2. Figure 4.3 gives these values when $u = 10^{-5}$ and $s = 0.2$. Notice that Δq_{mu} is about 10^{-5} throughout this low range in allelic frequencies, while Δq_s is 0.0 when $q = 0.0$ and becomes increasingly negative as q increases.

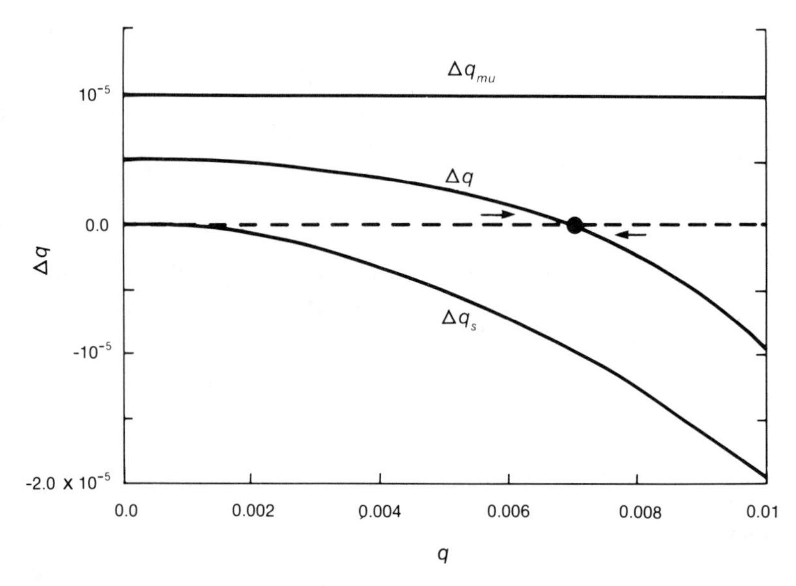

Figure 4.3. The change in allelic frequencies where both mutation to and selection against recessives occur. The line labeled Δq is the summation of the other two curves.

The equilibrium is

$$q_e = \left(\frac{10^{-5}}{0.2}\right)^{1/2} = 0.00707$$

and this is the point at which $\Delta q_{mu} + \Delta q_s = 0$. The two curves are summarized in the resultant Δq curve, and the solid circle indicates the equilibrium.

III. ESTIMATION OF MUTATION RATES

The estimation of mutation rates is a difficult problem, primarily because mutation is such a rare event. As a result, in order to obtain fairly accurate estimates of the mutation rate we must make an extremely large number of observations. Even when a new mutant is observed, it must be verified as a mutant rather than a contaminant from another source. Two basic approaches are generally used to estimate mutation rates: the direct approach, in which new mutants are counted, and the indirect approach, in which an equilibrium between mutation and selection is assumed.

Both the direct and indirect approaches have been widely used to estimate dominant mutation rates. The best approach is the direct method, in which the genotypes of both parents are known and neither of them carries a dominant mutant. Thus, where a dominant allele exists in an offspring, it must be the result of a new mutation. In this case, because each offspring receives two gametes that can potentially carry a mutant, the estimate of the mutation rate per gamete for this gene is

$$u = \frac{x}{2N}$$

where x is the number of mutant offspring and N is the total number of offspring examined. This procedure is used in Example 4.1 to estimate backward mutation rates for several loci simultaneously in mice and for a dominant human disorder.

The indirect approach has been used to estimate the mutation rate for recessive alleles that cause genetic diseases in humans. If a population is at a mutation–selection equilibrium, then the equation given above can be rearranged to give an estimation of the mutation rate in terms of the selective disadvantage and genotypic frequency as

$$u = sq_e^2.$$

Given that a population is at equilibrium and s and q_e^2 are known, then the mutation rate can be estimated (see Example 4.2).

Example 4.1. Because mutations at individual loci are generally rare events, large-scale experiments are necessary to estimate mutation rates adequately. One such massive study was undertaken to estimate both the forward and backward spontaneous mutation rates at five coat-color loci in mice (Schlager and Dickie, 1971). Over a period of six years, more than seven million mice at the Jackson Laboratory in Bar Harbor, Maine, were examined for spontaneous mutations at the *non-agouti*, *brown*, *albino*, *dilute*, and *leaden* loci. The estimated mutation rates for each locus and the overall values are given in Table 4.1. The overall forward mutation rate is 11.2×10^{-6} and the overall backward mutation rate was 2.5×10^{-6}, illustrating the general finding that the rate of forward mutations is higher than the rate of backward mutation.

An example of the use of the direct method to estimate the mutation rate for a dominant allele in humans was given by Stern (1973). Hospital records of 94,075 children in Lying-In Hospital in Copenhagen were examined for cases of achondroplasia, a dominant form of dwarfism. Ten of these newborns were achondroplastic dwarfs, but two had an affected parent; thus, only eight offspring appeared to be new mutants. Therefore, the estimate of the mutation rate was

$$u = \frac{8}{(2)(94{,}075)} = 4.2 \times 10^{-5}.$$

Table 4.1
Spontaneous mutation rates at five specific coat-color loci in mice (from Schlager and Dickie, 1971).

Locus	*Number of gametes tested*	*Number of mutations*	*Mutation rate per gamete* ($\times 10^{-6}$)
	Mutations from wild type (forward)		
non-agouti	67,395	3	44.5
brown	919,699	3	3.3
albino	150,391	5	33.2
dilute	839,447	10	11.9
leaden	243,444	4	16.4
Total	2,220,376	25	11.2
	Dominant mutations (backward)		
non-agouti	8,167,854	34	4.2
brown	3,092,806	0	0
albino	3,423,724	0	0
dilute	2,423,724	9	3.9
leaden	266,122	0	0
Total	17,236,978	43	2.5

IV. THE INCIDENCE OF GENETIC DISEASE

As discussed above, the general incidence of genetic disease is determined by the balance of the production of detrimental alleles by mutation and by selective elimination of these alleles. However, a number of factors can influence the incidence of genetic diseases in a variety of ways. The amount of inbreeding, population structure, level of medical treatment, changes in the mutation rate, and genetic counseling are examples of factors that can affect the incidence of genetic disease. In this section, we will briefly examine how changes in medical care or the mutation rate can affect the incidence of recessive genetic diseases.

The frequency of genetic disease may be increased either by improved medical care (relaxed selection) for individuals with genetic diseases, enabling

---

Example 4.2 Albinism is a recessive condition caused by a biochemical error that results in a lack of the pigment melanin. Individuals with albinism often have poor eyesight and are sensitive to sunlight, but their fitness is not greatly reduced relative to that in individuals with many other genetic diseases. One estimate (based on the relative viabilities) is that the fitness of albinos is about 0.9 that of unaffected individuals, making $s = 0.1$. The frequency of albinism is generally quite low, about 1 in 20,000 in many populations. Using the above expression, then

$$u = (0.1)(0.00005) = 5 \times 10^{-6}$$

a value not greatly different from mutation rates estimated using the direct approach.

However, the indirect approach may lead to some rather aberrant results, suggesting that the conditions necessary for its proper application may not be always present. For example, cystic fibrosis is a recessive disease that is generally fatal before reproduction because individuals with the disease easily contract respiratory infections. As a result, $s = 1.0$ except in the past few years during which better medical care has resulted in a greater survival rate. The incidence of cystic fibrosis is about 1 in 2,500 newborn Caucasians, making it the most common recessive disease in this population. Using the indirect approach then

$$u = (1.0)(0.0004) = 4 \times 10^{-4}.$$

This mutation-rate estimate is extremely high compared to other such estimates and suggests that factors besides those considered here are contributing to the frequency of cystic fibrosis. For example, some researchers have suggested that there may be several different genes that cause cystic fibrosis or that individuals heterozygous for the cystic fibrosis allele are at a selective advantage.

---

them to reproduce, or by an increase in mutation rate, resulting in the production of more mutant alleles. Both factors are described by the expression

$$q_e^2 = \frac{u}{s}$$

which notes that the equilibrium genotypic frequency is due to a balance between selection and mutation for a recessive disease. When there is a decrease in the selective disadvantage against the disease because of improved medical care, so that s' is the new selective disadvantage, then the new equilibrium becomes

$$(q_e^2)' = \frac{u}{s'}.$$

The ratio of these values is

$$\frac{(q_e^2)'}{q_e^2} = \frac{\frac{u}{s'}}{\frac{u}{s}} = \frac{s}{s'}.$$

In other words, the equilibrium is increased by the amount determined by the ratio of selective values. For example, if a disease that was formerly lethal before reproduction ($s = 1.0$) and now, with better medical care, is only slightly disadvantageous (say $s' = 0.1$), then its equilibrium frequency would be expected to increase tenfold.

Likewise, an increase in the mutation rate will also increase the equilibrium value. Assume that because of environmental pollutants, food additives, or some other factor there is an elevated mutation rate, u'. Then the new equilibrium is

$$(q_e^2)' = \frac{u'}{s}$$

and the ratio over the old equilibrium is

$$\frac{(q_e^2)'}{q_e^2} = \frac{\frac{u'}{s}}{\frac{u}{s}} = \frac{u'}{u}.$$

In other words, a tenfold increase in mutation rate will lead to a tenfold increase in the genotypic equilibrium value. The effect of an increase in mutation rate could be widespread because elevated mutation rates would probably occur simultaneously at a large number of loci.

However, the increase in genotypic frequency to the new equilibrium may take an extremely long time, particularly for a decrease in selection. For example, assume that initially $u = 10^{-6}$ and $s = 1.0$ and that either the amount of selection is reduced to $s' = 0.1$ or the mutation rate is increased to $u' = 10^{-5}$. When selection is decreased, it takes more than 1,000 generations to change halfway to the new equilibrium, while when the mutation rate is increased it takes more than 100 generations, still an extremely long time, assuming that a human generation is approximately 25 years.

SUMMARY

Mutation causes slow changes in allelic frequencies. It appears that many detrimental mutants are maintained at low frequencies by balance between mutation and selection. Estimates of mutation rates are low, about 10^{-5} per gamete for a given gene. The incidence of genetic disease may be slowly increased by either an increase in mutation rates or a decrease in the intensity of selection (through better medical care).

PROBLEMS

1. If $u = 10^{-6}$ and $v = 10^{-7}$, what is Δq for $q = 0.0, 0.25, 0.5, 0.75$, and 1.0? Plot these values on a Δq versus q curve. What is the equilibrium frequency of A_2 in this case?
2. If $u = 10^{-5}$ and $v = 0.0$, how much will the frequency of A_1 decline in ten generations?
3. If $u = 10^{-5}$, $v = 0.0$, and the selection coefficient against a recessive is 0.6, what is the expected equilibrium frequency of genotype A_2A_2 and allele A_2? If u is doubled, how much are the equilibrium genotypic and allelic frequencies changed?
4. If 3 dominant mutations in 50,000 offspring of normal individuals are observed, what is the estimated mutation rate?
5. Assume that a recessive disease is lethal and has a frequency of 0.0001. What is the estimated mutation rate?

CHAPTER 5

Gene Flow

> It is the continual movement of migrant individuals that genetically unites the many local populations of a species; these individuals are the links which cause all members of a species to share ultimately a common gene pool.
>
> BRUCE WALLACE

Previously, we have considered only a single population in which there was random mating throughout. In many species, however, populations are often subdivided into smaller units because of ecological or behavioral factors. For example, the populations of fish in tidepools, deciduous trees in farmlands, and insects on host plants are subdivided because the suitable habitat for the species is not continuous. Troops of primates and social-insect colonies are examples of population subdivision resulting from behavioral factors. Some organisms probably fit at any point along the spectrum of population structure between a random-mating population and a sharply subdivided population.

When a population is subdivided, differing amounts of genetic connectedness can exist among the parts of the population. This genetic connection depends primarily on the amount of *gene flow*—that is, genetically effective migration—among the subgroups. It is important to realize that some types of migration, such as the movement of birds that seasonally fly to the tropics, do not necessarily result in the exchange of genes among subpopulations. Clearly, movement from one area to another for food that involves no mating or reproduction is not genetically effective migration. In Chapters 9 and 12 we will consider the importance of movement from one area to another, *dispersal*, in determining the geographic distribution and population regulation, respectively, of an organism. Here in using the term gene flow (or migration), we assume that genetic exchange results from the movement. In this chapter we discuss several simple models of population structure in order to evaluate the potential effect of gene flow on genetic variation.

I. THE CONTINENT–ISLAND MODEL

The simplest model of migration is gene flow into a single population from an outside source. Such effectively unidirectional gene flow occurs, for example, in an island population that receives migrants from a continental source population. This description basically applies to species with populations on land islands and nearby large land masses, aquatic species in ponds with lakes as the sources of migration, and peripheral populations of any species that are constantly replenished by individuals from the main part of the species range. This model is used as well to describe admixture in some human populations—for instance, the American black population, discussed in Example 5.2.

To examine the effect of this type of population structure, let us assume that an island population receives migrants from a large source (continent) population, as shown in Figure 5.1. Although there may be reciprocal gene flow, we will assume that it has a negligible effect on the allelic frequency in the source population. Let the proportion of migrants moving into the island population each generation be m, so that the proportion of nonmigrants (residents) in the island population is $1 - m$. If the frequency of A_2 in the migrants is q_m and the frequency of A_2 on the island before migration is q, the allelic frequency after migration is

$$\begin{aligned} q' &= (1 - m)q + mq_m \\ &= q - m(q - q_m). \end{aligned}$$

The change in allelic frequency after one generation of gene flow is then

$$\begin{aligned} \Delta q &= q' - q \\ &= -m(q - q_m). \end{aligned}$$

From these expressions, it is obvious that there will be no change in allelic frequency if $m = 0$ or if $q = q_m$. Only the second alternative is of interest, because a value of $m = 0$ indicates there is no gene flow. Remember that q_m and m are assumed to be constant over time and have values between zero and unity. If q is smaller than q_m, the frequency of A_2 increases on the island, or Δq

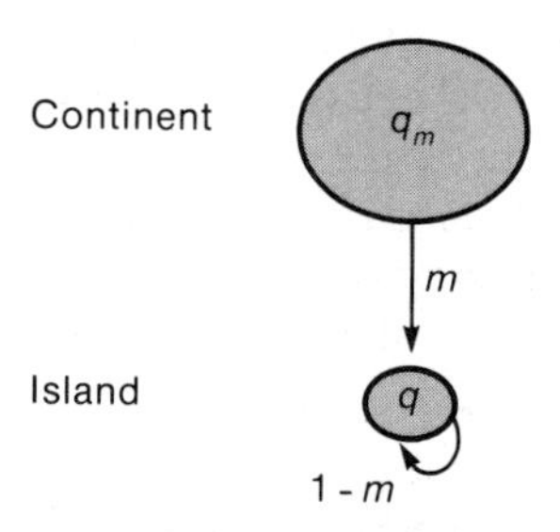

Figure 5.1. A model of gene flow from a continental population to an island population.

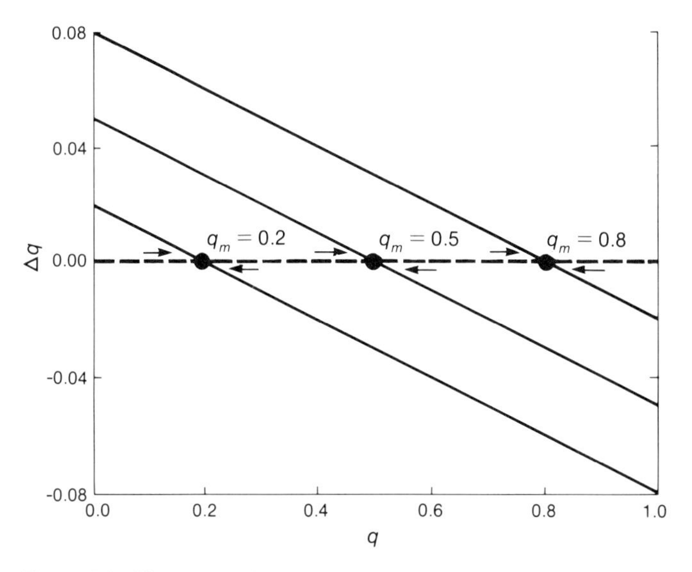

Figure 5.2. The expected change in allelic frequency in one generation due to gene flow when $m = 0.1$ and q_m is 0.2, 0.5, or 0.8.

is positive, and if q is greater than q_m, the frequency of A_2 will decrease, or Δq is negative. As a result, there is a stable equilibrium frequency of A_2 at $q_m = q$.

The effect of different allelic frequencies in the migrants on Δq when $m = 0.1$ can be seen in Figure 5.2. Note that the change in allelic frequency increases linearly as the frequency moves away from the equilibrium value and reaches an absolute maximum at either zero or unity, depending upon the value of q_m. The change in allelic frequency due to gene flow obviously can be important—potentially the magnitude of the change due to selection.

In order to examine the effect of gene flow on the frequency of A_2 over time, let us use subscripts in the first equation above to indicate generations, so that

$$q_0 = (1 - m)q_1 + mq_m$$

where q_0 is the initial frequency on the island. The general solution relating the frequency of A_2 in generation t to that in the initial generation is

$$q_t = (1 - m)^t q_0 + [1 - (1 - m)^t]q_m.$$

One can see from this expression that as t increases, the term $(1 - m)^t$ approaches zero because $1 - m$ is smaller than one. As a result, q_t approaches the equilibrium value q_m. The allelic frequency approaches the equilibrium asymptotically, as shown in Figure 5.3 for two values of q_0, 0.1 and 0.9, when $q_m = 0.4$ and $m = 0.1$. As we can expect from considering the values in the Δq curve, the allelic frequency initially changes quickly and then changes more slowly as the equilibrium is asymptotically approached. Because gene flow is

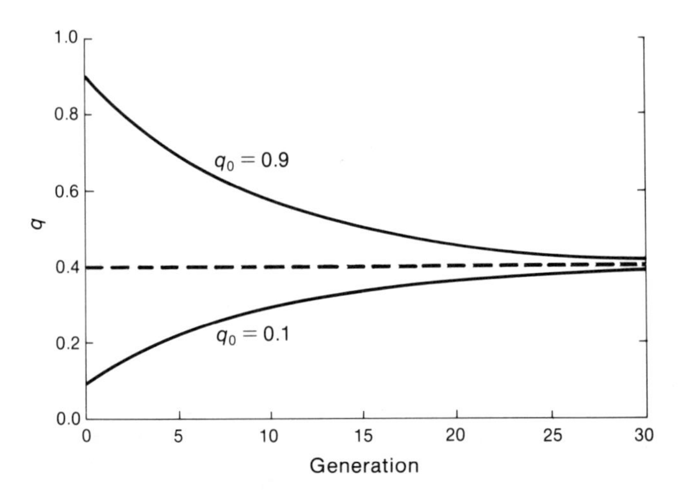

Figure 5.3. The change in allelic frequency over time when $q_m = 0.4$ and $m = 0.1$.

unidirectional in this model, the island population is eventually composed of individuals that have all descended from migrants. As a result, the allelic frequency on the island approaches that of the continent as more and more of the island population are descendants of migrants.

II. GENERAL MODEL

The continent–island model describes the allelic frequency change only in the island population and assumes that only gene flow to the island is important. A more general model assumes that gene flow can occur among all parts of a substructured population. As an example, let us briefly consider the situation in which there are three subpopulations in a population and that migration can occur among them in any direction (see Figure 5.4). The model becomes much more complex here, because nine migration rates (including the three values of nonmigration) and three allelic frequencies must be accounted for. For example, q_a is the frequency of A_2 in subpopulation a and m_{ab} is the proportion of migrants from subpopulation b to subpopulation a in each generation. Using these values, the allelic frequency in each subpopulation and in the total population can be calculated (see Example 5.1).

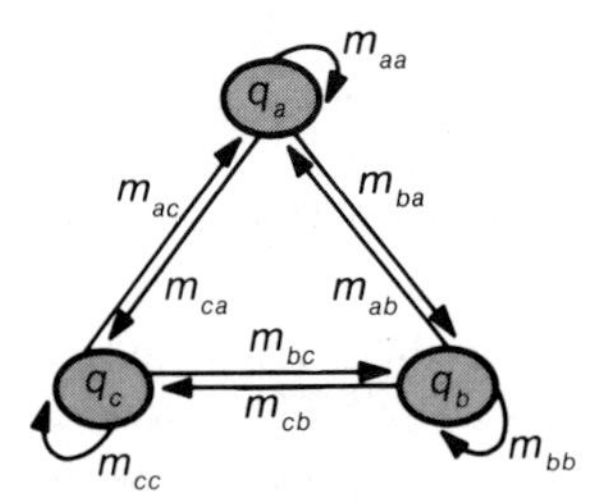

Figure 5.4. A model in which gene flow occurs among three subpopulations.

Example 5.1. Gene flow has been estimated in a number of human populations because of the availability of information on individual movement from generation to generation. One example of a population divided into three subpopulations comes from Sudan (Roberts and Hiorns, 1962). The proportion of migration among these subpopulations, as given in Table 5.1, shows that a high proportion of these individuals does not migrate; for example, $m_{aa} = 0.985$, or 98.5 percent of the Nuer individuals do not migrate to another group from one generation to the next. The estimated frequencies of the blood group allele *M* for the Nuer, Dinka, and Shilluk subpopulations are 0.575, 0.567, and 0.505, respectively. Using the migration rates in Table 5.1, we can predict the allelic frequencies for the three groups in future generations, as illustrated in Figure 5.5 (the equations are given in Hedrick, 1983). Notice that there is a very slow change in allelic frequency because the proportion of nonmigrants is large. In this case, the frequency of *M* is expected to approach an equilibrium value of 0.546 for the set of three subpopulations.

Table 5.1
The proportion of migration among the three parts of a human population in Sudan (after Roberts and Hiorns, 1962).

	Source population		
Recipient population	*a (Nuer)*	*b (Dinka)*	*c (Shilluk)*
a (Nuer)	0.9850	0.0125	0.0025
b (Dinka)	0.0138	0.9775	0.0087
c (Shilluk)	0.0000	0.0098	0.9902

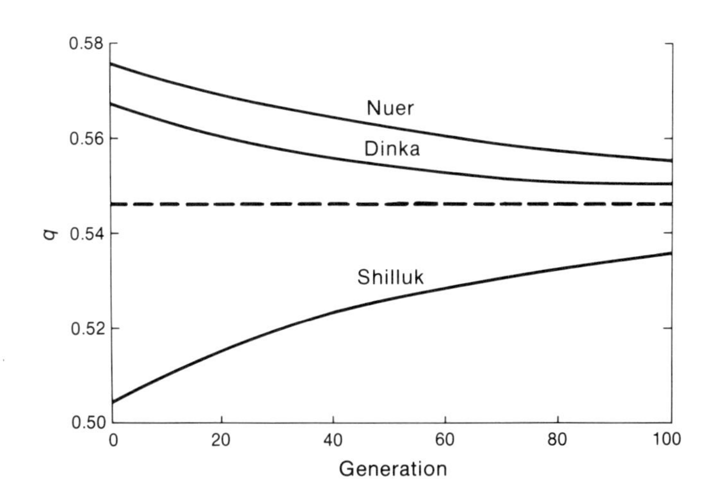

Figure 5.5. The expected change in allelic frequency over time for the *N* allele in three human populations from Sudan (from Roberts and Hiorns, 1962).

III. ESTIMATING ADMIXTURE

Many populations are composed of individuals that are descendants of different populations. If we assume that only two ancestral populations have

Example 5.2. The amount of admixture has been estimated in a number of human populations with ancestry from two populations. To estimate accurately the proportion of admixture, a locus is needed that has very different allelic frequencies in the ancestral populations and for which there is little selection. The American black population, a population that includes some white ancestry because of interracial matings, has been extensively examined. A locus that fits these conditions in these populations is the red blood cell locus *Duffy* because allele Fy^a at this gene is virtually absent in Africans and occurs with considerable frequency in American whites.

Table 5.2 gives the frequency for this allele in five different American black populations and two American white populations (Reed, 1969). Assuming that the frequency of Fy^a in Africans and whites is 0.0 and 0.429, respectively, the estimate of admixture in these populations from the expression below is given in the last column of Table 5.2. These estimates vary from 0.260 in Detroit to a low of 0.037 in Charleston, indicating that there appears to be a much lower proportion of white ancestry in these southern black populations. Remember that these estimates are based on only one locus and depend upon a number of assumptions, such as no past selection and accurate estimation of the allelic frequencies.

Table 5.2
The frequency of the *Duffy* allele Fy^a and estimates of white admixture in five American black populations (after Reed, 1969). The frequency of this allele in Africa (q_a) is nearly 0.0.

	Fy^a frequency		
Locality	*Blacks (q_h)*	*Caucasians (q_b)*	*M*
Nonsouthern			
New York City	0.081	—*	0.189
Detroit	0.111	—*	0.260
Oakland	0.094	0.429	0.220
Southern			
Charleston, S.C.	0.016	—*	0.037
Evans and Bullock Counties, Ga.	0.045	0.422	0.106

* The frequency was not estimated for these populations and a value of 0.429 was assumed.

contributed to the gene pool in a given population (for clarity we will call this population the *hybrid population*), then we can easily measure the amount of *admixture*, the proportion of the gene pool descended from one of the ancestral populations. We achieve this estimate by rearranging the first equation in the chapter,

$$q' = (1 - m)q + mq_m$$

and solving for m so that

$$m = \frac{q - q'}{q - q_m}.$$

However, the admixture that results between two ancestral populations generally occurs over an extended period of time, and the above expression is for one generation only. Assuming that the admixture has occurred over a number of generations, we can change the symbols in this expression to reflect this condition as

$$M = \frac{q_a - q_h}{q_a - q_b}$$

where M is the estimate of admixture of individuals from ancestral population b in the hybrid population h ($1 - M$ would be the proportion from ancestral population a). The frequencies of A_2 in populations a, b, and h are q_a, q_b, and q_h, respectively. Of course, this estimate of admixture assumes that other factors such as selection and genetic drift are negligible and result in little allelic frequency change relative to that caused by gene flow (see Example 5.2).

IV. DIRECT OBSERVATION OF GENE FLOW

The amount of gene flow can be measured directly in organisms where individuals can be identified. Because it is impossible for humans to identify individuals in most other species, direct observation of gene flow is generally possible only in humans when birth records or other information are available or in other organisms when individual identification marks are used. Many approaches have been employed to mark individuals differentially, such as toe clipping in rodents, banding in birds, or group marking with fluorescent dust in insects. Remember that movement of individuals from one area to another may not result in gene flow and that estimates of migration using these techniques should, for example, be contingent on successful breeding in the new area.

When the population is subdivided into discrete colonies or groups, then the probability of gene flow among groups may be estimated. In human populations, generally the birthplace of the parents and their offspring are used to establish such probabilities (as in Example 5.1). Because birth records in some populations are available for many generations, it is possible to evaluate the effect of migration patterns between villages or regions over long periods. In most other species, at best only short-term data are available, although the data for the marmot population discussed in Example 5.3 are the average for several generations.

Example 5.3. A population of the yellow-bellied marmot in Colorado has been observed over a period of 15 years (Schwartz and Armitage, 1980). Marmots in the study area are marked for individual recognition so that movement and other aspects of their behavior can be recorded each summer. In this population there were at most eight breeding colonies, each of which usually consisted of one adult male, two or three females, and juveniles of both sexes. The observed migration patterns among these colonies are given in Table 5.3 for females. The migrants that originated from outside the study area are classified as of unknown origin. On the average, 49 percent of the females did not migrate and potentially remained to breed in their natal colony. On the other hand, males in nearly every case migrated from their natal colony, with only 5 percent not migrating. As a result, even though the individual colonies have small population sizes, the continual migration of individuals, particularly males, among colonies and from other areas appears important in maintaining genetic variation in the population.

Table 5.3
Migration for females among eight marmot colonies (after Schwartz and Armitage, 1980).

Resident colony	Source colony								
	a	*b*	*c*	*d*	*e*	*f*	*g*	*h*	*Unknown*
a	0.62	0.02	0	0	0	0	0	0	0.36
b	0.07	0.57	0	0	0	0	0	0	0.36
c	0	0	0.60	0	0	0	0	0	0.40
d	0	0	0	0.79	0	0	0	0	0.21
e	0	0	0.05	0	0.05	0.11	0.15	0.11	0.53
f	0	0	0	0	0.04	0.15	0	0.07	0.74
g	0	0	0	0	0	0.13	0.37	0	0.50
h	0	0	0	0	0	0	0	0.74	0.26

V. GENE FLOW AND SELECTION

Gene flow and selection may frequently be important in the same population, making it necessary to consider their joint effect on allelic frequencies. Both migration and selection may be quite diverse in their effects on allelic frequencies, and the two factors together may result in even more complex potential outcomes. Here we will examine a simple combination of these two factors.

Let us again consider the continent–island model and investigate how selection affects the allelic frequency in the island population. Assume that the total change in allelic frequency on the island is the joint effect of gene flow and selection—that is,

$$\Delta q = \Delta q_s + \Delta q_{mi}$$

where Δq_s and Δq_{mi} are the changes in frequency due to selection and gene flow, respectively. If there is selection against a homozygote so that the relative fitnesses of A_1A_1, A_1A_2, and A_2A_2 are 1, 1, and $1 - s$, respectively, then

$$\Delta q = -spq^2 - m(q - q_m)$$

assuming that the denominator of the Δq_s expression is unity (as we did before when considering mutation and selection jointly).

In order to understand the joint effects of selection and migration, let us consider the particular situation in which $q_m = 1$—that is, where the source population is completely homozygous for allele A_2. Such a situation may result when a wingless variant or some other trait determined by the dominant A_1 allele is advantageous on the island but disadvantageous and, therefore, in very low frequency in the mainland population. Because selection on the island counterbalances the reduction in allelic frequency due to gene flow, there should be a polymorphic equilibrium. If we wish to calculate the equilibrium between these two factors, we can set the expression for Δq above equal to zero, let $q_m = 1$, and solve for q, or

$$0 = -spq^2 - m(q - 1)$$

$$spq^2 = mp$$

$$q_e = \left(\frac{m}{s}\right)^{1/2}.$$

In other words, the equilibrium frequency is a function of the ratio of migration rate and selective disadvantage, a result similar to that for a mutation–selection balance. However, m in general is much larger than u, so that the equilibrium for a gene flow–selection balance should not be as close to zero as that for a mutation–selection balance.

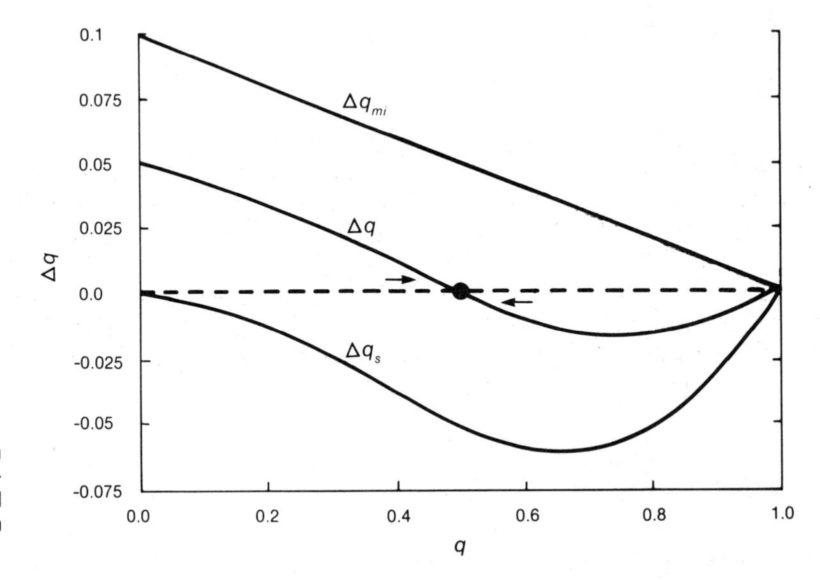

Figure 5.6. The change in allelic frequency in one generation due to gene flow and selection separately and then combined.

Example 5.4. The allelic frequencies of several different genetic disease alleles change over arrays of subpopulations. An example is the thalassemia allele over a linear series of towns in central Sardinia (solid line in Figure 5.7; from Livingstone, 1969). Notice that the frequency of the allele is highest on the coasts and lowest in the mountains at the center of the island. These frequencies correspond to the past presence of malaria along the coasts for which the thalassemia allele appears to confer some resistance.

A model combining selection and migration patterns was used to simulate this example. Gene flow was assumed to occur among a linear series of 40 populations mostly to adjacent populations (80 percent) and other nearby populations. In all subpopulations, the homozygotes with thalassemia, A_2A_2, were assumed to have a relative fitness of zero, and the other homozygotes, A_1A_1, a fitness of 1.0 (see Table 5.4). The heterozygote fitness varied in different populations, mimicking the advantage for the heterozygotes in malarial environments. The heterozygote

Table 5.4
The relative fitnesses used in a model to explain the frequency of the thalassemia allele in Sardinia.

Populations	A_1A_1	A_1A_2	A_2A_2
Coast populations with malaria (1–8, 33–40)	1.0	1.25	0.0
Mountain populations with no malaria (16–25)	1.0	1.0	0.0

As an example, let us assume that $m = 0.1$ and $s = 0.4$. Figure 5.6 gives the expected change for gene flow and selection separately and then sums them in the Δq curve. Notice that, at low frequencies of A_2, gene flow increasing the frequency of A_2 has the overriding effect and that, at high frequencies, selection decreasing the frequency of A_2 has the major effect. The equilibrium occurs at

$$q_e = \left(\frac{0.1}{0.4}\right)^{1/2} = 0.5$$

where the two forces balance each other.

VI. CLINES

The patterns of both gene flow and selection can be much more complex than suggested in the previous section. For example, there can be an array of

---

fitness was varied over the array from 1.25 in the coastal populations (1–8 and 33–40) and decreased in several increments to 1.0 in the mountain populations (16–25). The results of this model, calculated by computer simulation, are given as the broken line in Figure 5.7. Although these may not be the exact gene flow and selection patterns in this population, the simulated results are consistent with the distributions observed over the Sardinian population.

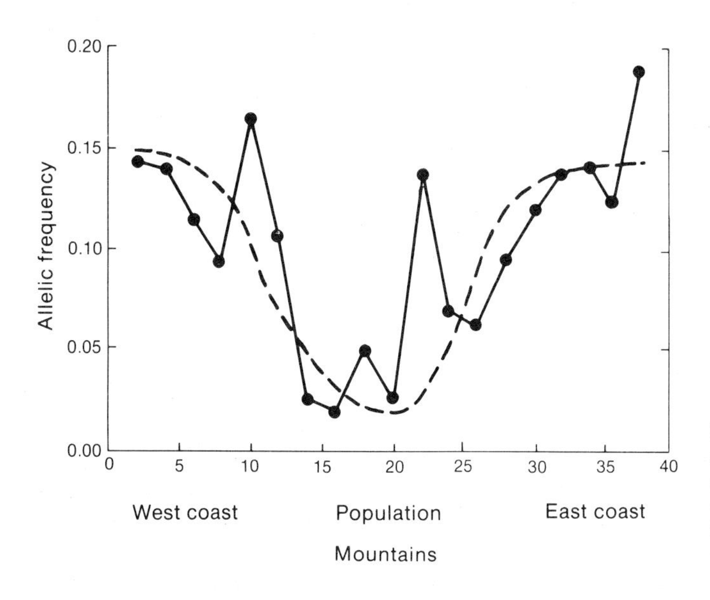

Figure 5.7. The observed frequency of the thalassemia allele across Sardinia (solid line) and a simulation (broken line) incorporating both gene flow and selection (after Livingstone, 1969).

subpopulations with gene flow between them, as in the general model of population structure discussed earlier. However, in this section we will only consider the situation in which the subpopulations are in a linear array, as might occur for species in tidepools along a shoreline or along a valley surrounded by sharply contrasting habitats.

Over such a linear array of subpopulations, selective factors may vary. As we will discuss in more detail in Chapter 15, selection may vary spatially owing to differences in the environment relating, for example, to temperature, soil, or the presence of other species such as parasites or competitors. These environmental factors may form a spatial gradient or may abruptly change from one area to another.

There are a number of examples of changes in allelic frequencies over a linear array of populations, a phenomenon called a *cline*. Such a trend is often thought to be caused by a change in the environmental selective pressures over the array and gene flow limited to adjacent or nearby subpopulations. We should note, however, that a cline can be unstable and could, for example, be the consequence of a historical difference in allelic frequencies caused by genetic drift in different subpopulations. Example 5.4 discusses the interaction of gene flow and selection, a combination that serves as a good explanation of a cline in the frequency of a disease allele.

SUMMARY

Two simple models of gene flow, the continent–island model and a general model of gene flow between subpopulations, are introduced. Gene flow can change the allelic frequency fairly quickly for both these models. The rate of gene flow can be estimated indirectly from admixture proportions or directly from observation of migration. The combination of gene flow and selection can lead to a polymorphism or, over a linear array of populations, may result in a cline of allelic frequencies.

PROBLEMS

1. Assume $m = 0.05$ and $q_m = 0.1$. What will the frequency be of A_2 after one and two generations if $q_0 = 0.6$? What will the frequency of A_2 approach after many generations?
2. If $q_a = 0.0$ and $q_b = 0.5$, what is M if $q_h = 0.1, 0.4$, or 0.49?
3. Assume that there is selection against a recessive on an island and $s = 0.2$. If $m = 0.1$ and $q_m = 1$, what is the equilibrium frequency of A_2? What is Δq if $q = 0.2$ and $q = 0.8$?
4. Compare the relative effects of selection, mutation, and gene flow on genetic variation.

5. Assume that a cline in the frequency of a color morph was observed. How would you design an experiment to evaluate gene flow and selection over this array? How would you test whether your estimates of these factors could adequately explain the cline?

CHAPTER 6

Genetic Drift

> Variation neither useful nor injurious would not be affected by natural selection, and would be left either a fluctuating element, as perhaps we see in certain polymorphic species or would ultimately become fixed, owing to the nature of the organism and the nature of the conditions.
>
> CHARLES DARWIN

Since the beginning of population genetics, there has been controversy concerning the importance of chance changes in allelic frequencies due to small population size. In part, this controversy stems from the fact that many natural populations are so large that chance effects appear to be unlikely in comparison with the effects of selection or migration. However, in certain circumstances in which population size is small, chance effects may be significant even for loci undergoing some selection.

First, some populations may be continuously small for relatively long periods of time owing to limited resources in the populated area, low movement between suitable habitats, or strong territoriality among individuals. For example, lizard numbers may be limited by perch sites and territoriality, bird populations by nesting sites and territoriality, and the number of colonizing plants by open habitat and movement between habitats. Isolated populations, whether they be land animals or plants on an island, vertebrates or invertebrates in a lake, or other groups living in a circumscribed area, may also be consistently small.

Second, some populations may have intermittent small population sizes. Examples are the overwintering loss of population numbers that occurs in many invertebrates, periodic crashes of populations in small rodents such as lemmings and voles, epidemics that periodically decimate populations of both plants and animals, and losses in many species stemming from the seasonal desiccation of ponds. These examples of population fluctuations generate population *bottlenecks*, periods during which only a few individuals survive to continue the existence of the population. A classic example of periodic oscillations is given in Figure 6.1, where estimates of the relative density are given (see Chapter 10 for measures of density) for the lynx and snowshoe hare. Both species show fairly regular changes, with the population density fluctuating by an order of magnitude or more. As a result, in periods of low density, individuals of both species often become exceedingly rare.

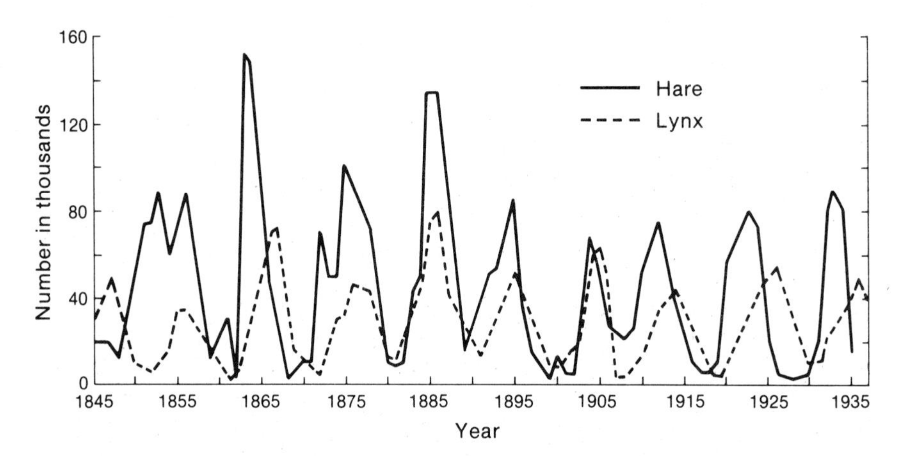

Figure 6.1. Oscillations in the density of the snowshoe hare and the lynx (after Odum, 1971).

When a population starts from a few individuals, the *founder effect*, which causes chance changes in allelic frequencies, may be important. For example, many island populations appear to have started from a very small number of individuals. If a population is founded by a single fertilized female, then the two alleles present in the female and the two alleles from the male that fertilized

---

Example 6.1. Some isolated religious groups have relatively high incidences of particular rare diseases. In fact, a number of diseases were originally described in populations such as the Amish in the United States. Often such groups were established by a small number of founders and keep their community closed to immigrants. As a result, it appears that genetic drift and subsequent inbreeding (see Chapter 7) are primary factors accounting for the abnormally high incidence of some genetic diseases in these populations.

The Amish population of Lancaster County, Pennsylvania, has a high incidence of a recessive disorder known as six-fingered dwarfism, or Ellis-van Creveld (EvC) syndrome (McKusick, 1978). In this population of about 13,000, 82 affected individuals in 40 affected sibships have been diagnosed. When pedigrees of the 80 parents in these 40 sibships were examined, it was found that all trace their ancestry to Samuel King and his wife, early members of this community. From this pedigree information, it appears quite certain that the high incidence EvC syndrome can be primarily attributed to the founder effect—that is, either Samuel King or his wife carried this allele and, because many individuals in the population are descended from them, the incidence of this disease is high.

---

her may constitute all the genetic variation at a locus in a new population. In plants, a whole population may be initiated from a single seed, containing only two alleles if self-fertilization occurs. Such initial restrictions in the number of founders also appear to be important in some human populations. For example, some religious isolates in North America such as the Hutterites, Amish, and others were initiated by small numbers of migrants from Europe (see Example 6.1).

Another situation in which small population size may be of great significance is in populations of many endangered and vanishing species. For example, only 20 whooping cranes were alive in 1920 because of hunting and destruction of their habitat. Their number has grown slightly due to protection and great care, and in 1977 there were 71 whooping cranes (see Example 14.9). Furthermore, some rare animals, such as certain lemurs, Siberian tigers, and a number of other species, exist primarily in zoos, resulting in small breeding populations.

I. THE EFFECT OF GENETIC DRIFT

All of the above examples of restricted population size can have the same general genetic consequence—that is, a small population size causes chance alterations in allelic frequencies. The chance effect that results from the sampling of gametes from generation to generation is called *genetic drift*. In a large population, on the average only a small chance change in the allelic frequency will be due to genetic drift. On the other hand, if the population size is small, then the allelic frequency can undergo large fluctuations in different generations in a seemingly unpredictable pattern.

An example will illustrate the type of allelic frequency change to be expected in a small population. Assume that a diploid population has five individuals (ten gametes) and that half initially are A_1 and half are A_2. Assume that the ten gametes for the next generation are randomly chosen from this gametic pool. The process is analogous to flipping a coin ten times and counting the number of heads and tails. Sometimes the proportion of A_1 is greater than A_2, sometimes they are equal, and sometimes the proportion of A_2 is greater than A_1. If a large number of populations, each containing ten gametes, are randomly chosen from a population with equal numbers of A_1 and A_2, then the proportion of populations with different frequencies of A_2 is given in Figure 6.2. For example, a frequency of 0.4 indicates that four of the gametes were A_2. Notice that in most of the populations frequencies were close to the initial frequency of 0.5, but that there is significant variation around that value.

On the other hand, if the population size is large, then very little change in allelic frequencies should occur, owing to genetic drift. For example, let us consider a large number of populations, each with 1,000 gametes instead of 10 gametes. In this case, almost all of the populations will have frequencies between 0.45 and 0.55 after one generation, a much narrower range than for the smaller population size.

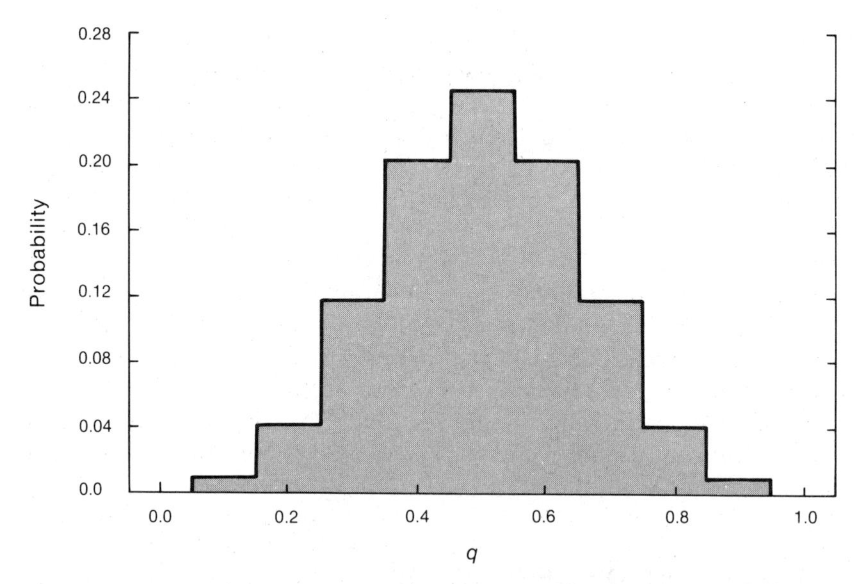

Figure 6.2. The probability that a population with ten gametes will have a given allelic frequency after one generation.

To illustrate the potential for cumulative genetic change over time, let us follow four particular populations or replicates of populations of size 20 (40 gametes) over a number of generations (Figure 6.3). The solid lines here indicate the four replicates, and the broken line is the mean frequency of allele A_2 over the four replicates. All of the replicates were initiated with the frequency of A_2 equal to 0.5. In the first generation, the frequencies are 0.625, 0.55, 0.55, and 0.475 because of the effect of genetic drift. By generation 15 they are more widely dispersed and are 0.8, 0.5, 0.475, and 0.3. In one of the replicates, the frequency of A_2 went to 1.0 in generation 19, and in another replicate the A_2 allele frequency went to 0.0 in generation 28. The other two replicates were still segregating for both alleles at the end of 30 generations. The mean of the four replicates varied from 0.625 in generation 19 to 0.475 in generation 30 but was generally near the initial frequency of 0.5. From this example it is obvious that genetic drift can cause large and erratic changes in allelic frequency of individual populations in a rather short time and can lead to differentiation among populations. On the other hand, if the populations were large over the same time span, there would be little change in allelic frequency and consequently little cumulative differentiation among populations.

However, if there are enough replicate populations, then no change would occur in the mean allelic frequency over all populations because the increases in allelic frequency in some populations are cancelled by reductions in allelic frequency in other populations. Eventually populations either go to fixation for $A_2 (q = 1)$ or to loss of $A_2 (q = 0.0)$ (see Example 6.2). The probability that a population will eventually become fixed for A_2 is equal to

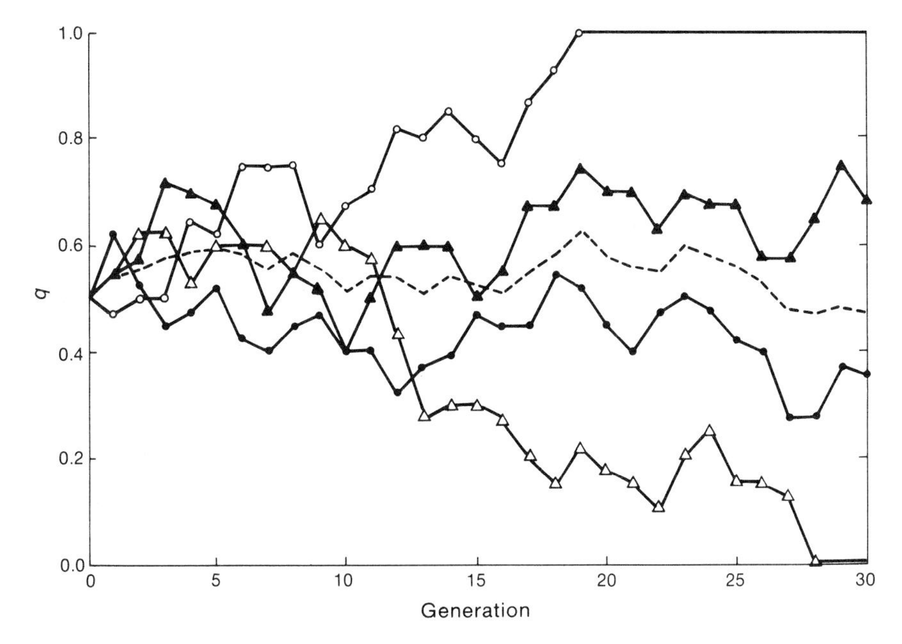

Figure 6.3. The allelic frequencies over 30 generations for four replicates of size 20 and their mean (broken line).

the initial frequency of A_2. For example, if $q_0 = 0.2$, then the probability that the population will be fixed for A_2 is 0.2 (the probability that it will lose the A_2 allele is 0.8). We can understand this point intuitively by realizing that, for A_2 to become fixed, it must change by chance from 0.2 to 1.0, a much bigger change than for the loss of A_2.

Because the mean frequency over replicate populations does not change and because the distribution of the allelic frequencies over replicate populations does, the overall effect of drift is best understood by examining either the variance of the allelic frequency or the heterozygosity over replicate populations. First, the expected variance in allelic frequency due to one generation of genetic drift is

$$V_q = \frac{pq}{2N}.$$

In a large population, V_q is small, indicating only a minor effect of genetic drift. To establish a general yardstick to measure the effect of genetic drift on allelic frequency relative to other factors, we note that the square root of variance is approximately equal to the average absolute value of the allelic frequency change, or

$$|\Delta q| \cong (V_q)^{1/2}.$$

For example, if $p = q = 0.5$ and $N = 50$, then

$$V_q = \frac{(0.5)(0.5)}{2(50)} = 0.0025$$

and

$$|\Delta q| \cong 0.05$$

Obviously, if N is relatively small, then the effect of genetic drift on allelic frequency may be substantial, comparable with that of selection or gene flow.

Second, we can use the expected heterozygosity to understand the effect of genetic drift. It is clear from the numerical example in Figure 6.3 that a finite population will eventually become homozygous by chance—that is, the

---

Example 6.2. A classic demonstration that illustrates how finite population size affects allelic frequency was provided by Buri (1956). He looked at the frequency of two alleles at the *brown* locus, which affects eye color in *Drosophila melanogaster*, in randomly selected populations of size 16. He chose the alleles bw^{75} and *bw* because they appeared to be nearly neutral with respect to each other. As a result, the effect of finite population size on allelic frequency could be examined almost independently of the effect of selection. Some data from his study are presented in Figure 6.4 as the number of the 107 replicate populations that had from 0 to 32 bw^{75} genes in different generations.

The histograms in Figure 6.4 illustrate that the distribution of the allelic frequencies over populations has a greater and greater spread with time. Although they are not given in Figure 6.4, the number of populations homozygous for one of the two alleles increased at nearly a linear rate after generation 4, and in generation 19 nearly equal numbers of the lines were fixed for the two alleles, 30 for *bw* and 28 for bw^{75}.

As expected, the mean allelic frequency over all populations remained close to 0.5 for the duration of the experiment, varying only between 0.47 and 0.51. The variance in allelic frequency increased from 0.0 in the zero generation to 0.16 by generation 19. Furthermore, the average heterozygosity declined from an initial value of 0.5 to approximately 0.10 in generation 19.

Overall, then, our predictions made concerning changes in the mean allelic frequency, variance in allelic frequency, and heterozygosity due to genetic drift are consistent with these observations.

---

heterozygosity will approach zero. The rate of approach is given by the following expression:

$$H_t = \left(1 - \frac{1}{2N}\right)^t H_0$$

where H_t is the heterozygosity in generation t. For example, if the initial heterozygosity is 0.5, in a population of size ten after two generations, the expected heterozygosity is

$$H_2 = \left(1 - \frac{1}{20}\right)^2 0.5 = 0.451.$$

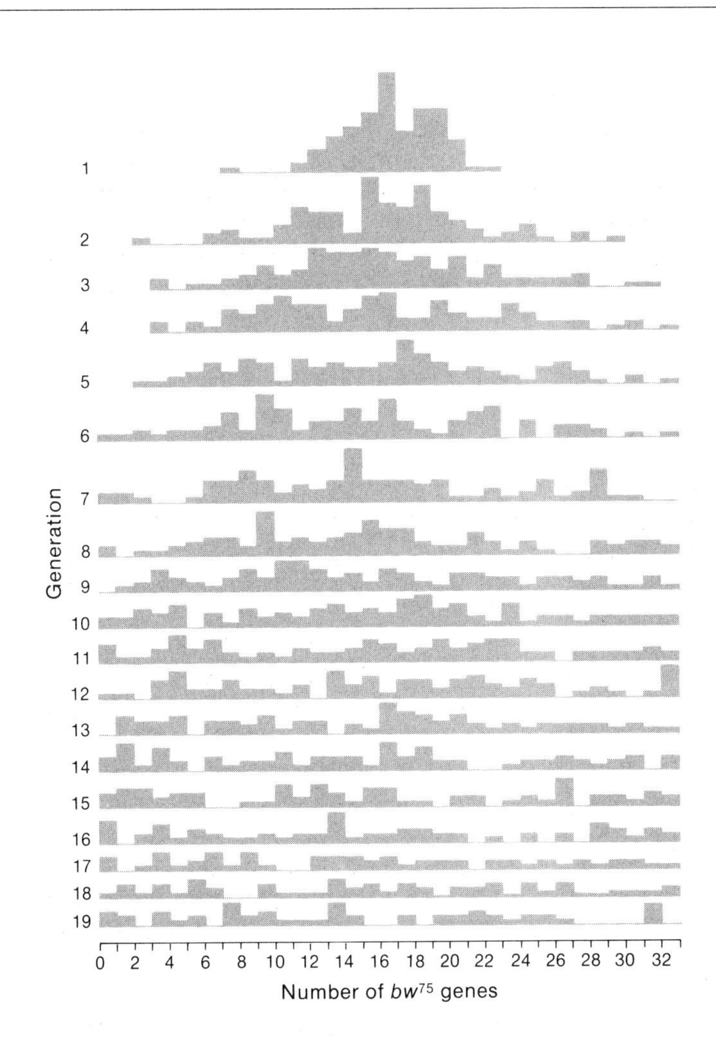

Figure 6.4. The number of bw^{75} alleles out of 32 in replicate populations over 19 generations (from Buri, 1956).

Obviously, because $[1 - 1/(2N)]$ is less than one, the expected heterozygosity will always decline and will approach zero over time at a rate dependent on the size of the population. In fact, the time until 50 percent of the heterozygosity is lost is $1.39\,N$. When $N = 10$, it takes approximately 14 generations for half of the heterozygosity to be lost. This loss of heterozygosity is due to changes in allelic frequency that, on the average, result in individual populations with frequencies approaching either zero or unity. Of course, the heterozygosity in a given population may temporarily increase due to chance, but the average effect is a loss of heterozygosity. Genetic drift generally does not result in a significant deficiency of heterozygotes relative to Hardy-Weinberg proportions within a population.

II. EFFECTIVE POPULATION SIZE

The total number of individuals in a population may not be indicative of the population size that is appropriate for evolutionary studies. For example, only a small proportion of the individuals may actually be reproductive at any one time. Additionally, some reproductive individuals may contribute more offspring than others, as when a single male fertilizes a number of females. A general estimate of the breeding population size does not take into account such variations in the genetic contribution to the next generation. As a result, it is often useful to consider the *effective population size*, a value that allows consideration of an ideal population of size N in which all parents have an equal expectation of being the parents of any progeny individual. In other words, the gametes are drawn randomly from all breeding individuals, with the probability of each adult producing a particular gamete equal to $1/N$. Populations that deviate from this ideal may then be compared with such an ideal population.

Let us consider an organism in which half the gametes come from individuals of one sex (females) and half from individuals of another sex (males). In this case, the effective population size, N_e, is

$$N_e = \frac{4N_f N_m}{N_f + N_m}$$

where N_f and N_m are the number of breeding females and males, respectively. If there are equal numbers of females and males, $N_f = N_m = \frac{1}{2}N$, then $N_e = N$, as we would expect.

In a number of species the number of breeding males and females are unequal. Frequently the number of breeding males is smaller than the number of breeding females ($N_m < N_f$), because some males mate more than once. However, the opposite is true in species in which the female mates with a number of males. Let us assume that one male mates with all the females in a colony or harem ($N_m = 1$), as appears to occur in some social vertebrate populations. In this case, the above expression becomes

Example 6.3. As discussed in Example I.2, the northern elephant seal was nearly hunted to extinction at the end of the last century. In fact, as few as 20 individuals on Isla Guadalupe may have been the ancestors of the present population of over 60,000. In addition to this bottleneck effect, the population size is also affected by social structure, in which a single, socially dominant male may mate with 40 or more females (see Chapter 18 for a discussion of social behavior). Therefore, in the bottleneck generation on Isla Guadalupe, let us assume that one male and ten females were reproductive. In keeping with the above expression, the estimate of N_e in this generation is

$$N_e = \frac{4(10)}{10 + 1}$$
$$= 3.6.$$

The heterozygosity in the southern elephant seal was estimated for a number of electrophoretic loci as 0.028. However, no heterozygotes were found in a sample of 159 northern elephant seals in an examination of 24 electrophoretic loci (Bonnell and Selander, 1974). If they had found an individual heterozygous for one locus, the average heterozygosity would have been 0.00026, two orders of magnitude below that observed in southern elephant seals. If we assume that the heterozygosity in northern elephant seals before overhunting was similar to that in southern elephant seals, we can ask how many generations it would take to reduce the heterozygosity to this low level. (Notice that heterozygosity here is used for a number of loci in a single population rather than for a single locus in a number of populations.) If we let $H_0 = 0.028$, $H_t = 0.00026$, and if we take the most extreme population size for the total period, $N_e = N = 3.6$, then

$$H_t = \left(1 - \frac{1}{2N}\right)^t H_0$$
$$0.00026 = \left[1 - \frac{1}{2(3.6)}\right]^t 0.028$$
$$t = 31.3.$$

In other words, it would take 31.3 generations, or approximately 250 years (assuming that a generation is approximately 8 years in length), for the heterozygosity to decrease this much owing to genetic drift alone. For the observed difference in heterozygosity between the two species of elephant seal to be caused by a bottleneck alone, it would have had to be about 31 generations in length (or a somewhat longer period if the population size were larger), making it likely that some other factors are important in determining the amount of heterozygosity.

$$N_e = \frac{4N_f}{N_f + 1}.$$

Notice that, as N_f becomes larger, this expression approaches a value of 4.0. In other words, because each sex must contribute half of the genes to the progeny, restricting the number of breeding individuals of one sex can greatly reduce the effective population size (see Example 6.3).

III. MUTATION IN A FINITE POPULATION

Although genetic drift has only a small effect in a large population, if there are many generations the cumulative effect may be substantial. Likewise, the mutation rate in any one generation is small, but the likelihood of a mutation over a number of generations is also substantial. Taken together, the joint effect of genetic drift and mutation may be important over long periods. Though we introduce this topic here, in Chapter 17 we will consider its application to molecular variation.

In order to examine mutation in a finite population, we need to assume that selection is not important—that is, that the alleles have no selective advantage or disadvantage, a concept called *neutrality*. For example, alleles A_1 and A_2 would be neutral with respect to each other if $w_{11} = w_{12} = w_{22}$. As discussed above, the probability of a new mutation becoming fixed in a population is equal to its initial frequency. Let us assume that a new mutant A_2 occurs in a population of size N that otherwise consists only of A_1 alleles. The initial frequency of the mutant is then

$$q_0 = \frac{1}{2N}.$$

Assuming that the two alleles are neutral with respect to each other, then the probability of fixation of the new mutant is equal to its initial frequency $1/(2N)$. The probability that the new mutant will be lost (equal to the fixation of the original allele) is $1 - 1/(2N)$. In other words, unless the population is quite small, a new neutral mutant will nearly always be lost from the population. If $N = 50$, the probability of fixation of a new mutant is only 0.01.

The two alternatives, fixation or loss of the new mutant, take quite different amounts of time to occur. Because the change of frequency necessary when loss occurs is small, from $1/(2N)$ to zero, the average amount of time for this to occur is short. On the other hand, fixation of the new mutant requires a substantial change in allelic frequency, from $1/(2N)$ to unity, and the time required is much larger. Figure 6.5 illustrates the time necessary for fixation and loss to occur for five separate mutants. In this example, one mutant, the first one, eventually becomes fixed, while the others are lost from the population due to genetic drift.

If a large number of potential alleles exists at a locus, then mutation will increase the number of alleles and genetic drift will reduce the number of

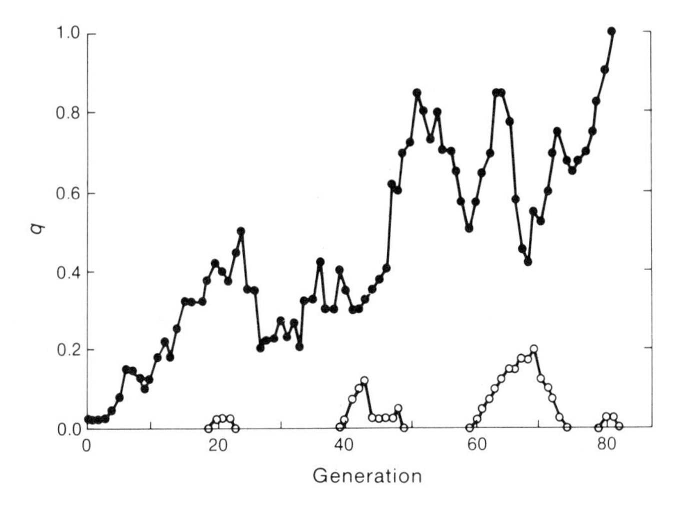

Figure 6.5. The frequency of five new mutants over time in a finite population.

alleles. The equilibrium heterozygosity resulting from the balance of these two factors for this model—called the infinite-allele model because each mutation is assumed to be to a new, unique allele—has been given by Kimura and Crow (1964). In this case, the equilibrium heterozygosity is

$$H_e = \frac{4N_e u}{4N_e u + 1}$$

where N_e is the effective population size and u is the mutation rate. The expected heterozygosity is greater than 0.5 if $4N_e u$ is larger than unity. If it is assumed that the mutation rate is between 10^{-5} and 10^{-7}, then the expected range of heterozygosity at equilibrium can be calculated for different effective population sizes (Figure 6.6). For example, if N_e is 10^4, then the equilibrium heterozygosity would be 0.0 when $u = 10^{-7}$ and 0.286 when $u = 10^{-5}$. In general, if $N_e u$ is much greater than one, then the mutation rate determines for the most part the amount of heterozygosity, so that H is quite high. On the other hand, if $N_e u$ is much less than one, then genetic drift becomes the major factor and H is low. If a population initially has no genetic variation, then mutation will introduce variation and the heterozygosity will increase over time. Eventually finite population size will limit the amount of heterozygosity and an equilibrium will be reached. However, even in an ideal situation, the time until equilibrium may be thousands of generations.

As we discussed in Chapter 1, there is extensive variation for electrophoretic variants. One hypothesis, called the neutrality theory, suggests that molecular variation is the result of the joint effects of genetic drift and neutrality as introduced here. Another hypothesis suggests that molecular variation is the result of balancing selection, such as heterozygous advantage (see Chapter 15). These views and related information are discussed in more detail in Chapter 17.

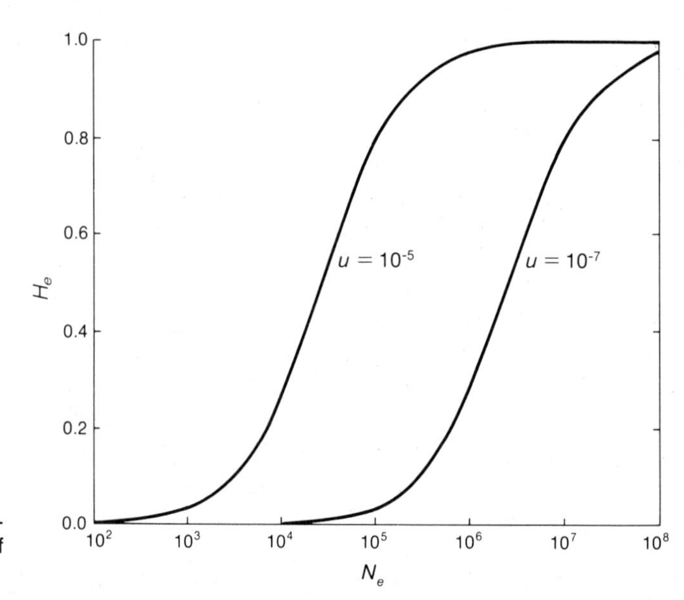

Figure 6.6. The equilibrium heterozygosity expected from a balance of mutation and finite population size.

SUMMARY

Small population size resulting from population bottlenecks, founder effects, or limitations of resources can lead to chance changes in allelic frequencies, or genetic drift. The effect of genetic drift can be evaluated using the variance of allelic frequency or heterozygosity measured over populations of the same size. In determining the potential importance of genetic drift in natural populations, the concept of effective population size is useful. The joint effects of genetic drift and mutation on heterozygosity are discussed as they relate to the concept of neutrality.

PROBLEMS

1. If the initial frequency in a population is 0.3 for A_2, what is the probability that the population will become fixed for A_2? For A_1?
2. In a population of 100 individuals with $q = 0.4$, what is V_q and the approximate value of $|\Delta q|$ after one generation? Assuming initial Hardy-Weinberg proportions, what is H_1 and H_2? If $N = 10$, what would H_1 and H_2 be?
3. What is the effective population size if there are two breeding males and six breeding females in a population?
4. What is the equilibrium heterozygosity if the effective population size is 10,000 and $u = 10^{-5}$? How does this compare to the heterozygosity estimated for electrophoretic variation?
5. In what ways is the effect on allelic frequency from genetic drift different from that due to mutation, selection and gene flow?

CHAPTER 7

Nonrandom Mating

What do I think about sex? Oh, I think it's here to stay.

MARILYN MONROE

Until now we have considered populations in which mating was random, an ideal that appears to be nearly met for many organisms. There are obvious exceptions, such as in highly self-fertilizing plant species, plant and animal species that can reproduce vegetatively or parthenogenetically, and human and other animal populations in which inbreeding occurs. Inbreeding is of particular importance in humans because many individuals who have recessive diseases are the products of matings between relatives.

Nonrandom mating with respect to genotype occurs in populations where the mating individuals are either more closely or less closely related than those drawn by chance from the population. The results of these two types of matings are called *inbreeding* and *outbreeding*, respectively. Neither inbreeding nor outbreeding can cause a change in allelic frequency but both cause a reorganization of the alleles into genotypes. In a population that is inbred, the frequency of homozygotes is increased and the frequency of heterozygotes reduced relative to random-mating (Hardy-Weinberg) proportions. With outbreeding, the opposite occurs, and the frequency of heterozygotes is increased and that of homozygotes reduced relative to random-mating proportions. These genotypic changes affect *all* loci in the genome. However, such changes may be quite ephemeral if the mating system changes. For example, the high frequency of homozygotes resulting from self-fertilization can be eliminated completely in one generation of random mating.

Nonrandom mating based on phenotypes rather than genotypes may also occur. If the mated pairs in a population are composed of individuals with the same phenotype more often or less often than expected by chance, then *positive assortative mating* or *negative assortative mating*, respectively, has occurred. Assortative mating generally affects the genotypic frequencies of only those loci involved in determining the phenotypes for mate selection. There is positive assortative mating in humans for some traits, such as height and eye color. Because positive assortative mating has effects similar to those of inbreeding, except that it generally affects only one or a few loci, in the following discussion we will consider only inbreeding.

I. COMPLETE SELF-FERTILIZATION

The most extreme type of inbreeding, assuming that there is sexual reproduction, is self-fertilization. In this case, both the pollen and egg or sperm and egg are produced by the same individual. With complete self-fertilization, a population is divided into a series of lines that quickly become highly homozygous. As shown in Table 7.1, where self-fertilization alone occurs there are just three mating types, and they are found in the relative proportions of the genotypes in the population. Allowing for segregation in the heterozygote, the proportions of the genotypes in the progeny are

$$\begin{aligned} P_1 &= P_0 + \tfrac{1}{4}H_0 \\ H_1 &= \tfrac{1}{2}H_0 \\ Q_1 &= Q_0 + \tfrac{1}{4}H_0. \end{aligned}$$

In other words, the frequencies of the two homozygotes are increased by $\frac{1}{4}H_0$ and the heterozygote decreased by $\frac{1}{2}H_0$ through one generation of self-fertilization. For example, if the initial frequencies of the genotypes were 0.25, 0.5, and 0.25 for A_1A_1, A_1A_2, and A_2A_2, then the frequencies would be 0.375, 0.25, and 0.375 after one generation of complete selfing.

One of the most important aspects of inbreeding is that even though genotypic frequencies may be greatly altered, the allelic frequencies remain unchanged. For example, the frequency of allele A_1 in the next generation is,

$$\begin{aligned} p_1 &= P_1 + \tfrac{1}{2}H_1 \\ &= (P_0 + \tfrac{1}{4}H_0) + \tfrac{1}{4}H_0 \\ &= p_0. \end{aligned}$$

In other words, there is no change in allelic frequency due to self-fertilization from generation to generation.

To understand how inbreeding changes the genotypic proportions, let us examine the frequency of heterozygotes. The relationship for heterozygosity in

Table 7.1
The mating types and frequency of progeny where there is complete self-fertilization.

		Progeny		
Mating type	*Frequency*	A_1A_1	A_1A_2	A_2A_2
$A_1A_1 \times A_1A_1$	P_0	P_0	—	—
$A_1A_2 \times A_1A_2$	H_0	$\frac{1}{4}H_0$	$\frac{1}{2}H_0$	$\frac{1}{4}H_0$
$A_2A_2 \times A_2A_2$	Q_0	—	—	Q_0
	1.0	$P_0 + \frac{1}{4}H_0$	$\frac{1}{2}H_0$	$Q_0 + \frac{1}{4}H_0$

two succeeding generations is a general one—that is, for generations one and zero, it is

$$H_1 = \tfrac{1}{2}H_0.$$

Between generations two and one, the relationship is

$$H_2 = \tfrac{1}{2}H_1.$$

By substituting the value of H_1 from the first expression, then

$$H_2 = (\tfrac{1}{2})^2 H_0.$$

This expression can be generalized for any number of generations so that

$$H_t = (\tfrac{1}{2})^t H_0.$$

Because the quantity $(\frac{1}{2})^t$ is always smaller than unity and decreases toward zero rapidly as t increases, the heterozygosity is reduced rapidly over time and asymptotically approaches zero. Reduction is rapid, and even when $H_0 = 0.5$, the heterozygosity is reduced to less than 1 percent after six generations—that is, $H_6 = (\frac{1}{2})^6\,(0.5) = 0.0086$.

II. THE COEFFICIENT OF INBREEDING

In order to describe the effect of inbreeding on genotypic frequencies in a general way, we will use a measure called the *coefficient of inbreeding* (f), which is the probability that the two homologous alleles in an individual are identical by descent. Alleles are identical by descent when the two alleles in a diploid individual are derived from one particular allele in their ancestry.

A general formulation of the proportion of the three genotypes A_1A_1, A_1A_2, and A_2A_2 using the inbreeding coefficient is

$$P = p^2 + fpq$$
$$H = 2pq - 2fpq$$
$$Q = q^2 + fpq$$

where the first term is the Hardy-Weinberg proportion of the genotypic frequencies and the second is the deviation from that value. Note that an individual may be homozygous because two alleles are present that are identical by descent, the fpq terms above, or because two alleles are present that are identical in kind through random mating, the p^2 and q^2 terms. The size and sign of the coefficient of inbreeding reflect the deviation from Hardy-Weinberg proportions of the genotypes such that, when f is zero, the zygotes are in Hardy-Weinberg proportions and, when f is positive, there is deficiency of heterozygotes.

The inbreeding coefficient from one generation of self-fertilization is $\frac{1}{2}$. Assuming that $H_0 = 2pq$, then from above

$$\begin{aligned} H_1 &= 2pq - 2fpq \\ &= H_0 - \tfrac{1}{2}H_0 \\ &= \tfrac{1}{2}H_0 \end{aligned}$$

which is the heterozygosity we obtained in Table 7.1. If $f = 1$, its maximum value, then

$$\begin{aligned} P &= p^2 + pq = p \\ H &= 2pq - 2pq = 0 \\ Q &= q^2 + pq = q. \end{aligned}$$

Example 7.1. Many species of plants have mixed mating systems—that is, they reproduce both by self-fertilization and by outcrossing (often assumed to be random mating) with other individuals. If the proportion of self-fertilization is high, then nearly all the individuals in the population should be homozygotes. A sample of data from *Avena fatua*, a highly self-fertilizing species of wild oats, widespread in many areas around the Mediterranean Sea and introduced into California, is given in Table 7.2. These data are for a locus that determines the black or gray color of the lemma (part of the grass spikelet). The three different populations given here vary in the frequency of the A_2 allele from 0.22 to 0.42. Estimates of f using the above formula are high and similar in populations b and c and much lower in population a. This difference probably reflects a smaller amount of self-fertilization in population a, but it should be noted that the genotypic frequencies may also be affected by selection and gene flow—that is, this estimate of the inbreeding coefficient is confounded with these other factors.

Table 7.2
The genotypic frequencies in three populations of wild oats and the allelic frequency and f value calculated from the data (from Jain and Marshall, 1967).

	Genotype				
Population	A_1A_1	A_1A_2	A_2A_2	q	f
a	0.71	0.14	0.15	0.22	0.59
b	0.67	0.06	0.27	0.30	0.86
c	0.55	0.07	0.38	0.42	0.86

In other words, only homozygotes are present, and they occur in the proportion of the allelic frequencies.

The expression for the frequency of heterozygotes above can be solved for f as

$$H = 2pq - 2fpq$$

$$1 - f = \frac{H}{2pq}$$

$$f = 1 - \frac{H}{2pq}.$$

From this equation, it is clear that f is a function of the ratio of the observed heterozygosity H over the Hardy-Weinberg heterozygosity $2pq$ (see Example 19.2 for an application). With inbreeding, H is less than $2pq$, so that f is greater than 0. This expression for f can be used to obtain a general estimate of inbreeding (see Example 7.1).

Using the genotypic frequencies expressed in terms of the inbreeding coefficient, we can show for any value of f that the allelic frequency remains the same. If the genotypic frequencies as given above are substituted in the equation for calculating the frequency of A_1, then

$$\begin{aligned} p &= P + \tfrac{1}{2}H \\ &= (p^2 + fpq) + \tfrac{1}{2}(2pq - 2fpq) \\ &= p^2 + pq \\ &= p. \end{aligned}$$

This is a general demonstration of the fact that inbreeding affects only genotypic frequencies and not allelic frequencies.

III. THE INBREEDING COEFFICIENT FROM PEDIGREES

As stated earlier, the inbreeding coefficient is known as the probability of identity by descent, which implies that homozygosity is the result of the two alleles in an individual descended from the same ancestral allele. The probability of identity by descent varies, depending upon the relationship of the parents of the individual being examined. For example, if the parents are unrelated, then there is no possibility that the individual will be homozygous by descent. The other extreme (given sexual reproduction) is when the two parents are the same individual and self-fertilization occurs. In this case, the probability of an offspring having identical alleles by descent is 0.5. As an illustration, assume that the parent has the genotype A_1A_2. Progeny are produced in the proportions 0.25 A_1A_1, 0.5 A_1A_2, and 0.25 A_2A_2, so that half the progeny, the A_1A_1 and A_2A_2 progeny, will have alleles identical by descent. We can obtain the inbreeding coefficient, f, from a pedigree in which there is a

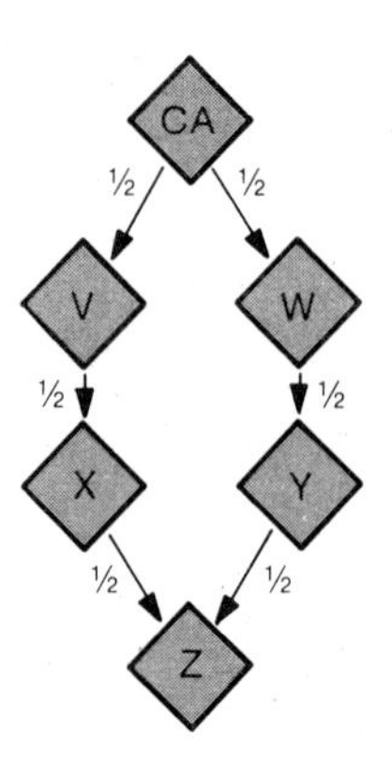

Figure 7.1. A pedigree illustrating a mating between two half first cousins, X and Y.

consanguineous mating, a mating between relatives, by calculating the probability of identity by descent.

As an example, Figure 7.1 gives a pedigree where two half first cousins (individuals who have one grandparent in common), X and Y, mate to produce an inbred offspring, Z. All unrelated individuals are omitted from the pedigree because they do not contribute to the inbreeding coefficient. The diamond symbols indicate that the individuals could be either males or females, and the double line connects the related, mating individuals. The parents of Z have a common ancestor CA, which in this case is either a grandmother or a grandfather with the genotype A_1A_2 (we assume here that CA is not inbred). In order to calculate the inbreeding coefficient, we must know the probability of Z being A_1A_1 or A_2A_2—that is, of having two alleles identical by descent from CA. These probabilities depend upon the probabilities of segregation of A_1 and A_2 into gametes between generations. For example, the probability that A_1 will be passed from CA to V is $\frac{1}{2}$.

The first alternative, Z being A_1A_1, can occur only if X contributes to Z a gamete containing A_1 and Y also contributes to Z a gamete containing A_1. The probability of an A_1 allele from X is equal to the probability of A_1 being passed on for three generations, CA to V, V to X, and X to Z, or

$$\Pr(A_1 \text{ from X}) = (\tfrac{1}{2})(\tfrac{1}{2})(\tfrac{1}{2}) = \tfrac{1}{8}.$$

Likewise, the probability that A_1 comes from CA to Z via Y is

$$\Pr(A_1 \text{ from Y}) = (\tfrac{1}{2})(\tfrac{1}{2})(\tfrac{1}{2}) = \tfrac{1}{8}.$$

Because A_1 must come from both parents to Z for identity, the probability of the identity of A_1 by descent is the product of the individual probabilities

$$\begin{aligned}\Pr(A_1A_1 \text{ in Z}) &= [\Pr(A_1 \text{ from X})][\Pr(A_1 \text{ from Y})]\\ &= (\tfrac{1}{8})(\tfrac{1}{8})\\ &= \tfrac{1}{64}.\end{aligned}$$

Likewise, the probability of identity of A_2 is

$$\Pr(A_2A_2 \text{ in Z}) = [\Pr(A_2 \text{ from X})][\Pr(A_2 \text{ from Y})] = \tfrac{1}{64}.$$

Because Z may be identical by descent and have either two A_1 or two A_2 alleles, the overall probability of identity by descent is then

$$\Pr(A_1A_1) + \Pr(A_2A_2) = \tfrac{1}{64} + \tfrac{1}{64} = \tfrac{1}{32}.$$

Therefore, the inbreeding coefficient of individual Z is $\frac{1}{32}$.

Example 7.2. Often in the breeding of dogs, horses, and other animals, complicated patterns of inbreeding occur because of the desire by breeders to incorporate favorable aspects of successful animals. By inbreeding, alleles that contribute to, say, racing ability or conformation, may be made homozygous. Figure 7.2 gives a hypothetical pedigree for an animal breeding program. In this case, the mating is between individuals that are first cousins once removed—that is, there are two generations on the left lineage and three on the right. Unlike the previous example, there are two common ancestors CA_1 and CA_2, from which alleles may originate and may be identical by descent. Using the chain-counting method, we find there are two chains, X-V-CA_1-U-W-Y and X-V-CA_2-U-W-Y. The overall inbreeding coefficient is the sum of these individual values, or

$$\begin{aligned} f &= f_1 + f_2 \\ &= (\tfrac{1}{2})^6 + (\tfrac{1}{2})^6 = \tfrac{1}{32}. \end{aligned}$$

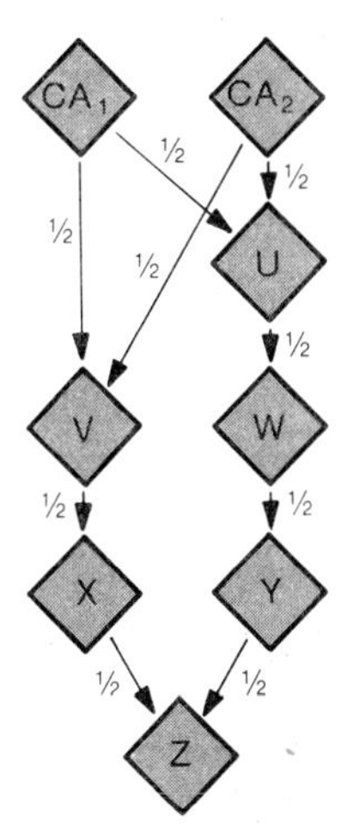

Figure 7.2. A pedigree illustrating a mating between two first cousins once removed.

Let us develop a straightforward approach often used to calculate the inbreeding coefficient from a pedigree, the *chain-counting technique*, so that we do not have to calculate all these probabilities. A chain for a given common ancestor starts with one parent of the inbred individual, goes up the pedigree to the common ancestor, and comes back down to the other parent. For the example in Figure 7.1, the chain would be X-V-CA-W-Y. The number of individuals in the chain, N, can be used in the following formula to calculate the expected inbreeding coefficient due to the presence of this common ancestor so that

$$f = (\tfrac{1}{2})^N.$$

In Figure 7.1, the chain for the pedigree includes five individuals (notice that the inbred offspring is not counted), so $f = \frac{1}{32}$, corresponding to our previous result (see Example 7.2).

IV. INBREEDING AND SELECTION

Inbreeding increases the frequency of homozygotes and decreases the frequency of heterozygotes. As a result of this deviation from Hardy-Weinberg proportions, the incidence of recessive diseases may increase and the mean fitness of a population may be reduced.

Let us first begin by examining the incidence of recessive genetic diseases among individuals from consanguineous matings. Inbreeding has the effect of increasing the frequency of homozygous recessives, which in this case increases the incidence of the recessive disease. The genetic basis for this effect becomes clear when the proportion of recessive homozygotes in a noninbred population ($Q = q^2$) is compared to that for a given inbreeding coefficient (Q_f). The ratio of these two quantities is

$$\frac{Q_f}{Q} = \frac{q^2 + fpq}{q^2}$$

$$= \frac{q + fp}{q}$$

which is greater than unity for any allelic frequency but is very large for low allelic frequencies. This ratio is given for two levels of inbreeding and several allelic frequencies in Table 7.3. For example, when $f = \frac{1}{16}$ (the inbreeding coefficient that results from a first-cousin mating) and $q = 0.01$, there are about seven times as many affected individuals in the random-mating group as in the inbred group (see Example 7.3).

One potential effect of a high level of inbreeding is that the frequency of a recessive deleterious allele may decrease. This decline results because under such circumstances more deleterious alleles are exposed to selection in

Table 7.3
The ratio of the proportion of recessives with a given inbreeding coefficient (f) to the Hardy-Weinberg proportion of recessives for several allelic frequencies.

	f	
q	$\frac{1}{32}$	$\frac{1}{16}$
0.001	32.22	63.44
0.01	4.09	7.19
0.1	1.28	1.56

homozygotes—that is, fewer are hidden from selection by dominance in heterozygotes.

However, in many societies today, the rate of consanguineous marriages is rapidly declining owing to an increased mobility, a growing population size, and other factors. An immediate consequence of this reduction in inbreeding is a reduced incidence of recessive genetic diseases. For example, using the logic above and assuming that $f = 0.05$ in the past, that random mating exists today, and that $q = 0.005$, the incidence of such a disease would be only 0.09 what it had been. In the very long run, however, the frequency of the disease allele may slowly increase, because not as much selection exists in homozygotes (there are relatively fewer homozygotes) to counter mutational pressure.

Because individuals are more likely to be homozygous for recessive deleterious alleles in an inbred population, the mean fitness of the total population may decline as a result. *Inbreeding depression*–the decline of population fitness due to inbreeding—seems to be a nearly universal phenomenon. The extent of inbreeding depression varies among different organisms, and, for example, three generations of mating sibs in Japanese quail resulted in a complete loss of reproductive fitness (Sittman et al., 1966). In other species, such as mice and *Drosophila melanogaster*, some lines have been maintained by mating sibs for hundreds of generations. Many plant species that have partial self-fertilization generally show little decline of fitness when inbred.

Let us examine how inbreeding can affect the mean fitness of the population. Assume as before that the relative fitness values for genotypes A_1A_1, A_1A_2, and A_2A_2 are w_{11}, w_{12}, and w_{22}, respectively. If we use the genotypic frequencies above, then the mean fitness with inbreeding is

$$\begin{aligned}\bar{w}_f &= w_{11}(p^2 + fpq) + w_{12}(2pq - 2fpq) + w_{22}(q^2 + fpq) \\ &= \bar{w} + fpq(w_{11} + w_{22} - 2w_{12}) \\ &= \bar{w} - fpq(2w_{12} - w_{11} - w_{22})\end{aligned}$$

where the first term is the mean fitness

$$\bar{w} = p^2 w_{11} + 2pq w_{12} + q^2 w_{12}$$

as before, and the second term measures the change in fitness due to inbreeding. Inbreeding depression can be defined as the difference in fitness between an outbred and an inbred population, or

---

Example 7.3. In isolated human populations, there are often many consanguineous matings because of the limited number of mates available. A pedigree from an isolated island population off the coast of Yugoslavia is given in Figure 7.3 (from Vogel and Motulsky, 1979). Circles and squares indicate females and males, respectively, and the darkened symbols are affected individuals. The generations begin at the top and progress downward. The disease, Mal de Meleda, is an autosomal recessive disorder that results in a thickening of the skin on the hands and feet. In this pedigree, there are two first-cousin matings, between individuals a and b in generation 3 and between individuals a and b in generation 4, indicated by the double lines connecting these mates. Individuals 3a and 3b must have both been heterozygotes, because two of their seven offspring were affected. The other first-cousin mating was between heterozygote 4a and a diseased homozygote 4b and resulted in two of the three progeny becoming affected. (Although this disease is rare in other populations, its commonness in this population results in many heterozygotes.) The high proportion of consanguinity in this population has resulted in a much higher incidence of Mal de Meleda than would have occurred in a random-mating population (or than occurs on the mainland).

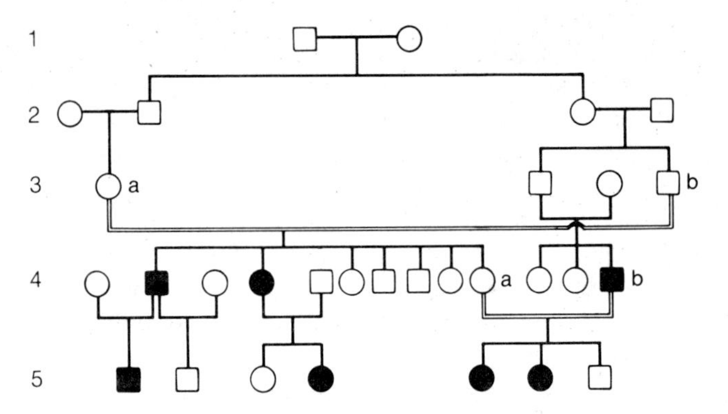

Figure 7.3. A pedigree with two first-cousin matings and a high incidence of a rare recessive disease (from Vogel and Motulsky, 1979).

$$\bar{w} - \bar{w}_f = fpq(2w_{12} - w_{11} - w_{22}).$$

Notice that the inbreeding depression is a function of the allelic frequencies and that the relationship between the fitnesses is a linear function of the inbreeding coefficient. If we assume that the fitnesses are 1, 1, and $1 - s$ for the three genotypes, then

$$\bar{w} - \bar{w}_f = fspq.$$

In other words, as f increases, then the inbreeding depression should increase a corresponding amount (see Example 7.4).

Example 7.4. A large body of data exists on the effects of inbreeding on viability in humans. Some of the best data are summarized in the report of Schull and Neel (1965) on the effects of the atomic bombs on people in and around Hiroshima and Nagasaki, Japan. In their study, they examined the effect of inbreeding on mortality in individuals not exposed to the effects of the bombs as a comparison to those who were exposed. A sample of these "control" data is given in Table 7.4 for infant mortality where the parents have different degrees of consanguinity. Here a trend appears, as expected, of increased mortality with an increasing amount of inbreeding.

Intense inbreeding may occur among many captive animals in zoos because unrelated individuals may be unavailable for mating. In a survey of the offspring of 16 species of captive ungulates, such as elephants, pygmy hippopotami, and gazelles, the juvenile mortality of 565 noninbred young was 23.2 percent, while that of inbred young, often with inbreeding coefficients greater than 0.25, was 44.5 percent (Ralls et al., 1979). These results suggest that the deleterious effects of inbreeding may be quite important in captive populations of exotic animals.

Table 7.4
Infant mortality in two Japanese populations where the parental consanguinity is known (from Schull and Neel, 1965).

	Parental relationship (f)			
Population	*Unrelated (0)*	*Second cousins ($\frac{1}{64}$)*	*$1\frac{1}{2}$ cousins ($\frac{1}{32}$)*	*First cousins ($\frac{1}{16}$)*
Hiroshima	3.55	4.43	7.18	6.12
Nagasaki	3.42	3.18	4.94	5.25

SUMMARY

The general effects of nonrandom mating are discussed. In particular, the effect of inbreeding is to reduce the frequency of heterozygotes at all loci without changing allelic frequencies. The effect of complete self-fertilization on genotypic frequencies is discussed. Estimation of the extent of inbreeding using genotypic frequencies and pedigree analysis is introduced. The joint effects of inbreeding and selection are considered to show how increased inbreeding results in an increase in disease incidence and a lowered mean fitness (inbreeding depression).

PROBLEMS

1. If the initial genotypic frequencies are 0.04, 0.32, and 0.64 for A_1A_1, A_1A_2, and A_2A_2, what are the frequencies after one generation of complete self-fertilization? What is the frequency of heterozygotes after five generations of complete self-fertilization?
2. Calculate the frequency of the three genotypes, given $f = 0.2$ and $q = 0.4$. Assuming that the frequencies of the three genotypes are 0.35, 0.1, and 0.55, what is f?
3. What is the expected f in a mating between an aunt and a nephew? Between two half sibs?
4. What is the expected inbreeding depression when $f = 0.25$, $s = 0.5$, and $q = 0.01$? If $q = 0.5$?
5. Discuss the general effects of inbreeding on genotypic frequencies, allelic frequencies, and mean fitness.

CHAPTER 8

Quantitative Traits

There is no security whatever that the individuals selected did not have excess of character owing to environmental and not to hereditary conditions.

KARL PEARSON

We have considered until this point traits for which the genetic basis is expressly known. For many traits of evolutionary importance, the genetic basis is not known precisely, and, in fact, it may not be feasible or even possible to determine the number of genes and their individual effects on a particular trait. The phenotypic value of traits such as size, weight, or shape can generally be measured on some quantitative scale. Although this phenotypic value may be the result of some general underlying genetic determination, for most quantitative traits there is no precise relationship between the phenotypic value and a particular genotype. In general, *quantitative traits* have a continuous distribution, appear to be affected by many genes—that is, they are polygenic—and are affected by environmental factors. Before we begin our discussion of quantitative traits, let us briefly examine these attributes. We will discuss natural selection on quantitative traits and its consequence, adaptation, in Chapter 16.

I. THE NATURE OF QUANTITATIVE TRAITS

To illustrate the polygenic nature of quantitative traits, let us assume that we can visualize the influence of multiple genes affecting a trait by comparing the phenotypic distribution resulting when one, two or four genes, each with two alleles, determine the phenotypic value of a trait (Figure 8.1). For simplicity, we assume that heterozygotes are exactly intermediate (that is, the effects of different loci additively determine the phenotype), that all alleles have a frequency of 0.5, and that the environment has no effect on the trait. The frequency of the polygenic genotypes is calculated as the product of the frequencies of the single-locus genotypes. From Figure 8.1, it is obvious that as the number of genes affecting the trait increases from one to four the distribution changes from one of discrete classes to one in which there is no discontinuity. In fact, the exact phenotype–genotype correspondence for all

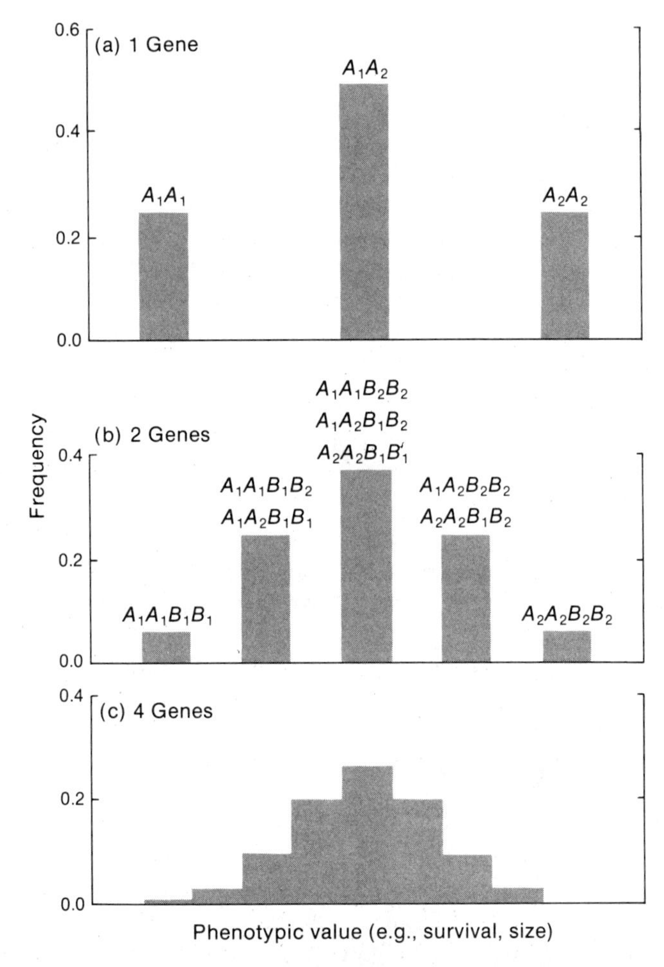

Figure 8.1. The determination of a quantitative trait by 1, 2, and 4 genes and the subsequent phenotypic distribution.

classes disappears when there are only two genes. For example, when two genes affect a trait, the second lowest phenotypic value can be the result of either genotype $A_1A_1B_1B_2$ or $A_1A_2B_1B_1$.

The effect of the environment on a quantitative trait can be illustrated by assuming that there is only one locus with two alleles affecting a trait in a particular population. The hypothetical phenotypic values resulting from the three different genotypes are given in Figure 8.2, where it is assumed that A_1A_1 is the best genotype on the average at lower environmental values, A_2A_2 is the best at higher values, and A_1A_2 is intermediate. If the average phenotypic value is calculated for different environmental values over all three genotypes (assuming for this illustration that all environments are equally likely), then the phenotypic distribution can be calculated for different frequencies of the genotypes. When the frequency of A_2 is 0.5 and there are Hardy-Weinberg

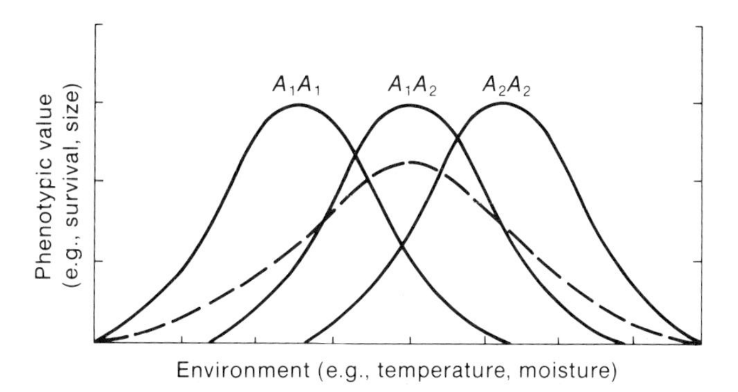

Figure 8.2. The phenotypic values for three genotypes and their mean (broken line) over an environmental spectrum.

proportions, the phenotypic distribution is given as the broken line in Figure 8.2. Notice that the distribution shows no discontinuities, even though only one gene with two alleles is responsible for determining the phenotype.

Typically, quantitative traits such as size and weight have a continuous distribution, given that the method of measurement is sufficiently accurate. The distribution of such traits often closely approximates a bell-shaped or normal distribution, with a high proportion of the individuals having some intermediate phenotypic value. Both the polygenic character of such traits and the influence of the environment contribute to determining the shape of such a distribution. A classical example of such a distribution is that for height in humans, such as given in Figure I.5 for the height of a company of soldiers.

However, some quantitative traits have only discrete classes, such as the number of eggs in a clutch for a bird, the number of scales on a reptile, the number of petals or leaves on a plant, or the number of vibrissae on the face of a mouse. Such discontinuous traits, sometimes called countable or meristic traits, are also assumed to be controlled by many genes and affected by environmental factors. At the extreme, there may be only two categories for a discontinuous trait such as the presence–absence situation that occurs for disease states (diseased, not diseased) or for reproductive states (fertile, sterile). In this case, one can imagine that the presence of one state or the other is the result of an underlying genetic determination. For example, the existence of a disease may be due to an underlying genetic determination of liability or susceptibility that has a continuous distribution (Figure 8.3). The presence of the disease in a particular individual is determined by a combination of the individual's underlying genetic liability and his or her particular environment. For example, Figure 8.3 gives the hypothetical threshold in a normal environment for a disease such as diabetes. Here a relatively small percentage of the population exceeds the threshold (the lightly shaded region of Fig. 8.3) and, as a result, exhibits diabetes. Given a different environment in which obesity, alcoholism, or other factors increase the likelihood of diabetes, the threshold is moved to the left (indicated as the threshold in a stress environment). This shift results in a higher incidence of the disease because

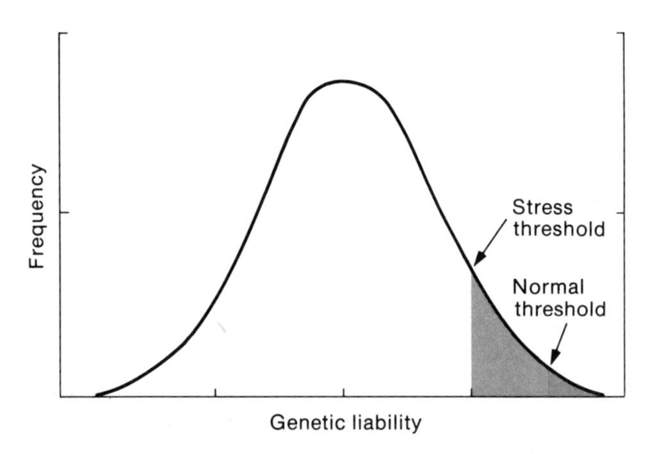

Figure 8.3. The genetic liability for a trait with thresholds indicating the value above which individuals have the disease phenotype.

some individuals, represented by the darkly shaded region, who would not contract the disease in a normal environment, now become afflicted.

It is important to realize that the genetic determination of quantitative traits may vary considerably among populations. For example, genes that are variable in one population may be invariant in another. Because selection can occur only on genes that are variable, different populations may have quite different potentials for selective responses. In addition, genetic drift or inbreeding may cause a differential loss of variants in different populations during the process of selective change. Such a varied response has occurred in different populations of insects exposed to various pesticides; the response may be primarily due to a single gene or to a number of genes that confer resistance to the action of a particular insecticide. In addition, the single genes and the method by which they confer resistance may differ among populations. For example, resistance to the insecticide DDT may occur because particular genes result in detoxification of the insecticide, the storage of DDT in fat bodies, or resistance to its diffusion into the circulatory system. Example 8.1 demonstrates that the selective response may be polygenically controlled and the result of a number of genes on different chromosomes.

It is generally assumed that genes that affect quantitative traits are essentially like genes (or are the same genes) that determine single-gene qualitative traits. In other words, there is no reason to assume that a different type of gene accounts for quantitative genetic variation, and, in fact, genes known because of their qualitative effects may also have effects on quantitative traits. It is assumed that genes affecting quantitative traits follow Mendelian patterns of inheritance and that they can have multiple alleles, can mutate, change in allelic frequency, and have dominance. Such assumptions allow us to formulate predictive models for quantitative traits that are in many ways extensions of the population genetic models we have already discussed. However, before we introduce these models, let us briefly mention some of the premises of quantitative genetic models.

It seems obvious that different quantitative traits would be affected by different numbers of genes. Very broadly defined traits, such as viability or size, are probably affected by large numbers of loci. For example, well over 1,000 loci cause genetic diseases and consequently affect survival in humans, and more than 400 loci in *D. melanogaster* are capable of mutating to lethals.

Example 8.1. Crow (1957) exposed a large population of *D. melanogaster* to DDT for a number of generations so that they acquired resistance to this insecticide over a period of time. He then crossed these flies with control (unexposed) flies, using a breeding scheme to construct flies with various combinations of control and resistance for the three major chromosomes. The survival of these flies was then tested in an environment containing DDT. From Figure 8.4, it is obvious that all three major chromosomes, the X, II, and III, contributed significantly to survival. In addition, detailed genetic analysis indicated that at least several genes on both the second and third chromosomes contributed to DDT resistance. In other words, resistance was controlled by a number of genes spread throughout the genome.

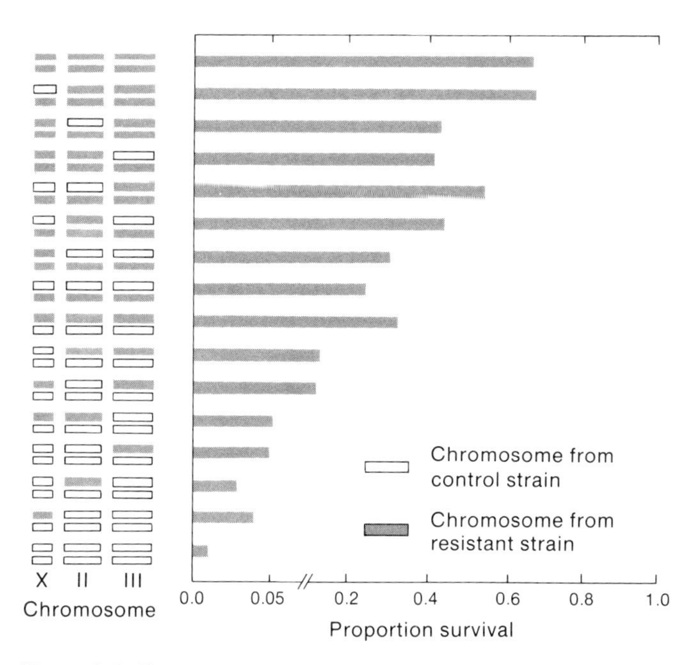

Figure 8.4. The survival of *D. melanogaster* composed of chromosomes of combinations of control and DDT-resistant strains in a DDT environment (Crow, 1957).

Of course, most of these extreme variants are rare, but it can be assumed that less severe allelic types at these loci are also segregating in populations, perhaps at even higher frequencies.

The simplest theory of quantitative genetics assumes that all genes affecting a trait contribute equally to the trait. However, when detailed genetic analysis has been carried out, it has been generally found that relatively few genes of large effect and many genes of small effect contribute to a trait (these are sometimes called major genes and minor genes, respectively). For example, about ten loci account for 75 percent of the genetic variation for the number of sternopleural bristles in populations of *D. melanogaster*, and several genes determine to a large extent heading time in wheat (see Example 8.2).

Example 8.2. Harrison and Owen (1964) examined the genetic determination of skin color differences between Caucasians and Africans by measuring light reflectance from the skin in a population from Liverpool, England. This population contained Africans and Caucasians, progeny from matings between Africans and Caucasians—or F_1 individuals—and backcross progeny from matings between F_1 individuals and both of the racial groups. The reflectance, as measured in a standard area of the body (underneath the upper arm), is given in Figure 8.5 for a range of different wave lengths. Because the F_1 value is almost intermediate between the two parental values and the backcrosses are almost intermediate between the F_1 and the respective parental values, it appears that there is no dominance for the genes affecting skin color. The number of genes that affect skin color can be estimated from these reflectance values and their variances. The estimate is that only three or four genes were responsible for the differences in skin color between the two parental groups.

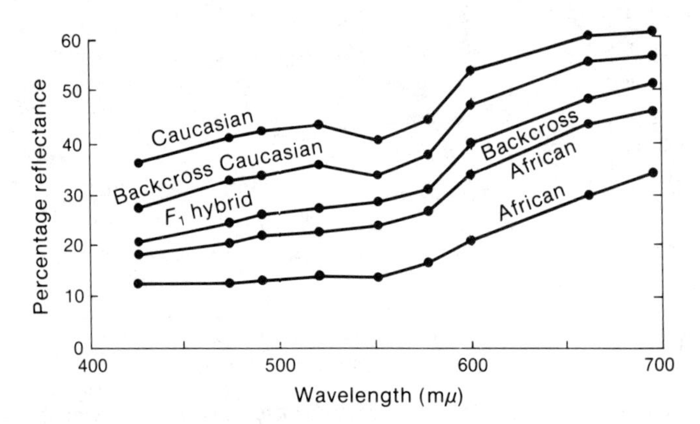

Figure 8.5. The reflectance from the skin of individuals of different ancestry (Harrison and Owen, 1964).

II. THE QUANTITATIVE GENETICS MODEL

In order to understand and examine the importance of quantitative traits, we need to construct a model that will allow us to break down phenotypic values into genetic and environmental components. We can accomplish this goal in a simple manner by symbolizing the phenotypic value for individual i in a particular environment j as

$$P_{ij} = G_i + E_j$$

where G_i is the genetic contribution of the ith genotype and E_j is the environmental deviation resulting from the jth environment (see Example 8.3). This simple model illustrates the basic genetic–environment or nature–nurture dichotomy often discussed in relationship to human intelligence and other traits. If a particular genotype is replicated, as plants can be by cloning or as animals can be by the use of an inbred line, then the average phenotypic value of that genotype over m different environments is

$$\bar{P}_i = G_i + \frac{1}{m}\sum_{j=1}^{m} E_j$$

$$= G_i.$$

In other words, the average phenotypic value becomes the genotypic value, because it is assumed that the environmental deviations over all environments cancel out one another.

In many cases, different populations have different mean phenotypic values. It is generally difficult to determine whether such a difference is the result of genetic factors, environmental factors, or a combination of both (see the discussion on race and IQ in section IV of this chapter). However, if the populations can be grown in the same environment (sometimes called a common-garden experiment), then we can make some inferences. For example, if individuals from different populations retain phenotypic differences in a common environment, then this result is consistent with the hypothesis that the populations are genetically different. If, on the other hand, the phenotypic differences disappear in the common environment, then this would indicate that environmental factors are important in determining the phenotypic differences. These conclusions are somewhat oversimplified, because, for example, such results may depend upon which of a range of possible common environmental conditions are used.

A particular genotype may do well in a specific environment. If such specific interactions occur between genotypes and environments, then the basic model above needs to be expanded to include a term for *genotype–environment interaction*, with the phenotypic value becoming

$$P_{ij} = G_i + E_j + GE_{ij}$$

Example 8.3. We can explore the relative importance of genetic and environmental factors by growing or raising individuals of known genotypes in different environments. Clausen et al. (1941) examined several plant species that ordinarily grow in a variety of environments from the warm, temperate California coast to the alpine areas of the Sierra Nevada. For example, they grew clones of *Potentialla glandulosa*, a member of the rose family, collected on the coast at an altitude of 100 meters, midway up the Sierras at 1,400 meters, and high in the Sierras at 3,040 meters, at each of the three locations. In other words, each genotype was grown in each of the three environmental conditions.

Figure 8.6 gives an example of these results, where each row represents a genotype from the different areas and each column indicates the environment in which these plants were grown. These results indicate that both genetic and environmental factors are important. For example, the coastal genotypes grow well at both low and intermediate altitudes but cannot even survive at the high altitude (first row), indicating the importance of environmental factors. On the other hand, the alpine genotype grows much better than the other genotypes at the high altitude (last column), indicating the importance of genetic factors.

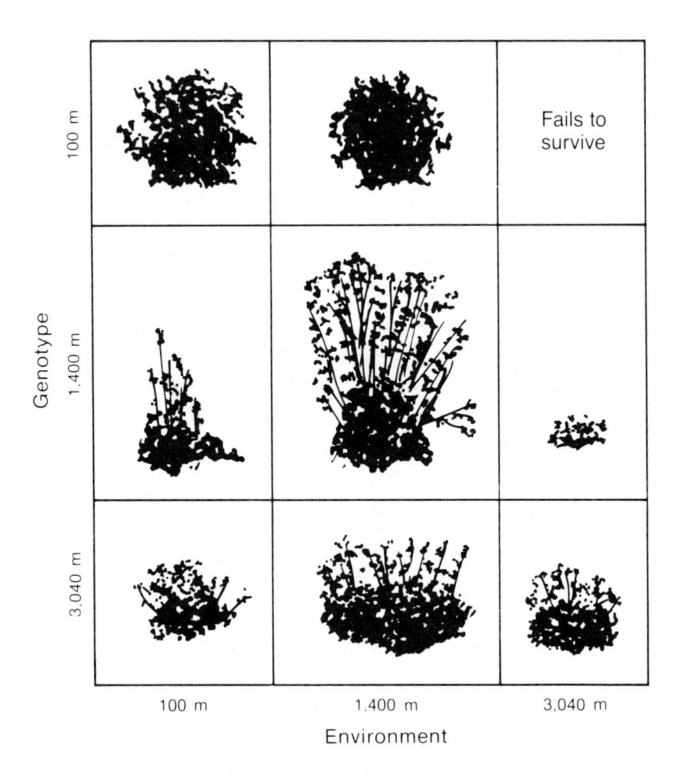

Figure 8.6. The phenotypes of plants from three different environments, each cloned and grown at the three environments (Clausen, 1941).

where GE_{ij} measures the interaction between the ith genotype and the jth environment. As with the environmental deviation, the genotype–environment interaction may be either positive or negative.

To illustrate the potential complexity introduced by genotype–environment interaction, consider the simple situation of two genotypes in two environments. It is possible that a reversal of the order of the phenotypic values of two genotypes may occur in two different environments, as for example with a domesticated strain and its wild progenitor of a species raised either in association with humans or in its natural environment. The wild progenitor will survive and reproduce better in the natural situation, while the domesticated strain will survive and reproduce more successfully in association with humans. As a result, the phenotypic value of, say, survival is primarily the result of the interaction between the specific genotype and the specific environment and cannot be predicted without information about both the genotype and the environment (see also Example 8.4).

The genetic model can be more precisely defined, with the value of the ith genotype being

$$G_i = A_i + D_i$$

where A_i and D_i are the additive and dominant components, respectively. In other words, the phenotypic value can be symbolized as

$$P_{ij} = A_i + D_i + E_j$$

if it is assumed that the genotype–environment interaction is small. The different types of genetic contributions can be illustrated by examining a one-locus model, assuming for the moment that there is no environmental effect on the phenotype. If all the genetic effects are additive, then the effect is as given in the example in Figure 8.8(a). Here, as an A_1 allele is added (actually substituted) as from A_2A_2 to A_1A_2 or from A_1A_2 to A_1A_1, the phenotypic value is increased by a given increment, $\frac{1}{2}x$ in this example. On the other hand, when there is dominance, then the effect at a locus is as given in Figure 8.8(b). In this case, the substitution of A_1 in A_2A_2 to give A_1A_2 does not increase the phenotypic value by the same increment as the substitution of A_1 in A_1A_2 to give A_1A_1. With dominance, the effect on the phenotype can be divided into an additive component, as given by the dotted line, and a deviation from this line due to dominance, as indicated by the arrows.

As pointed out earlier in this chapter, for quantitative traits there is no exact genotype-to-phenotype correspondence. In addition, because the average environmental deviation in a population must be zero, the environmental effect cannot be easily described using the model above. As a result, the above models are only generally useful in a heuristic sense. Instead, a more practical approach to examining quantitative traits is to partition the population variation for a particular phenotype into components due to genetic factors and those due to environmental factors. If the basic model given above is used to examine the phenotypic variance (with some

Example 8.4. Significant genotype–environment interactions may exist for traits involving behavior (see also Chapter 18). An example is the comparison in two strains of rats of the ability to complete a maze in three different environments: restricted, normal, and enriched (Cooper and Zubek, 1958). These environments are ranked on the amount of visual stimulus present during the maturation of the rats. The two strains examined had been selected in a normal environment for either increased or decreased maze-learning ability and are called the maze-bright and maze-dull lines, respectively. In the normal environment, animals from the maze-bright strain make, on the average, 50 fewer errors than animals from the maze-dull strain, reflecting the effectiveness of selection (see Figure 8.7). However, in both the restricted and the enriched environments, the two strains make a similar number of errors, primarily because the maze-bright strain is unable to run the maze well under restricted conditions and the maze-dull strain is able to perform well under enriched conditions. The implications of these observations are intriguing and suggest that differences in behavioral performance may only be present in a narrow range of environments and may not be generalizable to a diversity of environments.

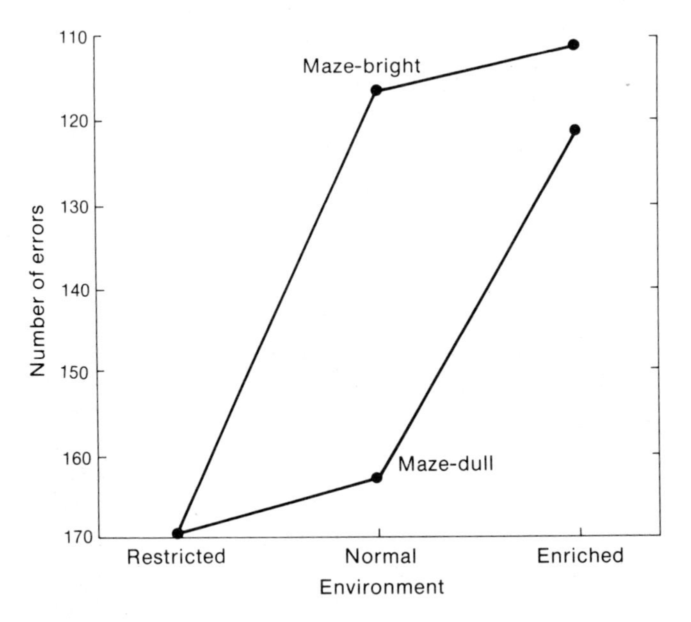

Figure 8.7. The numbers of errors in running a maze for two strains of rats, maze-bright and maze-dull, when raised in three different environments (Cooper and Zubek, 1958).

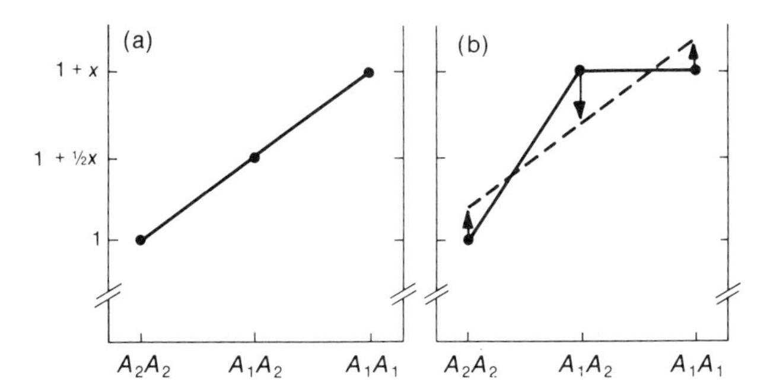

Figure 8.8. The phenotypic value for three different genotypes when there is (a) no dominance or (b) complete dominance.

assumptions), it becomes

$$V_P = V_G + V_E$$

where V_P, V_G, and V_E are the phenotypic, genetic, aand environmental variances, respectively. If both sides of this expression are divided by V_P, it becomes

$$\begin{aligned} 1 &= \frac{V_G}{V_P} + \frac{V_E}{V_P} \\ &= h^2 + e^2. \end{aligned}$$

In this expression h^2 and e^2 are the proportion of phenotypic variation due to genetic factors and environmental factors, respectively. The term h^2 is generally known as the *heritability in the broad sense* and can be signified as h_B^2—that is,

$$h_B^2 = \frac{V_G}{V_P}.$$

Adopting the same approach, if we use the expression with separate additive and dominant components to examine the phenotypic variance, then it becomes

$$V_P = V_A + V_D + V_E$$

where V_A and V_D are the variances due to additive and dominant effects, respectively. If this expression is divided by V_P, then we can obtain the proportion of phenotypic variance due to the different genetic components. The ratio of the additive genetic variance to the phenotypic variance is known as the *heritability in the narrow sense*, or

$$h_N^2 = \frac{V_A}{V_P}.$$

The magnitude of the heritability in the narrow sense is important in determining the rate and amount of response to directional selection and is often estimated by plant and animal breeders prior to initiating a selection program.

III. ESTIMATING GENETIC VARIANCE AND HERITABILITY

We can take several different approaches in estimating the amount of genetic variance and heritability. However, before we discuss these approaches, it is important to realize that such measures of genetic variance or heritability for a trait are specific to particular populations and environments. For example, the genes that are variable and that affect the trait may be different in different populations, making such measures population specific. In addition, the environmental factors that affect the phenotypic variance may be quite important in some environments and unimportant in others. As a result, estimates of heritability for a particular trait may vary from one population or environment to another. We will briefly discuss three approaches used to estimate genetic variation and heritability: those that reduce or eliminate one component of variance, the observed response to directional selection, and the resemblance between relatives.

In some organisms, it is possible by experimental manipulation or breeding either to reduce or to eliminate the environmental or genetic-variance components in order to estimate the complement variance component. In other words, if $V_G \cong 0$, then $V_P \cong V_E$ or if $V_E \cong 0$, then $V_P \cong V_G$. For example, the genetic variation is zero or near zero in inbred lines, clones, hybrids between inbred lines, and repeated measurements on the same individual. Environmental variance may approach zero if all important environmental factors, such as food, moisture, and temperature, are known and carefully controlled.

The phenotypic variance in both inbred lines and crosses between inbred lines should be nearly equal to the environmental variance, because little or no genetic variation exists within inbred lines or their hybrids. However, inbred lines often have very large amounts of phenotypic variation. The difference in phenotypic variation between inbred lines and hybrids is apparently due to the greater tendency of inbreds to reflect variation in the environment. An example of such data is given in Table 8.1 from Falconer (1981). In some of these comparisons, the variation in hybrids is less than 50 percent of that in inbreds. The exact basis for this phenomenon is not clear but it has been suggested that hybrids may have a greater developmental buffering against environmental influences.

The rate and amount of response to directional selection are a function of the heritability in the narrow sense of the trait under selection. (Selection that

Table 8.1
The phenotypic variance in inbred lines and a F_1 cross between inbreds for several traits (from Falconer, 1981).

Organism	*Trait*	*Inbreds*	*Hybrids*	*Hybrids/Inbreds*
D. melanogaster	Wing length	2.35	1.24	0.528
Mouse	Weight at birth	119	59	0.496
	Weight at 60 days	24	19	0.792
Rat	Weight at 90 days	522	170	0.326

is determined by human criteria, rather than natural factors, is called *artificial selection.*) This is quite obvious when h_N^2 is either 0 or 1. When $h_N^2 = 0$, then none of the phenotypic variation is from the additive-genetic-variance component, and, as a result, selecting even extreme individuals as parents would not change the phenotypic mean. On the other hand, if $h_N^2 = 1$, then all the phenotypic variation is from the additive-genetic-variance component, and phenotypically extreme individuals are that way because of their genotypes. In general, the *response*, the difference in the parental mean and the offspring mean, is

$$R = h_N^2 S$$

where the S is the *selectional differential*, the difference in the mean of the selected parents and the mean of all the individuals in the parental population. This equation may be rearranged so that an estimate of heritability in the narrow sense is

$$h_N^2 = \frac{R}{S}.$$

This expression is also called the *realized heritability* because it is the ratio of observed response over the total response possible.

The relationship between the level of realized heritability and the selection response is given in Figure 8.9 for a hypothetical quantitative trait. Here, the overall phenotypic distribution is indicated under the curve, while the shaded region in the top part of the figure is the segment of the parental population selected to be parents for the next generation. The selection differential and response are

$$S = \bar{P}_S - \bar{P}$$

and

$$R = \bar{P}' - \bar{P}$$

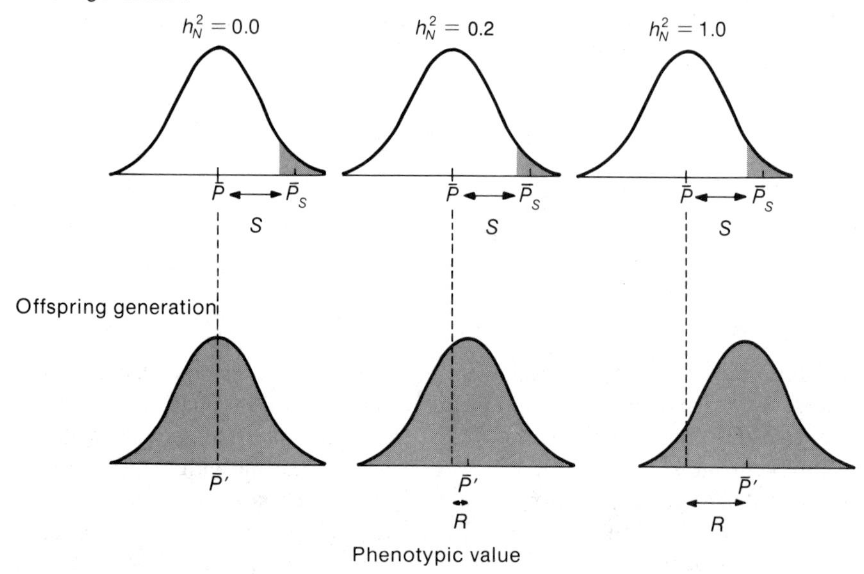

Figure 8.9. The change in phenotypic distribution and mean from one generation of selection when the heritability is 0.0, 0.2, or 1.0.

where $\bar{P}$, $\bar{P}_S$, and $\bar{P}'$ are the means of the parental population, the selected parents, and the offspring, respectively. When h_N^2 is 0 or 1, then $R = 0$ and $R = S$, respectively. If h_N^2 is smaller than 1—say $h_N^2 = 0.2$, as in the middle of Figure 8.9—then $R < S$. Such values occur because some of the individuals selected to be parents are extreme because of good environments and not because of genetic factors.

Lewontin (1974) reviews the type of traits that have been selected successfully in *Drosophilia melanogaster*. These include size, bristle number, developmental rate, fecundity, behavioral traits, mutant expression, environmental sensitivity, and resistance to insecticides. He concludes that "there appears to be no character—morphogenetic, behavioral, physiological or cytological—that cannot be selected in *Drosophila*." Furthermore, it seems that virtually any population has enough genetic variants in it for response to selection for any trait. Of course, some traits respond much more quickly than others, which is generally a reflection of the amount of genetic variation for that trait.

Many directional (or artificial) selection experiments give evidence of a change in phenotypic values over 10 to 40 generations, often reaching a maximum or plateau before the termination of the experiment. There is evidence now that selection response may continue for many generations if there is an adequate initial sample of genetic variation, a reasonable population size during the selection process, and an effective selection scheme (see Example 8.5).

Example 8.5. One of the longest selection experiments ever conducted is that still running at the University of Illinois for oil and protein percentage in corn. High and low directional selection has been carried out for 76 generations (Dudley, 1977) and progress is still continuing, although large changes in oil and protein percentage have already been made. The data for the high-oil (HO) and low-oil (LO) lines are given in Figure 8.10. The percentage of oil has increased nearly fourfold in the high line and has been reduced to below 1 percent in the low line. In generation 48, reverse selection lines were begun, and both the low selection from the high line (RHO) and the high selection from the low line (RLO) were successful in changing the mean values. These data, along with other estimates of genetic variability, indicate that genetic variation still exists in these lines, even after many generations of selection. This is even more remarkable because it is generally accepted that a relative lack of genetic diversity exists in major cultivated plant species (see discussion in Chapter 21).

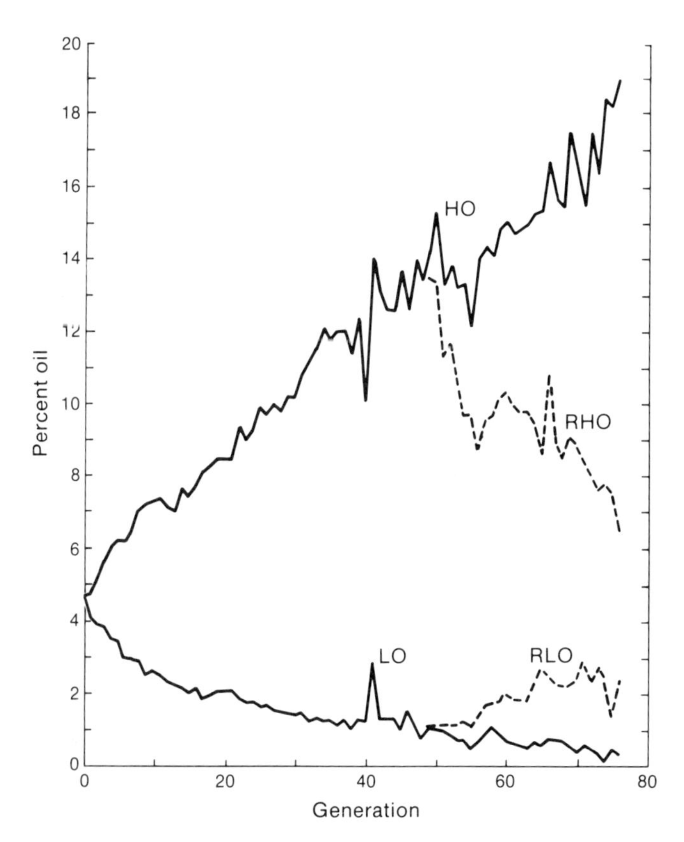

Figure 8.10. The percent of oil content in a long-term selection experiment in corn (Dudley, 1977).

Perhaps the most important approach used to measure components of genetic variation and heritability is based on the phenotypic resemblance between relatives. If genetic factors are important, then one would expect the offspring of parents with high phenotypic values to have high phenotypic values and the offspring of parents with low phenotypic values to have low phenotypic values. Obviously, close relatives will share more genes in common with each other than with nonrelatives or distant relatives, and thus should have greater phenotypic resemblance. For example, monozygotic twins will share all of their genes, on the average siblings will share one-half of their genes, and first cousins will share one-eighth of their genes. As a result, the phenotypic values between close relatives should be more highly correlated than those between more distant relatives (see Example 8.6). Estimates of heritability for a number of traits using this approach in different organisms

---

Example 8.6. Fingerprints have been widely used for identification purposes because the fingerprint patterns of an individual are unique. A general way of classifying fingerprints is to use a count of the number of ridges, the raised areas of skin, between certain aspects on the fingertips. There are three basic configurations of ridges on fingertips known as

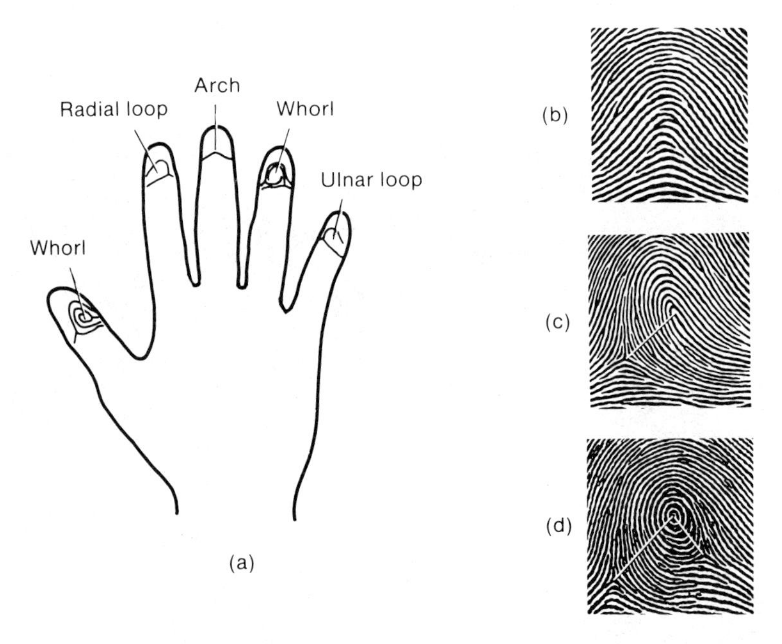

Figure 8.11. The position of fingerprints on a hand and examples of the three basic figureprint types (after Holt, 1961).

are given in Table 8.2. Notice that the traits most closely related to fitness, such as litter size and viability, appear to have lower heritabilities.

A number of diseases, such as diabetes, cleft palate, and spina bifida, have a strong genetic component but do not appear to have a single-gene mode of inheritance. For such diseases, it can be assumed that the genetic liability to the disease is polygenically controlled and that, consequently, close relatives of affected individuals should be more likely to become affected. Obviously, the closer the affected relative and the more relatives affected, the higher would be the genetic liability in an individual.

For such diseases, empirical recurrence risks of the disease are often calculated from a large number of case studies, although heritability measurements may also be used. Empirical risk values may be calculated where both parents are affected, where one parent is affected, where one sib is

---

the arch, loop, and whorl (Figure 8.11). The number of ridges is illustrated for fingerprints of each type in Figure 8.11(b), (c), and (d) (after Holt, 1961). For arches, the count is zero because the triradii—which are points where three ridges join and the landmarks used for counting ridges—are missing [Figure 8.11(b)]. For loops, the ridge count is the number of ridges from the triradius, the Y-shaped structure just left and below the center in Figure 8.11(c), to the center of the loop. In this case, the count is 13 ridges along the line from the triradius to the loop center. In the whorl pictured in Figure 8.11(d), the ridge count is 23, 15 from the left triradius to the center of the whorl and 8 from the right triradius to the center. Fingerprint ridge count seems to be almost entirely genetic, with little noticeable environmental effect. Table 8.3 gives, for different degrees of relationship, the observed correlation and theoretical correlation, assuming that the trait is completely genetically determined and that variation is only due to additive genes. Notice that the observed values are very close to those predicted in every case, indicating the high degree of genetic determination.

Table 8.3
The observed and theoretical correlation coefficient for the total fingerprint ridge count for different relationships (Holt, 1961).

Relationship	*Observed*	*Theoretical*
Monozygotic twins	0.95	1.00
Mother–child	0.48	0.50
Father–child	0.49	0.50
Dizygotic twins	0.49	0.50
Full sibs	0.50	0.50

Table 8.2
Approximate values of heritability of various characters in several species (after Falconer, 1981).

Organism	*Trait*	*Heritability*
Swine	Back fat thickness	0.55
	Body weight	0.3
	Litter size	0.15
Poultry	Egg weight	0.6
	Body weight	0.2
	Viability	0.1
Mice	Tail length	0.6
	Body weight	0.35
	Litter size	0.15
Drosophila melanogaster	Bristle number	0.5
	Body size	0.4
	Egg production	0.2
Corn	Plant height	0.70
	Yield	0.25
	Ear length	0.17

affected, or in other situations. Because of the complex genetic and environmental factors that affect such diseases, empirical risk values can only be used as a guideline, since specific genetic or environmental factors may either increase or reduce the risk in a particular family (see Example 8.7).

IV. RACE AND INTELLIGENCE

In the late nineteenth century, some British scientists suggested that the high reproductive rate among the "lower" classes was leading to an overall deterioration of the average intelligence of the British population. The discussion that followed was based on what became known as the nature–nurture controversy—that is, given that there were real differences in intellectual ability among social classes, was the basis of this difference due to genetic or environmental factors? In the United States, a similar controversy arose in the early part of this century resulting from concern over the effect of large numbers of immigrants.

The controversy was revived again in the late 1960s when it was claimed that the difference in intelligence quotient (IQ) scores among racial groups was genetically determined (Jensen, 1969) and, in particular, that the lower IQ scores among American blacks was the result of genetic factors. This view, called *the hereditarian school*, is premised on several factors: (1) there is a significant difference in the mean IQ scores between American blacks and whites; (2) IQ is a phenotypic measure and can be analyzed like any other quantitative trait; (3) compensatory education has not eliminated the difference in IQ between blacks and whites; and (4) there is a high heritability for IQ, approximately of 0.8 or larger. Based on these considerations, Jensen and others concluded that IQ differences were genetically determined and recommended various social and educational programs. A number of scientists, educators, and others reacted strongly to these views and proposals. Because the hereditarian view is based on genetic differences, it is instructive to note that many geneticists were critical of it. For example, a resolution adopted by the Genetics Society of America in 1975 states that "in our view, there is no convincing evidence of genetic difference in intelligence between races."

One basic question concerns the validity of the measurement of intelligence by standardized tests. Obviously, mental ability is a complex entity that can be expressed in a number of different forms, some of which may not be determined by standardized IQ tests. For example, artistic creativity, a mental ability held in great esteem by most people, is not evaluated by IQ tests. One of the most serious objections to many IQ tests is that they may be to

---

Example 8.7. Pyloric stenosis is a disease in which the pyloric valve, the sphincter valve between the stomach and the small intestine, is not completely opened. Before corrective surgery was introduced to open this valve in the 1920s, most individuals with the disease died in infancy because little or no food was passed on to the small intestine. After corrective surgery, individuals with pyloric stenosis have normal digestion and generally no other related problems. Risk values for this disease are given in Table 8.4 with respect to the disease state of the parents and the number of affected sibs. The risk, as intuitively expected, increases as either the number of affected sibs or affected parents increases. For example, with no affected relatives, the risk is only 0.3 percent, while with two affected sibs and two affected parents the risk is greater than 40 percent.

Table 8.4
Empirical risks for congenital pyloric stenosis for the English population (after Bonaiti-Pellié and Smith, 1974).

Number of affected sibs	*Neither parent affected*	*Father affected*	*Mother affected*	*Both parents affected*
0	0.003	0.037	0.051	0.298
1	0.030	0.115	0.135	0.365
2	0.119	0.216	0.231	0.431

some extent culturally bound—that is, they may include questions that imply particular cultural experiences limited to mainstream culture and, therefore, discriminate against individuals brought up in minority cultures. These and other concerns about the tests and testing procedures constitute obvious problems with the basic data.

Many researchers have suggested that environmental factors may have a substantial effect on IQ scores, citing a range of environmental factors, including such obvious elements as nutrition (prenatal, during childhood, and at the time of testing), socioeconomic status of parents (see Example 8.8), and educational and testing experience. Of course, more subtle factors may be

---

Example 8.8. One concern in the comparison of IQ scores of blacks and whites was that these groups differed not only by ethnic group but also by education, urban versus rural upbringing, and other socioeconomic factors. An obvious approach to evaluating the effect of socioeconomic factors was to compare blacks and whites within the same socioeconomic category. Nichols and Anderson (1973) attempted to do this for two samples, one of relatively high overall socioeconomic status (SES) from Boston and one of relatively low overall socioeconomic status from Baltimore and Philadelphia. The index of socioeconomic status referred to parental occupation, education, and income data. Their results for the scores of seven-year-olds in the two samples are given in Figure 8.12. Obviously, the distributions have large amounts of overlap and their mean values are quite close—the mean values for Boston whites and blacks were 104.2 and 100.0, respectively, and for the Baltimore–Philadelphia sample they were 95.3 and 91.2, respectively. These differences, about four points in both cases, indicate that the differences cited by hereditarians (and larger than those in this study) are probably due in large part to socioeconomic differences between blacks and whites.

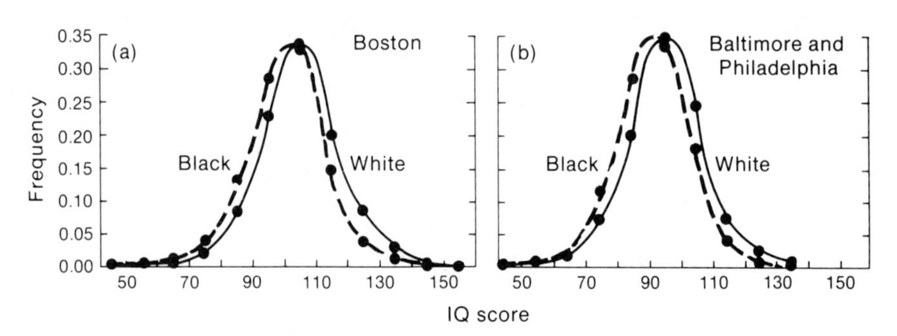

Figure 8.12. The IQ scores of high SES black and white populations in Boston and low SES in Baltimore and Philadelphia (Nichols and Anderson, 1973).

---

significant in many cases. One important finding is that IQ scores are subject to change with changes in environmental factors.

If whites have higher IQ scores than blacks because of genetic factors, then one might expect that blacks with higher proportions of white ancestry would have higher IQ scores. Of course, a confounding environmental factor exists if blacks with high-white ancestry experience a different environment from blacks with low-white ancestry. For example, the offspring of black–white matings should have 50 percent black and 50 percent white ancestry and, in a hereditarian view, should have IQs intermediate between those of blacks and of whites, assuming no dominance exists. But several studies have shown that the IQ scores of such mixed-race children instead do not differ significantly from their white peers. For example, Tizard (1974) examined mixed-race children in England who had spent part of their early life in institutions and found that they had slightly higher scores than their white peers. And a study of illegitimate offspring of American servicemen and German women in World War II provides another comparison in which no overall difference in average IQ was found between the children of black or white servicemen.

Implicit in the hereditarian argument is the assumption that, when heritability is high, then differences in IQ scores between groups must be genetically based. However, such reasoning may be completely fallacious, since the cause of variation within a group and the cause of mean difference between groups may not be the same at all. There is certainly no a priori reason to assume a very high positive correlation between these factors—that is, a large genetic causation within groups (high heritability) does not necessarily suggest that the difference in the means of two groups is genetically caused (see Example 8.9).

The problems of establishing the relative genetic and environmental contributions to variance in IQ are substantial. In any case, this information is not useful in identifying individuals of higher or lower innate ability, because variance estimates are relevant only to populations and not to individuals within a population. Because some hereditarians have suggested that individuals of different racial groups be separated in schools and occupations, there has been a strong reaction both from minorities and from individuals who feel that equal opportunity is a fundamental right. This concern is well stated in the following quotation from Goldsby (1971).

> It is likely that public policies based on the belief that differences in the environment account for the Black–White difference would differ from policies based on the alternative genetic hypothesis. Subscription to the environmental view suggests that improvement of the environment, extension of opportunity, and efforts to compensate for obvious social, educational, and economic disadvantages, if sufficiently massive and continuous, will narrow and eliminate the gap. Such a hypothesis leads to the conclusion that there are few American Black plumbers, executives and stock brokers because social factors have operated to exclude them. It leads to a policy of actively opening up opportunities and, if it is properly applied, to adequately training capable people for them. On the other hand, a plausible extension of the genetic hypothesis

suggests that the underrepresentation of Blacks in some areas of the society is as one might expect because the pool of able individuals is proportionately lower in that population. On the basis of a genetic hypothesis, low achievement scores by whole groups or throughout an entire school or even school system may be no cause for concern about the quality of the educational experience: The students are assumed to be inherently less able and hence can be expected to do less well. Under such a hypothesis the solution to the problem of low achievement scores can be to expect less.

With these considerations in mind it is not extreme to suggest that whichever hypothesis is correct, a belief in the environmental one is likely to lead to social policies aimed at the expansion of opportunities and the inclusion of racial minorities whereas subscription to the genetic one could produce a policy of contraction and exclusion.

SUMMARY

Quantitative traits are affected by many genes and by the environment. A quantitative genetics model is introduced that considers genetic and environmental components and heritability. We also introduced estimation of genetic

---

Example 8.9. Perhaps the best way to comprehend the fallaciousness of the hereditarian logic is to consider the following situation (after Lewontin, 1970). Assume that we are examining a phenotypic character for which the major environmental factors are known and can be controlled—for example, height in a crop plant such as corn. First, let us consider two highly inbred strains of corn, A and B, which, because of long-term inbreeding, are assumed homozygous but for some different alleles. As a result, when the lines are grown under the same conditions, they have different height distributions and means as shown in Figure 8.13(a). The phenotypic variation within each inbred line is entirely the result of environmental factors, and the heritability within each line is thus nearly zero. However, the cause of the mean phenotypic difference between the lines is entirely genetic, because different alleles have been fixed in the two inbred lines.

Now let us consider an outbred line, C, which has a large amount of genetic variation, and separate it into two groups. Both of these groups will be grown in a tightly controlled environment, and no variation will therefore occur in humidity, temperature, soil, or other factors known to affect plant growth. However, the soil for one group will contain sufficient amounts of all trace elements, while the soil for the other group will lack an essential trace element, zinc. Zinc is essential for normal growth at the very low concentration of six parts per million. Because of the tight environmental control within each group, there will virtually be no within-group environmental variation, and, consequently, the heritability will be near unity. However, there will be a difference in mean plant height

variance and heritability by elimination of one variance component, selection response, and phenotypic resemblance between relatives. These principles were applied to the race–intelligence controversy.

PROBLEMS

1. Discuss the characteristics of quantitative traits and distinguish them from single-gene traits. Explain how disease susceptibility may be a quantitative trait.
2. Given that V_p in an inbred line is 0.022, and V_p in an outbred line is 0.033 for a particular trait, what is h_B^2?
3. If the phenotypic mean in a parental group is 22, the mean of the selected parents is 30, and the mean in the progeny is 24, what is an estimate of h_N^2?
4. Discuss how selection, mutation, gene flow, and genetic drift would affect the phenotypic and genetic variation for a quantitative trait.
5. What are the major determinants of IQ? Explain, using some examples or data to support your answer.

---

between the groups, and this difference will be entirely environmental, resulting from a subtle environmental difference, the presence or absence of zinc [Figure 8.13(b)].

This example illustrates that there need not be a positive association between the extent of genetic determination within a population and causation of phenotypic differences between the populations. Furthermore, the result shown in Figure 8.13(b) may be analogous to the measurement of IQ in blacks and whites.

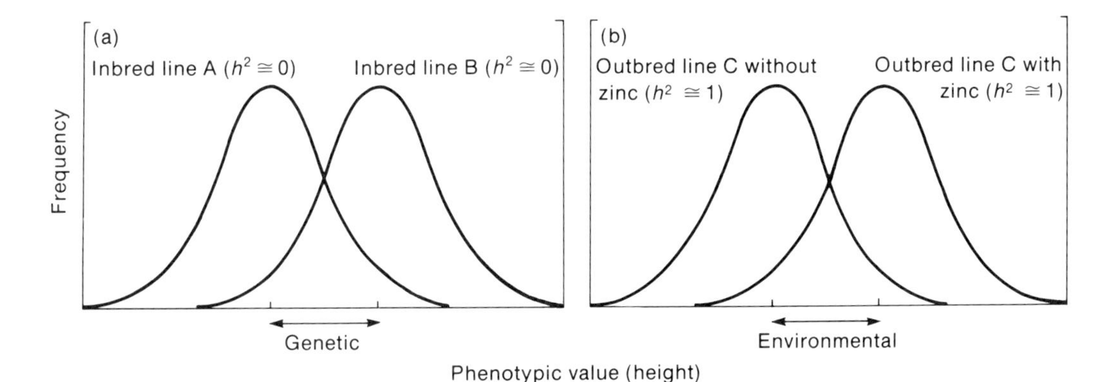

Figure 8.13. A hypothetical example in which (a) the difference between two inbred lines is entirely genetic and $h^2 \cong 0$, and (b) the difference between two samples of an outbred line with $h^2 \cong 1$ is entirely environmental.

CONCLUSIONS TO PART 1

A large amount of genetic variation exists in most populations, as indicated by studies of electrophoretic variation and selection experiments for quantitative traits. A variety of factors can influence the extent of genetic variation (see Table C.1). Some of these may have different effects in different situations as selection that can either reduce or maintain the amount of genetic variation in a population. Expressions for the rate of change in the frequency of an allele, Δq equations, are useful in determing the relative importance these factors potentially have.

In many populations more than one of these factors may be important, and, of course, factors can act together to further reduce the amount of genetic variation, as may the combination of gene flow and selection. Of particular importance are the instances in which factors counteract the effects of each other and result in maintenance of genetic variation (see Table C.1). Such equilibria may occur due to the joint effects of mutation and selection, gene flow and selection, and genetic drift and mutation.

Table C.1
A summary of the factors that can influence the amount of genetic variation and their general effect either singly or jointly.

	Effect on genetic variation		
Factor or factors	*Reduce*	*Maintain*	*Increase*
Single			
Selection	Yes	Yes	—
Mutation	—	—	Yes
Gene flow	Yes (q_m = 0 or 1)	Yes ($q_m \neq$ 0 or 1)	Yes ($q_m \neq$ 0 or 1)
Genetic drift	Yes	—	—
Nonrandom mating	Yes (inbreeding)	—	Yes (outbreeding)
Joint			
Mutation–selection	—	Yes	—
Gene flow–selection	Yes	Yes	Yes
Genetic drift-mutation	—	Yes	—
Inbreeding–selection	Yes	—	—

PART 2

Population Ecology

The ecology of populations is concerned primarily with the factors responsible for the *distribution* of organisms geographically and the *abundance* of these organisms in areas where they exist. In many cases, the important factors affecting distribution and abundance are the same. For example, temperature can affect both the geographical extent of the distribution of an organism and the abundance in areas in which it exists. In this section, we will consider primarily the ecological, or immediate, factors in the environment, such as temperature or predation, that affect distribution and abundance. The ecological factors may be divided into abiotic (or physical) factors and biotic factors. Biotic factors include the effects of other species, such as competitors and predators, and intrinsic aspects such as dispersal, birth rates, and death rates. In Part 3 we will consider some of the evolutionary factors that affect the distribution and abundance of organisms.

The ecological factors that determine distribution and abundance are important in a number of applications. For example, an understanding of these factors may be used in fish, wildlife, and forest management, permitting the maintenance of appropriate densities for harvesting. For rare or endangered species, if we know the important factors affecting their numbers, we may be able to prevent their extinction. In addition, an understanding of the interactions among species and their effects on abundance is a fundamental aspect of pest control by the use of predators, competitors, or parasites. Perhaps the most important application is in determining the consequences of the human population on other species and the potential effects of such factors as disease, food scarcity, and human numbers on the human population itself.

CHAPTER 9

Factors Affecting Distribution and Abundance

In solving ecological problems we are concerned with *what animals do* in their capacity as whole, living animals, not as dead animals or as a series of parts of animals. We have next to study the circumstances under which they do these things, and, most important of all, the limiting factors which prevent them from doing certain other things. By solving these questions it is possible to discover the reasons for the *distribution and numbers of animals in nature.*

CHARLES ELTON

In this chapter, we will focus on the abiotic and biotic factors that can potentially affect the distribution and abundance of an organism over an area (abundance will be discussed in more detail in following chapters). In this section, we are assuming that a species does not change genetically in response to the presence of either abiotic or biotic factors. In Part 3, we will consider adaptation, life-history evolution, coevolution, and related topics that deal specifically with genetically based responses to environmental factors.

In the distribution of an organism, there are often areas of high abundance (or density, a measure of abundance that we will discuss in Chapter 10) in which all the factors necessary for the continued existence of the species appear to be present in amounts well suited for the species. As the value of an environmental factor declines from an optimum level, then the density of the species should decline. Figure 9.1 illustrates this for an environmental factor that can be measured on a linear scale, such as temperature or rainfall, and that ranges from a low to a high value. Toward the environmental extremes, the environment becomes so stringent that the species can no longer successfully reproduce. For example, at very high or very low temperatures, individuals can no longer exist, the density becomes zero, and the boundary of the species distribution occurs. Figure 9.2 gives a schematic example of a species distribution indicating different densities resulting from the changes in an environmental factor over space. In this example, an area of high density is surrounded by an area of moderate density. As the conditions become more severe—say, because of low moisture—the density decreases, and eventually the species is absent altogether. The edge of the distribution of the species is indicated by the solid line on the left.

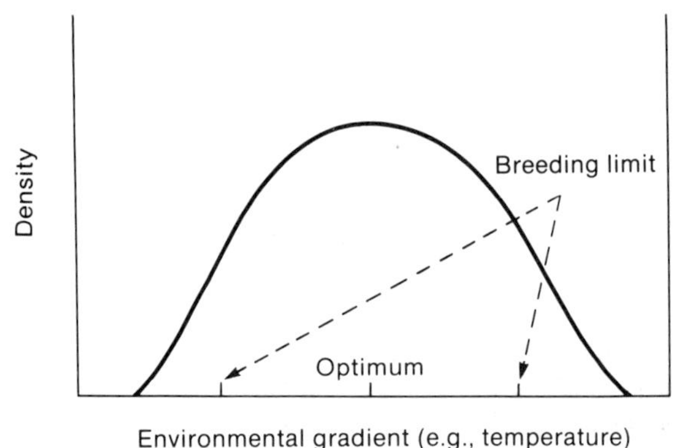

Figure 9.1. The effect of an environmental factor on the density of an organism.

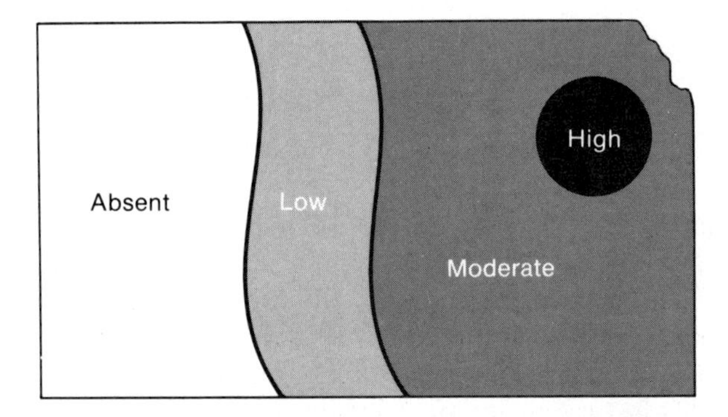

Figure 9.2. A schematic representation of a species distribution indicating areas of low, moderate, and high density.

I. ECOLOGICAL NICHES

For any given species to survive and reproduce in an area, the environmental conditions must fall within a given range. One way to imagine these limits is to consider two factors simultaneously, such as temperature and moisture. Figure 9.3(a) illustrates a range of temperature and moisture in which a hypothetical organism can exist (shaded region). Beyond these limits—for example, at hot temperatures or low moisture levels—the species cannot exist. A laboratory example of these limits is given in Figure 9.3(b) for two species of grain beetles that occur in Australia (after Birch, 1953). The limits indicate the environmental extremes in which the species can maintain a population in the laboratory. As would be predicted from these laboratory experiments, *Rhizopertha* occurs in more arid areas of Australia.

In reality, many environmental factors may be simultaneously important to the existence of a species. In considering all such factors, we can define the

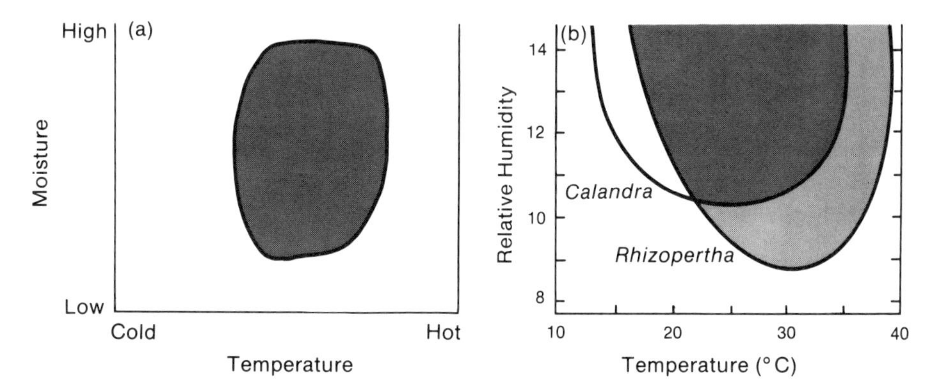

Figure 9.3. (a) The limits in temperature and moisture within which a hypothetical species can exist. (b) The moisture and temperature conditions in which two species of grain beetles can persist (Birch, 1953). The darkly shaded region indicates the conditions in which both can exist.

ecological niche as the total set of environmental conditions necessary for the persistence of a species (for a historical review of the niche concept, see Vandemeer, 1972). In theory, this can be imagined as a volume with as many dimensions as there are important factors. In other words, the ecological niche is that domain in multidimensional space defined by the range of each environmental factor that allows the organism persist. The niche defined in this way is the *fundamental niche* of a species—that is, all the potential combinations of environmental conditions in which a particular species may exist. In general, the different niche dimensions denote *resources* necessary for the species, such as food type, habitat type, and moisture. Along a resource dimension, such as habitat type, only a certain array of conditions will be suitable for and utilized by the species. In many cases, the species may not actually exist in all environments that could potentially support it. The conditions in which it does actually exist are termed the *realized niche* of a species. For example, a species may be excluded by a competitor from an area in which it could exist, making the realized niche smaller than the fundamental niche.

In many cases, the fundamental niche and the distribution of an organism are closely related. Obviously, a species can only exist in areas that fit the environmental constraints of its fundamental niche. On the other hand, the distribution, or range, of a species may not include all areas that fit its fundamental niche conditions, because, for example, the suitable area is occupied by a competing species or the site is so remote that it cannot be easily colonized. In addition, over the dimensions of the fundamental niche, the density of a species may vary. The region of environmental combinations for which the species has the greatest density is called the *preferred niche* of the species.

In some cases, particularly on islands or in isolated areas, it appears that no organisms fill a particular environmental combination of factors—that is,

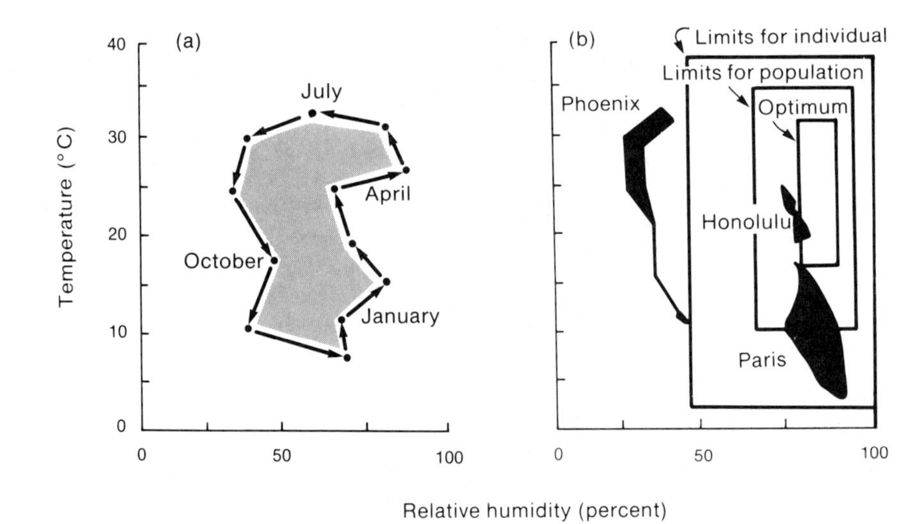

Figure 9.4. The limits of mean temperature and humidity over a year for (a) a hypothetical area and (b) three areas along with the biological limits for the Mediterranean fruit fly (from Ricklefs, 1975).

there is an *empty niche*. However, the vacancy may only be an apparent one, and it may be experimentally examined by the introduction of an appropriate species. If the new species successfully colonizes and does not affect other species, then it would appear that the niche was indeed previously empty (see the discussion in next section).

In some species, only a few factors largely determine the fundamental niche. For example, the temperature and relative humidity (or rainfall) patterns over a year may be the most significant factors. The seasonal variation in two such factors can be diagrammed, using measures such as the monthly mean temperature and the relative humidity, in a *climatograph* as in Figure 9.4(a). Here the hypothetical values enclose a polygon circumscribing the range of environmental combinations for a particular area.

Figure 9.4(b) gives an application of this approach to represent the seasonal variation in temperature and humidity as indicated for three areas—Phoenix, Honolulu, and Paris (after Ricklefs, 1976). The rectangular areas indicate the general ranges of these environmental factors tolerated by the Mediterranean fruit fly, a pest on many types of fruit. Obviously, Honolulu is an excellent habitat for this species, which has in fact colonized the Hawaiian Islands and is now well established there. Miami has a seasonal range of temperature and humidity similar to those of Honolulu. The Mediterranean fruit fly has invaded Florida several times, and only extreme control measures have kept it out (recall a similar event in California in 1981). On the other hand, the relative humidity in Phoenix is so low that colonization would appear to be impossible. Although the environmental conditions in Paris for part of the year are suitable, the winters are so cold that a population could not persist.

The breadth of tolerance, or the niche requirements, may vary consider-

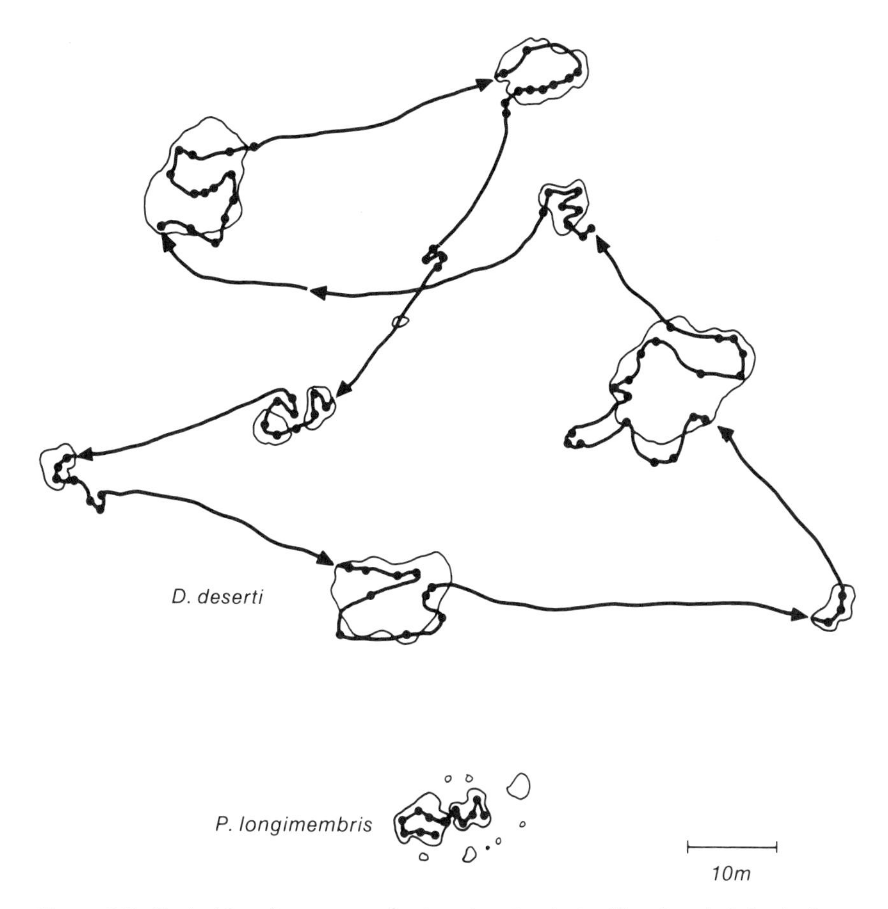

Figure 9.5. Typical foraging patterns for two desert rodents. The closed circles indicate locations of foraging activity, and the irregular shapes represent the outlines of shrubs (after Thompson, 1982).

ably among species, and is best described in terms of the *niche width* or *niche breadth* of the species (see also Chapter 13). In a theoretical sense, the niche width is the proportion of the multidimensional niche space utilized by a particular species, and similarly, the niche breadth is the range of a particular resource—say, prey size—used by a species. For example, some species may have a narrow niche width—that is, a limited tolerance or specific requirements along one or more niche axes—and are called *specialists*. Other species, having much broader niche widths, are called *generalists* or *opportunists*. For example, some birds, such as hummingbirds, are specialist feeders on nectar-producing flowers of a given morphology, and coyotes are generalist feeders on organisms of different sizes, such as insects, rodents, or deer, depending upon the relative availability of these prey (see Chapter 14).

The effects of the temporal pattern of the environment may be linked to the generation length of an organism. A general way of classifying environmental variation is to distinguish between environmental variation that occurs within a generation of an organism—*fine-grained variation*—and environmental variation in different generations—*coarse-grained variation.* The same pattern of environmental variation, such as the seasonal distribution of rainfall, may be fine-grained for a tree and coarse-grained for an annual plant. In addition, different animals may experience the same habitat in different fashions owing to mobility differences. For example, the foraging patterns of desert rodents may be quite different in the same area. Figure 9.5 gives examples of the foraging patterns for two seed-eating rodents, the bipedal *Dipodomys deserti* and the quadrupedal *Perognathus longimembris* (from Thompson, 1982). *D. deserti* utilizes shrubs in a fine-grained manner, foraging in different shrubs for large seeds, while *P. longimembris* forages in a coarse-grained fashion within a single or a few shrubs.

II. CHANGES IN DISTRIBUTION

Before we consider the factors affecting distribution patterns, let us first consider the dynamic aspects of a species distribution. First, the extent of a species distribution may vary from year to year. During a sequence of several environmentally good (optimal in some sense) years, the distribution may become temporarily extended, though it subsequently diminishes when the climate is not so favorable. For example, prairie grasses may expand their range in moist years into normally dry areas, only to have their distribution subsequently reduced by a periodic drought. As a result of such temporal fluctuations, it is often best when identifying the boundary of the species distribution to consider not isolated individuals but some measure based on a minimum viable population size or the extent of reproductive populations, measures that may not fluctuate so erratically.

The distribution of a species may change, either expanding or contracting, over relatively long periods of time. The distribution may increase over time when a species is invading a new area or decrease when the favorable habitat for the species disappears. In recent times, such changes have often been influenced either by human introduction of a species or human destruction of a suitable habitat. Human predation or human habitat alteration may lead to an extreme reduction of a species range and have led to the extinction or near extinction of a number of species (see Chapter 14).

Figure 9.6 gives an example of the range expansion of the osage orange, a tree with large and heavy fruit, often 10 cm in diameter. The natural range in recent centuries of the osage orange is represented by the shaded area, mainly in the state of Texas (Little, 1971). It is assumed that glaciation initially restricted the distribution of osage orange southward and that the tree's heavy fruit limited its expansion. Because the wood of this tree is extremely strong and rot resistant, the species was introduced into a number of more northern states for use as fence posts and as windbreaks (the wood is also used for longbows in archery). The distribution in these areas is indicated by the dots in different counties (McGregor and Barkley, 1977). Such "dot maps" are widely used to indicate the maximum distribution of a species and probably include some areas in which the organism in question does not reproduce.

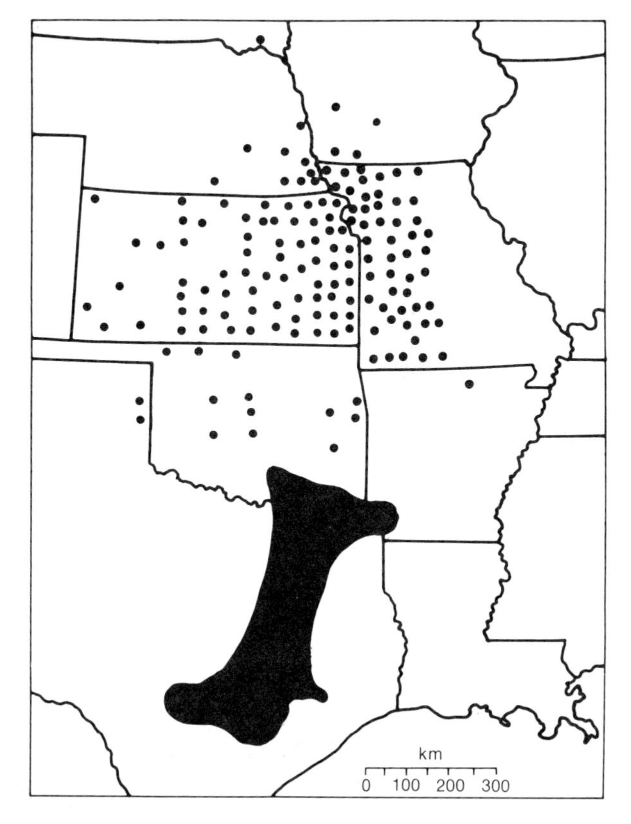

Figure 9.6. The distribution of the osage orange. The dark area indicates its natural range and the dots the extension by humans (after Little, 1971, and McGregor and Barkley, 1977).

Expansion of the range of a species has often occurred when humans have either purposely or accidentally introduced a new species from one continent to another or to an isolated area such as an island (see Carlquist, 1974, for numerous examples). This may occur for plants—examples are wild oats from Spain and eucalyptus from Australia, which have colonized large areas of California; plant diseases—the chestnut blight and Dutch elm disease introduced from Europe; insects—the spruce budworm from Europe; or vertebrates—the mongoose introduced into Hawaii from Asia or the rabbit into Australia from Europe. The rate of range expansion of an introduced species depends upon a number of factors, but the dispersal ability appears to be most important. For example, chestnut blight, a fungus that is wind dispersed, upon its introduction in 1904 spread to include the total natural range of the American chestnut by 1950. On the other hand, Dutch elm disease, which is insect dispersed by a bark beetle and which was introduced in the 1930s, has not yet infected many American elms.

A number of bird species were introduced from Europe into North America in the last half of the nineteenth century. Two of these species, the

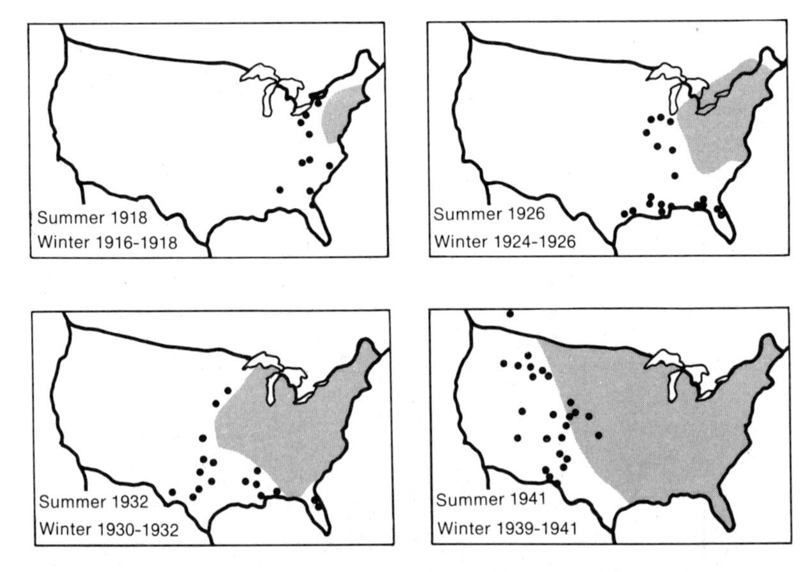

Figure 9.7. The expansion of the range of the starling where the shaded area is the summer breeding range and the dots indicate winter occurrences (after Kessel, 1953).

English sparrow and the starling, have spread extensively and they are now among the most common North American birds. After their introductions, both species spread rapidly, and now the English sparrow is found in Central America and in many parts of Canada. The starling was introduced in New York City in 1890, and within 50 years it had spread nearly across the continent (see Figure 9.7). In fact, in some places, the high density of flocking starlings before the fall migration has become a local nuisance.

Long-term environmental change may also lead to the reduction of a species distribution. For example, the bristlecone pine, a tree that may live to be more than 4,000 years old, now only occurs on the highest mountaintops in the western United States (Figure 9.8; Wells, 1983). However, middens of wood rats in which vegetation has been preserved indicate that the range of the bristlecone pine was much larger until 10,000 years ago, the approximate end of the last ice age. Since then, warming temperatures and subsequent colonization by more temperate conifers have led to a great reduction in the range of bristlecones.

If a particular area is isolated or small, then the probability that it is part of the species distribution is a function of the dispersal ability of the organism and the ability of the species to persist in the area. If both the dispersal ability and the ability to persist are low, then it is unlikely at any given time that the area will be occupied by the species. On the other extreme, if the dispersal ability and persistence of the organism are high, then the probability of the area being occupied is high. Such considerations allow an understanding of the size of the distribution of a species over time and space, particularly when considering disjunct areas such as islands.

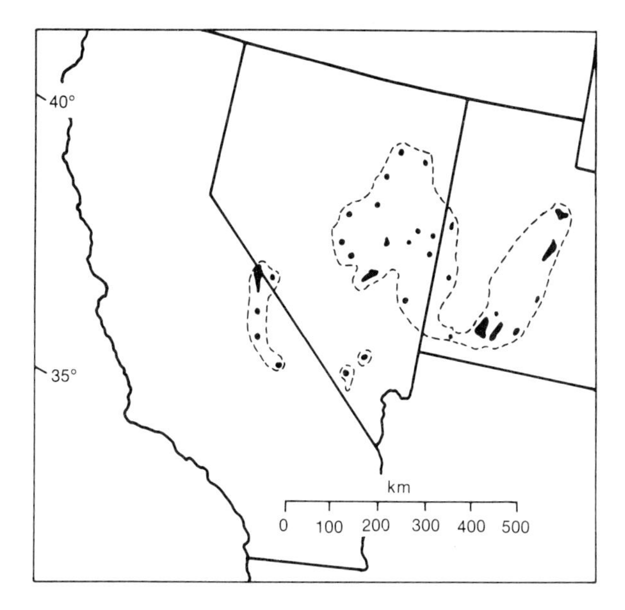

Figure 9.8. The existing distribution (darkened areas) of bristlecone pine and the approximate distribution (broken lines) during the last ice age, based on fossil evidence (from Wells, 1983).

If this logic is extended to many species simultaneously, it forms the basis for the prediction of the equilibrium number of species present on an island, a central concept in the study of *island biogeography* (see MacArthur and Wilson, 1967). In other words, the expected number of species on an island should be a function of the size (and habitat complexity) of the island and the dispersal ability and persistence of the species involved. A natural example of this process is found on the island of Krakatau, between Sumatra and Java, after it blew up because of a volcanic eruption in 1883. The entire remaining area was covered with volcanic debris and the fauna and flora were completely destroyed. However, by 1920 184 species of plants and 27 species of land and freshwater birds inhabited the island. A census of birds in 1934 also found 27 species, but 5 of those found in 1920 had disappeared, they had been replaced by 5 new species, suggesting that the number of bird species had reached an equilibrium, while the composition changed over time owing to species establishment and loss.

III. FACTORS AFFECTING DISTRIBUTION

In analyzing the distribution of an organism, the basic question is: Why is the species present in some areas and not in others? The factors involved may be numerous and many have been thoroughly investigated. Such factors may

Table 9.1
A general classification of factors affecting the distributions of organisms and some examples in each category.

Factors	*Examples*
Abiotic	
Physical	Temperature, light, fire, currents
Chemical	Moisture, salinity, soil nutrients, oxygen
Biotic	
Other species	Competition, predation, disease
Intrinsic factors	Dispersal, habitat selection

be classified into several general categories, as in Table 9.1. The first two categories, physical and chemical factors, are factors in the abiotic environment, while the last two are the aspects of the biotic environment. There are, of course, many different potential factors in the abiotic environment, and their importance depends upon various characteristics of the organism—for example, whether it is acquatic or terrestrial, sessile or mobile.

It is often difficult to pinpoint a single factor that limits the distribution of an organism, and in many cases the species boundary may result from a combination of factors. For example, a combination of increasing temperature and decreasing rainfall may serve to make an area inhospitable to an organism. On the other hand, one factor may compensate for an extreme in another factor. For example, plants may exist on relatively mesic north-facing slopes in areas out of their general range because they are not exposed to the desiccating effect of warm afternoon sun.

By utilizing information concerning the physical and chemical environment, we can predict the potential or natural vegetation undisturbed by human activity for a given area. General vegetation patterns, such as temperate deciduous forest, coniferous forest, desert, or grassland, appear to be determined primarily by a combination of temperature and moisture (see Figure 9.9). For example, prairies or grassland occur in areas of relatively low rainfall and moderate temperatures. More detailed vegetation patterns or types—for example, different types of prairie, such as tall grass, short grass, or sand prairie—also depend on other factors, such as soil type and photoperiod. Furthermore, the vegetation type is not just the result of climatic and soil conditions, since the vegetation can itself modify various aspects of the environment, such as moisture level and soil fertility.

In theory, it is possible to determine whether intrinsic biotic factors are responsible for limiting a distribution. If an area is not occupied because it is inaccessible due to the limited dispersal of an organism, then a transplant to that area should become established. (The same logic should apply in examining habitat selection if the area is isolated enough from the preferred habitat.) Such a hypothetical experiment is illustrated in Figure 9.10, in which the shaded area indicates the actual distribution and the adjacent area

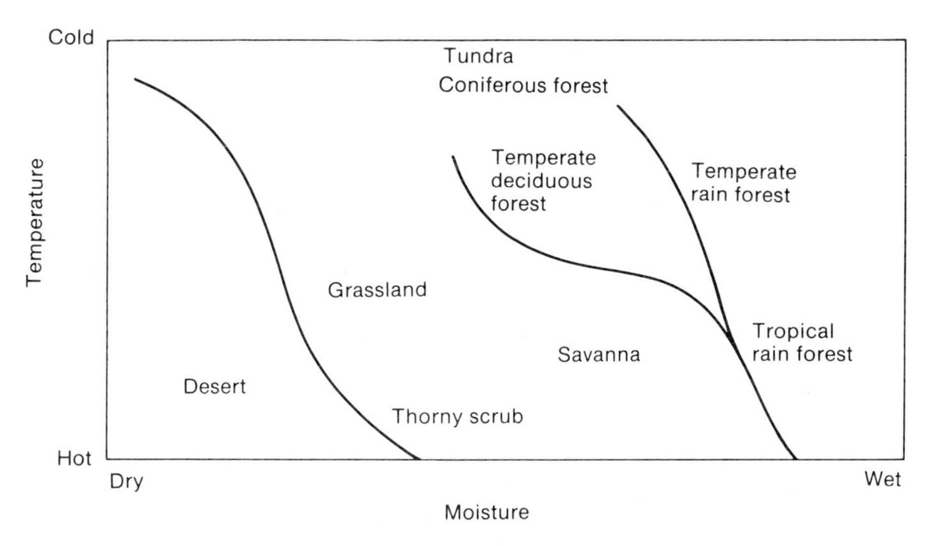

Figure 9.9. General associations of vegetation patterns with moisture and temperature (after Ellseth and Baumgardner, 1980).

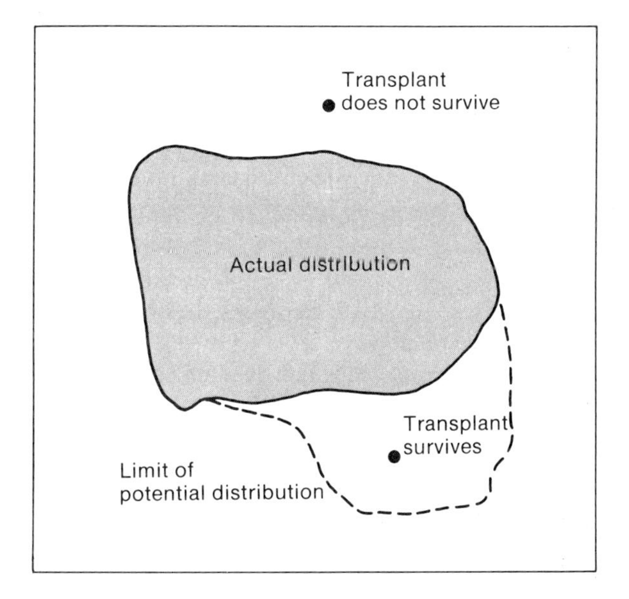

Figure 9.10. A diagram illustrating the determination of the distribution of an organism using transplant experiments.

indicates the geographic area in which the organism should be able to live (based on environmental conditions) but does not. A transplant to this area should survive, while a transplant to an area in which there is present some extrinsic factor, such as a competitor or an extreme physical aspect of the environment, does not result in establishment. Therefore, a transplant

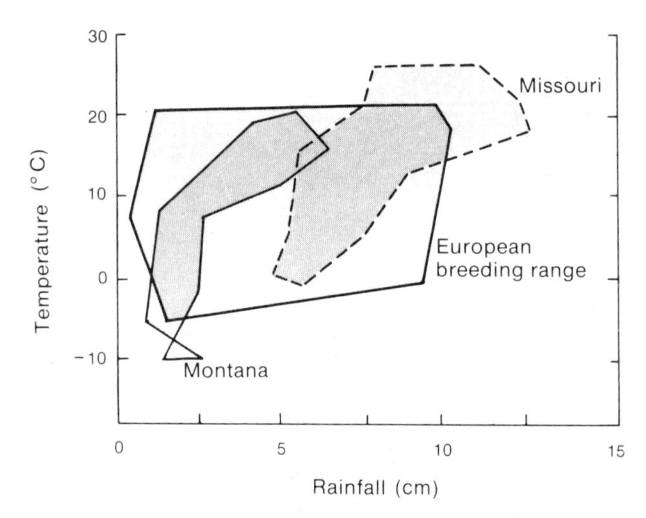

Figure 9.11. Climatographs for the European breeding range of the Hungarian partridge (heavy line) and two regions, Missouri and Montana, in which it was introduced (light lines) (after Ellseth and Baumgardner, 1980).

experiment can distinguish between intrinsic biotic factors and all other factors. An example of the limitation of a distribution because of low dispersal is the osage orange (Figure 9.6). Here the fruit has very low dispersal, and only by human introduction has its distribution expanded beyond its historical range.

The successful introduction of exotic species to new areas also allows examination of the factors important in determining geographic distributions. For example, the Hungarian partridge, *Perdix perdix*, is widespread in Europe and was introduced to several areas in the United States, including Montana, where the species became established, and Missouri, where the introduction failed. Figure 9.11 gives the climatographs for these regions and shows that neither Montana nor Missouri is completely enclosed within the European environmental conditions (after Ellseth and Baumgardner, 1980). It appears that the partridge can survive the slightly colder Montana winters but the hotter and wetter summers of Missouri prevent normal reproductive activity, resulting in its failure to establish there.

Below are given some examples in which specific factors appear to be responsible for the limit of the distribution of an organism either by increasing mortality or reducing reproduction. As discussed above, combinations of several factors may be involved simultaneously in limiting a distribution. The exact mechanisms by which physical and chemical factors limit a distribution are a function of physiological constraints. In fact, one important area of ecology, physiological ecology, considers the effects of abiotic factors on the survival of organisms. Here we will only briefly mention biotic factors because Chapters 13 and 14 are devoted to competition and predation, respectively.

a. Temperature

Extremes of temperature often appear to be important in limiting both the distribution and abundance of organisms. For many plants and animals, high temperatures will result in desiccation. On the other hand, low temperatures can limit the distribution of organisms that are intolerant of freezing temperatures. For example, the saguaro cactus, an extremely large succulent, cannot endure more than 36 hours of below-freezing weather. If a thaw occurs every day, then the saguaro can survive. The distribution of the saguaro in the Sonoran desert (see Figure 9.12) follows closely the area that does not have long freezes. Figure 9.12 also gives the mean number of days per year in which the temperature remains below freezing. Notice the correspondence between the distribution of the cactus and this measure of temperature.

Some organisms are tolerant of extreme temperatures. An example is the desert pupfish, a small fish that exists in a few streams and pools in Death Valley. These aquatic areas are the remnants of a larger habitat area that has dramatically shrunk since the last ice age. The pupfish are amazingly tolerant

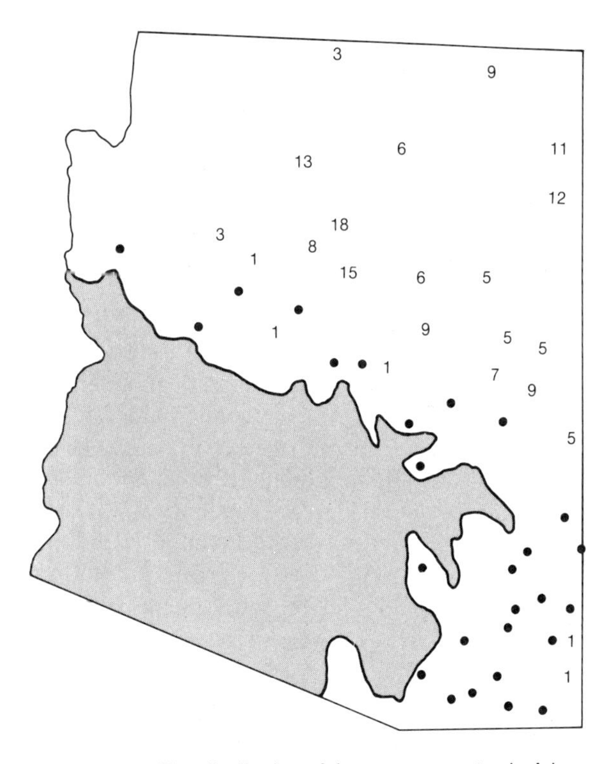

Figure 9.12. The distribution of the saguaro cactus in Arizona (shaded region). The mean number of days per year that the temperature remains below freezing is also indicated. The closed circles indicate sites where mean temperatures are below freezing less than half a day per year (after MacArthur, 1973).

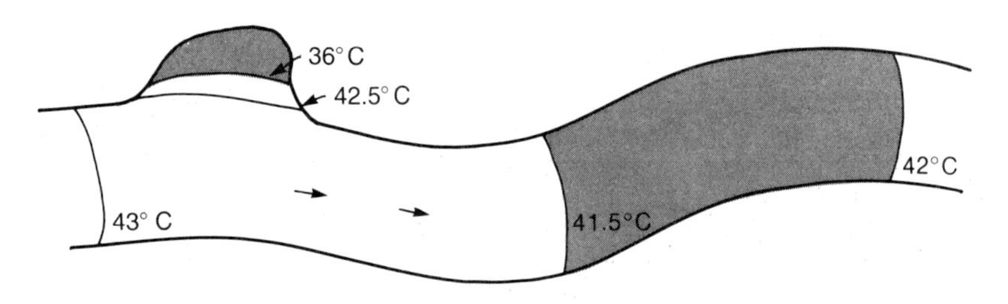

Figure 9.13. The distribution of the desert pupfish (shaded region) in a stream originating from a hot spring to the left (after Brown, 1976).

of extreme salinity and temperature conditions in these pools and exist in some pools with salt concentrations as high as sea water and in other areas with extremely high temperatures. Apparently, the maximum temperature that they can endure is 42° C. Figure 9.13 gives the water temperature and the distribution of pupfish in a stream that originates from a hot spring. The fish will swim at the edge of an area with a temperature greater than 42° C so that they can feed on blue-green algae growing in these areas but will avoid entering these areas, which are lethal to them.

b. Moisture

Obviously, the presence of water is the primary factor determining the distribution of aquatic plants and animals. In addition, moisture levels are of major importance in limiting the distribution of many terrestrial plants, invertebrates, and vertebrates. A number of phenotypic characteristics enable both plants and animals to help conserve moisture and thus to exist in arid areas. For example, plants may lose their leaves or have water-storage organs or deep roots; animals behaviorally avoid the desiccating sun, foraging during cooler periods of the day, and may use metabolic water for moisture.

Because temperature and rainfall interact to determine the amount of moisture available, classification of climate is based on the concept of *potential evapotranspiration*—that is, the amount of water that would be evaporated or transpired if it were available. Potential evapotranspiration is a function of temperature and sunlight and is greatest in hot, sunny areas or months (see Figure 9.14). In conjunction with rainfall patterns, the potential availability of water in different seasons can be evaluated. For example, in a desert area in California. [Figure 9.14(a)], a water deficit exists throughout the year, while in a temperate deciduous forest in Tennessee [Figure 9.14(b)], a water surplus exists except in the summer months.

Whittaker (1956) examined in detail the vegetation patterns along transects at three elevations in the Great Smoky Mountains. Figure 9.15 gives the percent of the stand that is composed of several of the major tree species at the lower (900 to 1500 m) and higher (2100 to 2700 m) transects. Notice that along the transect of 13 stations generally one species is sequentially replaced

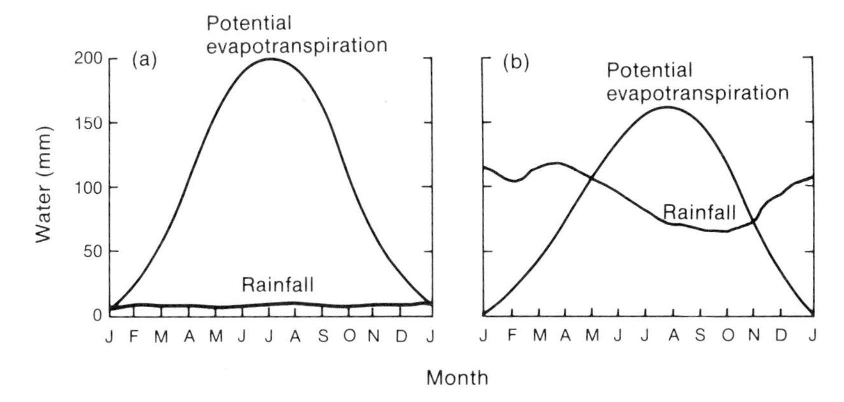

Figure 9.14. The seasonal variation in potential evapotranspiration and rainfall (a) at a desert site and (b) in a temperate deciduous forest (after Ellseth and Baumgardner, 1980).

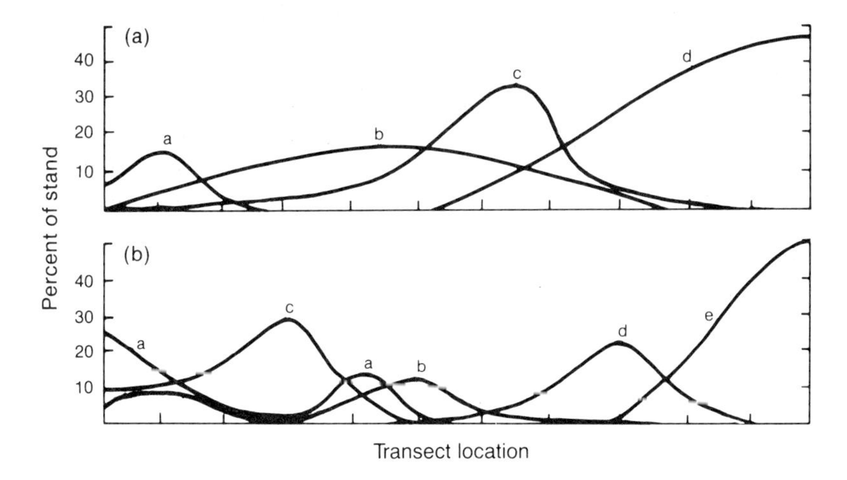

Figure 9.15. The percent of stand for the major tree species along (a) a lower-elevation transect and (b) a higher-elevation transect with gradients in moisture on both in the Great Smoky Mountains (after Whittaker, 1956).

by another. Although the species found on the two transects are different, the most xeric species on both transects are pines and the second most xeric species are oaks. Stations at different elevations have different potential evapotranspiration, with the cooler, high-elevation locations having lower values.

c. Soil Nutrients and Type

Many plants are limited in their distribution either by the lack of necessary soil nutrients or by the presence of certain chemicals in the soil that inhibit growth. For example, serpentine soils, which are primarily magnesium iron silicate and

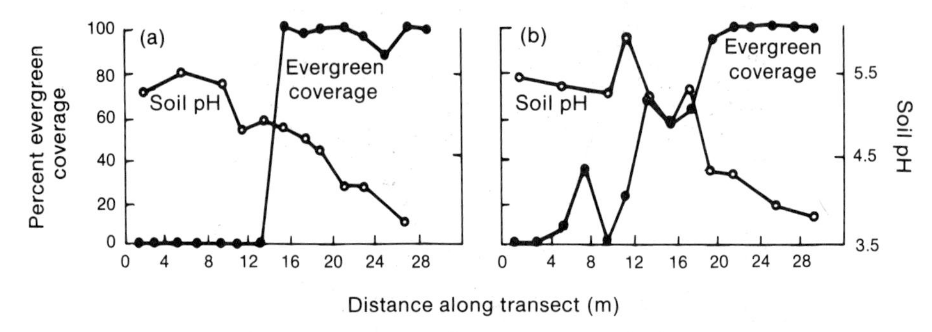

Figure 9.16. The percent of evergreen coverage on two transects of increasing soil pH at a site in northwestern Mexico (after Goldberg, 1982).

may contain other minerals, are scattered around the world. Even though such soils are lethal to many plants, some species are able to colonize these soils and utilize these areas. Other areas with particular soil types or aspects often have endemic flora; granite outcrops in the southeastern United States are an example.

The distribution of an organism can be determined by the joint effects of soil type and other factors. For example, the distribution of broad-leaved evergreens in some areas of northwestern Mexico appears to result from a combination of soil type and interspecific competition from deciduous trees (Goldberg, 1982). The evergreens occur on acid, infertile soils, while the deciduous vegetation dominates on more fertile soil. Figure 9.16 illustrates the change in soil pH associated with the generally abrupt boundary observed between the deciduous vegetation and the evergreen patches. The two transects in the figure, representing the most and the least abrupt transitions found, illustrate that, as the pH goes below 5.0 (soils become more infertile), the evergreen coverage quickly increases. These evergreens can exist in soils with a higher pH, but it appears that the higher growth rate of the deciduous plants competitively excludes them from these areas.

d. Competition

The presence of other species that have similar niche requirements may limit the distribution of a species (see Chapter 13 for an extended discussion). For example, several species may compete for food or space with one species, effectively excluding a second species from an area. As a result, two competing species may have nonoverlapping geographic distributions. An example of a group of related species that appear to have similar resource requirements are the pocket gophers of the western United States (Miller, 1964). These burrowing animals are highly territorial and often aggressive towards the pocket gophers of other species. Interactions and to some extent soil-type preferences among the species appear to result in sharply delineated distribu-

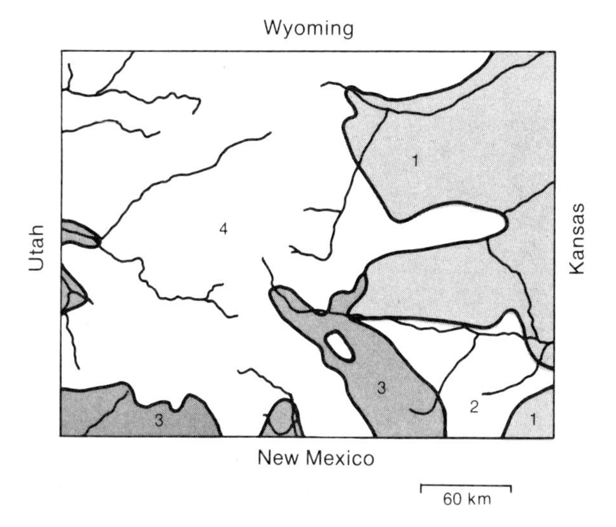

Figure 9.17. The distribution of four species of pocket gophers in the state of Colorado (after Miller, 1964).

tional boundaries. The distributions of the four species present in Colorado are given in Figure 9.17; the species in this state occupy mutually exclusive but adjacent ranges.

Plants can limit the distribution of other plants by producing toxic or allelopathic chemicals. One of the first observations of allelopathy was the effect of black walnut trees on nearby grass. Walnut trees were surrounded by areas of dead grass several times greater than the extent of their canopies. A substance from walnut roots was later identified and shown to be toxic to several other species. Another example of an allelopathic plant species is barley, a crop known to have value as a weed suppressor. Overland (1966) showed that barley inhibited the growth of chickweed in experiments in which there was no competition for water or light. Again, a chemical found in the roots was the allelopathic agent.

e. Predation

Predators can limit the distribution of their prey, and if a predator can only utilize a particular prey species, then the absence of prey may limit the distribution of a predator (see Chapter 14 for an extended discussion). A clearcut example of a predator's limitation of the distribution of a prey is the effect of sea urchins on the presence of algae (see also Example 14.7). Figure 9.18 illustrates the correspondence between the presence of a sea urchin and the absence of algae in a transect extending from the shore in Ireland. Both close to shore and away from the shore, where the sea urchins are absent the algal cover is nearly complete, while in the intermediate area the sea urchins eliminate the algae. A direct demonstration of the effect of the sea

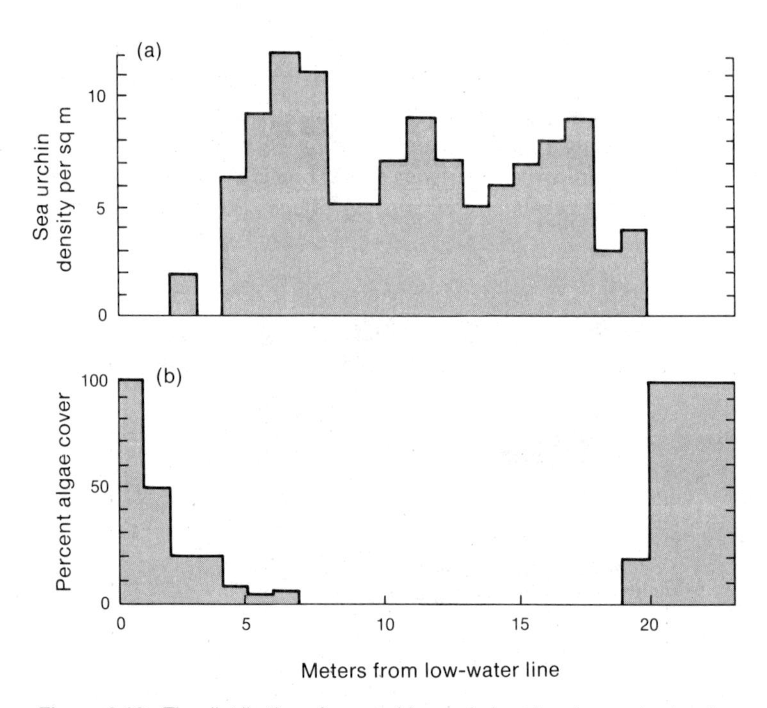

Figure 9.18. The distribution of sea urchins and algae in a transect extending from the seashore (after Kitching and Ebling, 1961).

urchin on the algae was given when sea urchins were removed from a section of the area, and, within a year, algae had completely covered it.

The distributions of specialized predator species are limited by the presence of their prey or resource species. One well-known specialist is the panda, which eats only bamboo, and large amounts of it. As a result of this specialization and other factors, its geographic distribution is limited to a small area of China. Such obligate associations may be used to identify a predator species that will control an introduced pest and not affect other related native species (see Chapter 21).

SUMMARY

The ecological niche of an organism is the total set of environmental conditions necessary for the persistence of a species. In general, the closeness of environmental conditions to the niche of a species determines both the species abundance and geographic distribution. Geographic distribution may change as the result of ecological changes such as climatic shifts or human activity. Both abiotic and biotic factors affect geographic distributions. Temperature, moisture, soil nutrients, dispersal, competition, and predation all have significant effects.

PROBLEMS

1. What are the differences between the fundamental, realized, and preferred niches of an organism? How would you determine these different niches for a particular species?
2. What is the relation between a species distribution and its density? How would you determine the edge of a species distribution and whether the distribution was constant over time?
3. Assume that the distribution of a particular plant species had an abrupt edge to its distribution. Design a series of experiments that could potentially identify the factor(s) determining the distribution of this plant.
4. It is thought that the distribution of a particular mouse species is limited either by competitors or predators. Give a series of observations that could help distinguish between these alternatives. What experiments would help distinguish between these alternatives?
5. The effect of human activity on the distributions of many plant and animal species have been enormous. Describe the effects you imagine the removal of all human activity in North America would have on the distribution of several specific species with which you are familiar.

CHAPTER 10

Population Density and Dispersion

Mathematics ... was repugnant to me ... [But] I have deeply regretted that I did not proceed far enough at least to understand something of the great leading principles of mathematics; for men thus endowed seem to have an extra sense.

CHARLES DARWIN

How many individuals are there in a population? What is the variation in population numbers temporally and spatially? To understand the influence of various factors on population numbers, first we must be able to answer these questions accurately by quantifying the numbers in a population. For sessile organisms, enumeration of a population is possible if the numbers are not too large. However, for many motile organisms, reliable estimates of density are often difficult to obtain.

In natural populations, the density of a population over time may follow a number of different temporal patterns. For example, the numbers may remain relatively constant over time, as shown in Figure 10.1 for the amount of

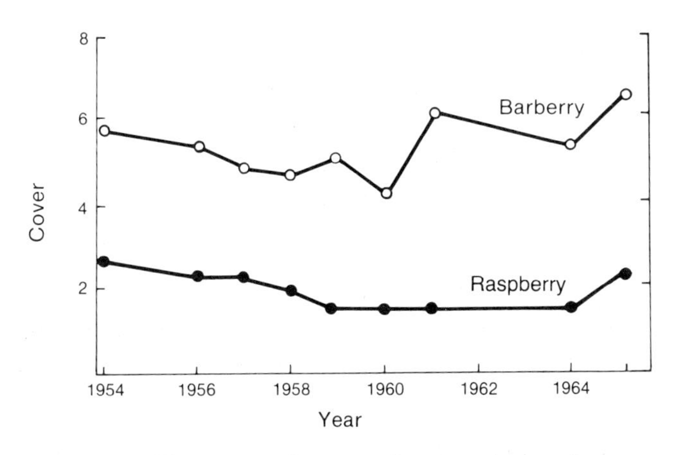

Figure 10.1. The amount of coverage in a forest by two shrubs, a barberry and a raspberry over a 12-year period (from Rosensweig, 1974).

cover, an indicator of density, for two shrubs in a forest over a 12-year period. Most populations show variation in density over time but remain within some broad limits that are well above extirpation and well below the overutilization of their resources. For example, Figure 10.2 gives the number of white storks in an area of Germany over 35 years. Here obvious fluctuations occur in the population density, with a peak in the late 1930s and smaller numbers in recent years, but the overall variation is confined to a relatively small range.

A few species appear to have more or less regular temporal density variations. One example was given for the lynx and the snowshoe hare in Chapter 6; often small rodents, such as voles and lemmings, exhibit marked and fairly regular fluctuations in density (see Example I.1). Some species have large and apparently erratic variations in densities that result in occasional outbreaks of extremely high densities. A classic example of such an organism is the desert locust, which periodically reaches incredible numbers. For example, one such outbreak, or plague, was estimated to be composed of more that 10^{10} individuals and to weigh more than 50,000 tons over its total area. During such outbreaks, the locusts destroy virtually all vegetation in the areas that they cover.

In addition, population density may vary considerably over space, resulting in a number of spatial patterns. These spatial patterns are called patterns of *dispersion* (note the difference between this and dispersal which concerns the movement of individuals). For example, some organisms may exist in groups in particular areas and not in other areas. The variation in density over space may be quite large for organisms that require particular ecological conditions or that form social groups, such as many marine mammals. On the other hand, some organisms may be quite regularly spaced. For example, the colonies of two species of groundforaging ants are quite evenly spaced (Figure 10.3). In this case, the pattern of regularly distributed colonies is consistent with the hypothesis that competition for food among

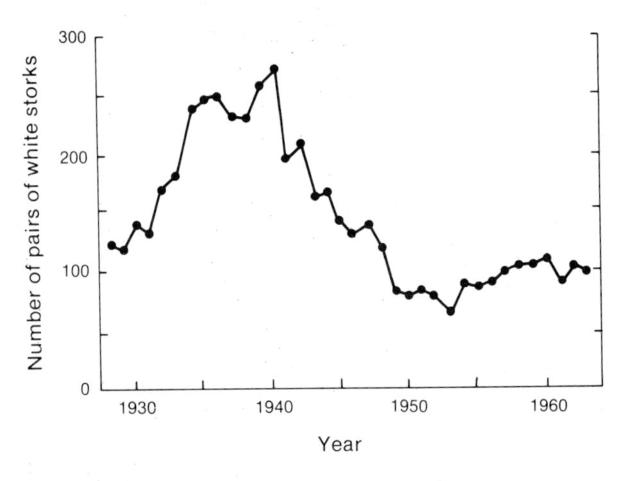

Figure 10.2. The number of pairs of white storks over 35 years in an area of Germany (from Lack, 1966).

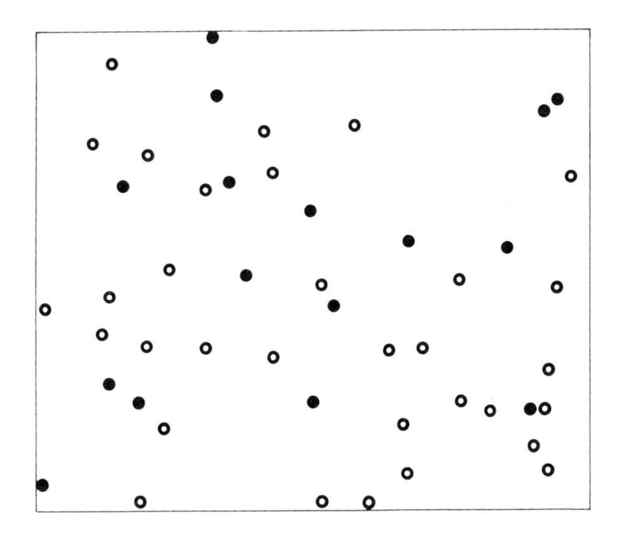

Figure 10.3. The dispersion of two species of ground-foraging ants in a sample plot in Panama (after Levings and Franks, 1982).

colonies of both species affects spacing among colonies (see Chapter 13 for a discussion of interspecific competition).

I. ABSOLUTE DENSITY

In general, when we are enumerating a population, the basic measure of abundance is the *density*–that is, the number of individuals per unit area (or volume in the case of small aquatic or subterranean organisms). Generally, population density is used synonomously with the abundance of the organism. In some cases, the term *population size* can be used in an ecological context as the number of individuals of a species in a given area. Remember that the number of individuals important for genetic considerations is related to the number of mating individuals (see Chapter 6), while the number important for ecological considerations, such as resources consumed, may be the total census number.

To estimate density accurately, one must be able to count precisely the organisms in an area that is the "appropriate" size. For example, counting the number of deer in an acre tells little about their density because deer move over areas much larger than an acre. On the other hand, it would be impossible to count the number of soil bacteria in an acre because the number would be too large. Owing to differences in such factors as the size and mobility of organisms and variation in resource availability, a wide range of densities exists in various organisms (see some examples in Table 10.1).

For some organisms, it is possible to obtain a total count of the organisms in a population. For example, censuses of human populations attempt to

Table 10.1
Estimates of the density of several different organisms (after Krebs, 1978).

Organism	*Density*
Humans	
(Netherlands)	375/sq km
(Canada)	2/sq km
Deer	10/sq mi
Woodland mice	5/acre
Soil arthropods	500,000/sq m
Trees	200/acre

count all the individuals in a population, although underenumeration of some groups is often a problem. All the trees of a particular species in a forest patch or the number of territorial birds in an area may be counted. In some cases, aerial photographs allow a complete census of migrating animals, such as caribou, or widely separated plants, such as some large cacti. For most organisms, however, total counts are impossible, and sampling of the population is the only way to estimate absolute density. Here we will discuss two simple sampling techniques that are used to estimate density, the quadrat method and capture–recapture method.

a. The Quadrat Method

In situations in which the density of an organism is extremely high, it is often only feasible to count the number of organisms in a *quadrat*, a small part of the area of interest of a given size. Given the number in such an area (or several areas) of known size, then one may extrapolate this density to the total area of interest. For example, if the average number of snails per quadrat over several square-meter quadrats is 12, then the estimated number in 100 square meters in 1200. In general, an estimate of the density is

$$N = k\bar{N}$$

where $\bar{N}$ is the average number counted in a quadrat and k is the number of quadrats contained in the area of interest. In the example above

$$N = (100)(12) = 1200/100 \text{ sq m}.$$

There are several important assumptions in this approach. First, one must be able to count accurately all the individuals in a given quadrat. In some cases, this may be difficult because the organism moves or is cryptic. Second, the area of each quadrat must be known. For example, when the terrain is uneven, it may be difficult to measure the exact area of quadrats on steep hillsides. Last, the quadrats should be representative of the whole area to be sampled. As a result, in most cases, a number of sample quadrats should be taken that are randomly chosen from the total area of interest (see Example 10.1).

In some cases, there may be large-scale environmental heterogeneity, and a random sample from the total area may thus include many empty quadrats. To avoid this problem, the investigator can select areas in which the organisms are known to exist and choose random quadrats within these areas; this technique is called *stratified random sampling*. For example, the density of periodical cicadas was estimated from random samples under trees from sites known to have only the particular species of interest (Karban, 1982).

Example 10.1. Quadrat sampling is often used for estimating the density of soil organisms. For example, estimates of the density of click beetle larvae is important because this species damages the roots of many crop plants. One approach used to estimate their density is to take core samples of the soil and count all the larvae in these cores (Salt and Hollick, 1944). The counting is somewhat tedious and involves breaking up the soil, using sieves for soil separation, and floating the larvae in a solution. In one example, $\bar{N}$, the average number of larvae in 240 cores, 4 inches in diameter and 6 inches deep, was 15.6. Because there are approximately 5.0×10^5 such cores in an acre, the estimate of the density per acre was then

$$N = (5.0 \times 10^5)(15.6)$$
$$= 7.8 \times 10^6/\text{acre}.$$

Quadrat sampling is often used in plant ecology to estimate the density of different species. Below is an example for several species of trees more than 25 cm tall in a hardwood forest (from Krebs, 1978). The quadrat in this case was the area within one meter of a transect (straight line) approximately 100 meters long in the forest. The average number of chestnut oaks for three transects was 22 and the average number of beeches was 14.7. Approximately 19 transects of this size are contained in an acre, so the estimates of the densities per acre are

$$N(\text{oak}) = (19)(22) = 418/\text{acre}$$

and

$$N(\text{beech}) = (19)(14.7) = 278.7/\text{acre}.$$

---

b. The Capture–Recapture Method

Many animals cannot be completely enumerated, even for a small area, because they move to avoid capture or observation. As a result, a different type of sampling technique is necessary, which depends on two (or more) subsequent attempts to capture or observe the organism. Below we will discuss the simplest such capture–recapture sampling technique, often called the Lincoln-Petersen method.

In the Lincoln-Petersen method, there are two different capture times. In the first instance, a sample of the population of size M is captured, marked in some manner, as with paint, tags, or toe-clipping, and then released. Later, a second sample of size n is taken. In the second sample, the number of marked individuals m, called the recaptures, is counted. If the density of the whole population during this time period is N, then the proportion of the marked individuals in the second sample should be equal to the proportion of total individuals captured in the first sample or

$$\frac{m}{n} = \frac{M}{N}.$$

This expression can be solved for an estimate of the population density N so that

$$N = \frac{Mn}{m}.$$

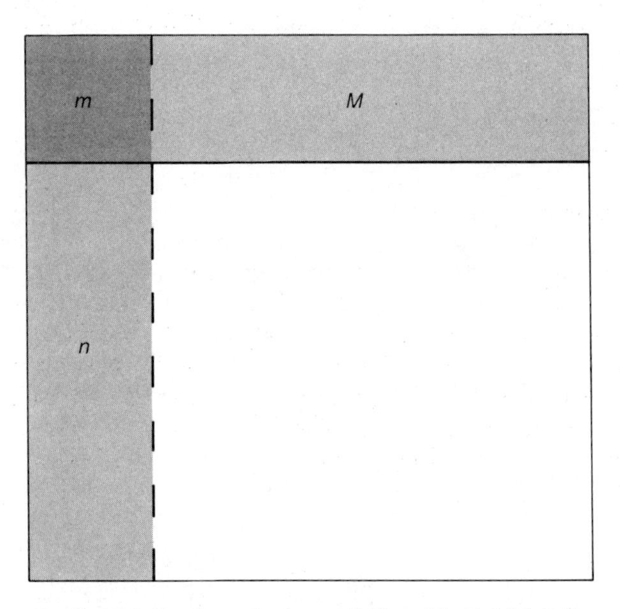

Figure 10.4. A schematic representation of the total density of a population illustrating the capture–recapture technique (see text for key to symbols).

The logic of this approach is illustrated in Figure 10.4. The total area of the square represents N, the total population density. The shaded region across the top indicates the number marked in the population. The recapture attempt results in catching both some marked individuals, the darkly shaded region, and some unmarked individuals, the lower shaded region. In theory, the proportion of marked individuals in this recapture attempt should be the same as the proportion of marked individuals in the total population. For example, if there are 1,000 total individuals (N) and 400 are marked (M), then one would expect that in a second attempt, in which 200 were caught (n), 80 (m) would be marked, or

$$N = \frac{(400)(200)}{80} = 1000$$

(see Example 10.2).

A number of assumptions are implicit in the capture–recapture approach. First, it is assumed that the marks are not lost before the time of the recapture attempt. For example, some organisms may remove tags or other markings, making it difficult to distinguish recaptured individuals. Second, marked individuals should not suffer more (or less) mortality than unmarked individuals because either the marking or handling affects survival. Obviously, such treatment as rough handling or keeping fish out of water during marking may increase mortality. Third, the marked and unmarked individuals should mix freely and their spatial distributions should be identical. For example, if captured individuals do not disperse, then they may be more readily caught in the second attempt. Last, it is assumed that in the second capture attempt all

---

Example 10.2. The density of fish populations is often estimated by capture–recapture techniques. For example, the population size of mature yellow perch in a one-acre pond was estimated using the above approach (from Pielou, 1974). In the first catch, the investigators captured 284 fish using a seine and marked them by clipping their dorsal fins. They handled these fish carefully and released them to prevent mortality and two days later made a second catch, of 1,392 fish, which included 86 marked fish—that is, $M = 284$, $n = 1,392$, and $m = 86$. Using the expression above, we find that the estimate of the population size is then

$$N = \frac{(284)(1,392)}{86} = 4,562.$$

In this case, the pond was later drained and the true number of fish was found to be 4,123. The estimation procedure came reasonably close to predicting the actual number of perch in the pond.

---

individuals, marked and unmarked, are equally likely to be caught. For example, some rodents may become "trap-shy" or "trap-happy" and, as a result, either avoid the traps or return specifically to the traps. When animals are trap-shy or trap-happy, then the probability of recapture is lower or higher, respectively, than a random expectation. At the extreme, there may be no recaptures, $m = 0$, if captured individuals in a population become extremely trap-shy. Related to these points, there may be sex or age differences in the catchability of many species.

Often capture–recapture studies are carried out for more than two time periods. The information from multiple recaptures can yield a better estimate of population density and can also be used to measure the increase in population density due to birth and immigration and to estimate population decreases resulting from death and emigration (e.g., see Blower et al., 1981, for a discussion of the various estimation procedures available).

II. RELATIVE DENSITY

The density of some organisms cannot easily be enumerated by either the quadrat or capture–recapture methods because it is difficult to count them directly. In some cases, the only feasible approach is an indirect one based on factors that indicate the presence or numbers of the organism (see Table 10.2). In general, these techniques can only be used relative to the same measure in another year. For example, if 1,000 lynx pelts were sold in one year and 100 in a subsequent year, then there appears to be a tenfold difference in density, though the actual density is not known. As a result, many of the relative density measures are often used as preliminary estimates of density with subsequent, more precise census data being necessary to measure accurately the population density. However, some of these measures, such as vocalization frequency in species in which nearly all males call, or artifacts that can be enumerated easily and accurately, as in some bird nests and the pupal cases of cicadas (see Example 14.6), may be nearly as good as direct measures of

Table 10.2
Several different approaches used to estimate relative density and examples of organism for which they have been used.

Techniques	*Organisms*
Number of fecal pellets	Rabbits, coyote
Pelt records	Beavers, lynx
Catch per unit effort	Fish, lobster
Trapping	Plankton, ground insects
Vocalization frequency	Frogs, pheasants
Number of artifacts	Birds (nests), cicadas (pupal cases)
Tracks in snow	Deer, mice
Amount of cover	Shrubs, algae

density. Even in these better situations, there will still be some unknown proportion of males that don't call or some unknown proportion of nests or pupal cases that aren't observed.

III. DISPERSION

In many cases, the density of a population may vary over time, and the techniques described above can be used to measure this variation. In addition, as we discussed earlier, the population density may vary over space. In this section, we will briefly describe two techniques that can be used to evaluate the distribution of organisms over space. Generally, measures of spatial distributions, dispersion, are most important and applicable to organisms that are either sessile or relatively sedentary (at least in part of their life cycle), such as plants or mollusks.

a. Index of Dispersion

One way to describe the variation in numbers over space is to count the number of individuals in a series of sample areas. The number of sample areas with different numbers of organisms can then be summarized; thus, for example, there may be 30 sample quadrats with 2 individuals, 20 with 3 individuals, and so on. Using these data, one can then calculate the mean number of individuals per quadrat and the variance of the number per quadrat. The assumptions of this approach are similar to those used to estimate absolute density by quadrat sampling—that is, one must be able to count all the individuals, know the area of the quadrat, and choose the quadrats randomly.

The distribution of individuals over an area can be categorized into three basic patterns (see Figure 10.5). First, the distribution is random if the

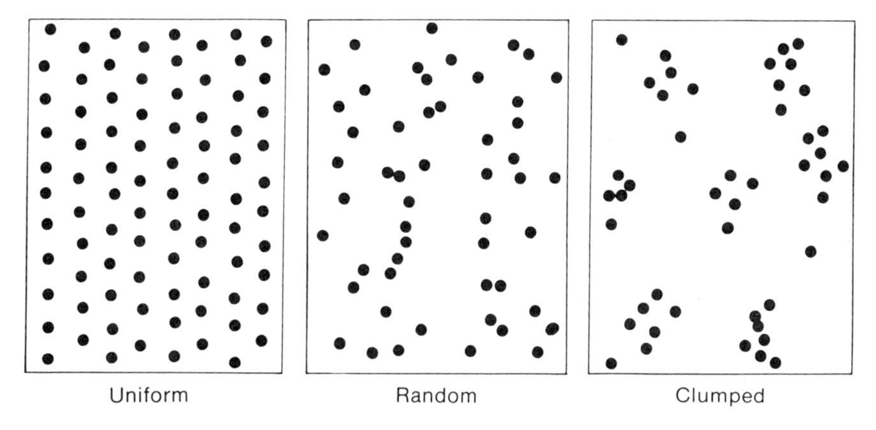

Figure 10.5. The dispersion of an organism over an area illustrating uniform, random, and clumped distributions.

probability of the presence of an individual in a quadrat is independent of the presence of other individuals. In this case, the mean should be equal to the variance (this is a property of the Poisson distribution; see below). Second, if individuals are clumped together in certain quadrats, then the variance

---

Example 10.3. Dispersion is particularly important in tree populations because the spatial patterns may vary from the random distribution and indicate particular ecological factors. For example, a uniform distribution may indicate strong competition for some resource such as water or light among individuals within a species. On the other hand, a clumped distribution may result from low seed dispersal or a patchy distribution of suitable habitat.

Pielou (1974) examined a population of second-growth lodgepole pine, in which nearly all the trees were approximately 40 to 50 years old. This population had grown up after a fire, and the trees did not appear to be so crowded that they interfered with each other. A sample of 100 random quadrats, each of 7.3 square meters, was established, and the number of trees in each was counted (Table 10.3). For example, 7 quadrats had no trees and 16 had only 1 tree. The mean number of trees per quadrat was

$$\bar{x} = \frac{1}{n}\sum_{i=1}^{100} x_i$$

$$= \frac{1}{100}(2860) = 2.86.$$

Using this mean value and the formula on page 172, we can calculate the expected number of quadrats with a given number of individuals. For example, the probability that a quadrat will have no individuals is

$$\text{Pr (0 individuals)} = e^{-\bar{x}} = e^{-2.86} = 0.057.$$

Because there were 100 quadrats, then the expected number of quadrats with no individuals is $(0.057)(100) = 5.7$. Likewise, the probability that a quadrat would have two individuals is

$$\text{Pr (2 individuals)} = \frac{2.86^2 e^{-2.86}}{2!} = 0.234$$

and the expected number of quadrats with two individuals is $0.234(100) = 23.4$. These and the other expected numbers are given in Table 10.3. Notice that the expected values are quite close to the

---

may be much larger than the mean. Last, if there is a very even distribution of individual numbers over quadrats, with each quadrat having similar numbers, then the variance will be quite small and smaller than the mean.

A measure commonly used to compare mean and variance and to

observed values. A χ^2 test may be calculated to determine if these observed values are statistically significantly different from the expected numbers. The χ^2 value is 0.99, the sum of the last column, and, with five degrees of freedom, the values are not significantly different from those expected (see Chapter 2 for a discussion of the χ^2 test).

The variance of these observed numbers is

$$V = \frac{1}{n} \sum_{i=1}^{100} (x_i - \bar{x})^2$$

$$= \frac{1}{100} \sum_{i=1}^{100} (x_i - 2.86)^2 = 2.84.$$

Therefore, the index of dispersion is

$$I = \frac{2.84}{2.86} = 0.99$$

extremely close to 1.0, the expected value if the population was randomly dispersed.

Table 10.3
The number of lodgepole pine trees observed in quadrats in a forest and the expected number based on random dispersion.

	Number of quadrats		
Number of trees per quadrat	*Observed*	*Expected*	$\frac{(O-E)^2}{E}$
0	7	5.7	0.30
1	16	16.4	0.01
2	20	23.4	0.49
3	24	22.3	0.13
4	17	16.0	0.06
5	9	9.1	0.00
6 and over	7	7.1	0.00
			0.99

describe the dispersion in a population is called the *index of dispersion.* This measure is equal to

$$I = \frac{V}{\bar{x}}$$

where $\bar{x}$ is the mean number in a quadrat and V is the variance in the number. When I is 1.0, then the organisms are distributed randomly over the quadrats. If I is smaller than or greater than unity, then the organisms are more evenly distributed or more clumped than random, respectively.

If the organisms are distributed at random over the habitat, then the probability that a quadrat has a given number of individuals can be calculated from

$$\text{Pr}(x \text{ individuals}) = \frac{\bar{x}^x e^{-\bar{x}}}{x!}$$

where e is the base of natural logarithms, 2.718. The distribution of probabilities obtained from this formula is called the Poisson distribution. For example, if we want to calculate the probability of no individuals in a quadrat, then

$$\text{Pr}(0 \text{ individuals}) = \frac{\bar{x}^0 e^{-\bar{x}}}{0!} = e^{-\bar{x}}$$

because both $\bar{x}^0$ and 0! are equal to unity. If $\bar{x} = 2$, then the probability of a quadrat having 0 individuals is

$$\text{Pr}(0 \text{ individuals}) = e^{-2} = 0.135.$$

Likewise, the probability that a quadrat will have three individuals is

$$\text{Pr}(3 \text{ individuals}) = \frac{\bar{x}^3 e^{-\bar{x}}}{3!} = \frac{\bar{x}^3 e^{-\bar{x}}}{6}.$$

Again, if $\bar{x} = 2$, then

$$\text{Pr}(3 \text{ individuals}) = \frac{2^3 e^{-2}}{6} = 0.180.$$

Example 10.3 illustrates how this approach can be applied to a population of lodgepole pines.

b. Nearest-Neighbor Distance

The type of dispersion pattern determined by the index of dispersion may be relative to the quadrat size. For example, Figure 10.6 gives a hypothetical distribution of an organism and two possible scales of quadrat size. If

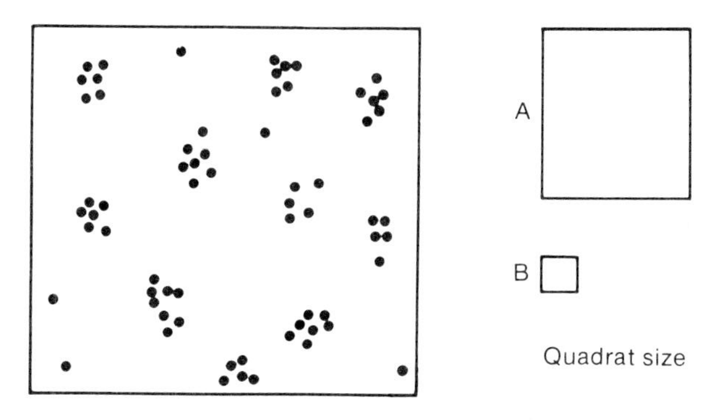

Figure 10.6. A hypothetical distribution of organisms over an area and quadrats of two different possible sizes.

quadrats of the larger scale, A, were used for the population, then the dispersion would appear to be uniform, because each of the four quadrats would have nearly the same number of individuals. On the other hand, if the quadrat of smaller scale B were used, then the measure of dispersion would indicate that the population is clumped, because many of the quadrats would be empty or have a number of individuals. Clearly, the scale of the quadrat size is critical in evaluating dispersion using the index of dispersion.

One approach that avoids this dependence on the scale of quadrat size is measuring the nearest-neighbor distance. If individuals are randomly dispersed, then the expected distance between a randomly chosen individual and its nearest neighbor is

$$d = \frac{1}{2N^{1/2}}$$

where N is the density of the population. In this case, the scale is implicit in the scaling of N. This expected value can be compared to the average observed nearest-neighbor distance, which is

$$\bar{d} = \frac{\sum_{i=1}^{n} d_i}{n}$$

where d_i is the nearest neighbor to the ith random individual and n is the number of observations.

The two values can be compared as the ratio of the observed and expected distance

$$J = \frac{\bar{d}}{d}$$

to give a measure of the dispersion in the population. When J is 1.0, then the organisms are distributed randomly over the area. If J is smaller than unity, then the organisms are more clumped than random. For example, with a very closely aggregated population, the average observed nearest-neighbor distance should be very small, making the ratio near zero. If the ratio is greater than unity, it indicates an evenly distributed distribution. Of course, this measure assumes that both the density and nearest-neighbor distance are accurately measured.

As a simple example, let us assume that the density is four individuals per square meter, so that the expected distance to the nearest neighbor is

$$d = \frac{1}{2(4 \text{ sq m})^{1/2}} = 0.25 \text{ m}.$$

If the nearest neighbors for five randomly chosen individuals are 0.10, 0.12, 0.16, 0.21, and 0.16 m away, then

$$\bar{d} = \tfrac{1}{5}(0.10 + 0.12 + 0.16 + 0.21 + 0.16) = 0.15 \text{ m}.$$

Therefore,

$$J = \frac{0.15}{0.25} = 0.6$$

Example 10.4. It is generally difficult to measure dispersion in animal populations because individuals may move. However, dispersion can be measured in some bird or mammal populations from aerial photographs. For example, Miller and Stephen (1966) examined the dispersion in flocks of the sandhill crane, related to the whooping crane (see Chapter 14) and a large bird that forms large flocks. In one flock, the density from a photograph was 0.0094 individuals per unit area. Therefore, the expected nearest-neighbor distance if the birds were randomly distributed was

$$d = \frac{1}{2(0.0094)^{1/2}} = 5.16 \text{ units}.$$

The observed distance between randomly selected nearest neighbors was actually $\bar{d} = 5.82$ units. Therefore, the ratio of the observed to the expected values was

$$J = \frac{5.82}{5.16} = 1.13.$$

This value is quite close to unity, so it appears that the cranes were randomly distributed over the area.

---

a value lower than unity, indicating clumping of individuals in the population (see Example 10.4).

SUMMARY

The density of a population may vary both over time and over space. Density in a given area can be measured using measures of absolute density, such as the quadrat or capture–recapture methods, or measures of relative density, such as the number of calls or animal artifacts. The distribution of individuals over space, dispersion, may be in general clumped, random, or even. These patterns of dispersions can be measured by the index of dispersion or the nearest-neighbor distance.

PROBLEMS

1. Assume that the average number of spiders under a square meter of litter is 30. What is the estimated number in 100 square meters? What assumptions are used in this estimate?
2. In a capture–recapture study, 100 individuals were captured initially and in the recapture 200 were caught, 20 of which were marked. What is the estimate of N and what assumptions are made?
3. What are the basic patterns of dispersion? If four quadrats had 10, 12, 8, and 10 individuals, what type of dispersion would this be (calculate I to illustrate)? Calculate the expected number of quadrats that contain 1, 3, and 5 individuals in Example 10.3.
4. In a population of a species of tropical tree, the density is approximately 16 per square kilometer and the observed nearest-neighbor distance is 0.07 kilometer. Calculate J and interpret your result.
5. Pick an organism with which you are relatively familiar and design a program to estimate its density over time and its dispersion. Try to anticipate any problems you might have and suggest ways to deal with them.

CHAPTER 11

Population Growth

> First, that food is necessary to the existence of man.
> Secondly, that the passion between the sexes is necessary and will remain nearly in its present state .... Assuming then, my postulata as granted, I say, that the power of population is indefinitely greater than the power in the earth to produce subsistence for man.
>
> CHARLES MALTHUS

The density of a population over time varies in a number of patterns, as discussed at the beginning of the last chapter. If the population density is relatively low, then in general the population will grow to some higher level at a rate determined by a number of factors. Some of these factors, such as birth and death rates, are intrinsic to the population under consideration; these we will discuss in this and the next chapter. Other factors are extrinsic and may be caused either by other interacting organisms such as competitors and predators (discussed in Chapters 13 and 14) or by physical aspects of the environment such as light, water, or temperature. If the influences of all these factors are known, then in theory we should be able to predict the general rate of population growth.

To understand the process that results in changes or maintenance of population density, let us examine a model that involves the intrinsic factors contributing to population growth. The numbers in a population are increased both by births and immigration; together these are known as *recruitment* to the population. On the other hand, population numbers are reduced both by death and emigration, together known as *loss*. These factors can be combined in a general way so that the change in population numbers from one time to the next is

$$N_{t+1} - N_t = B + I - D - E$$

where B, I, D, and E are the number of births, immigrants, deaths, and emigrants in a given time period and N_t is the number in the population at time t. Because here we will consider only a single population in which I and E are zero, the above expression becomes

$$N_{t+1} - N_t = B - D.$$

The levels of dispersal in a species (I and E values) are important in other situations—for example, in considerations of geographic distributions of a species (see Chapter 9) or regulation of population density (see Chapter 12).

The total numbers of births and deaths are both a function of the number of individuals in the population. They can be expressed as

$$B = bN_t$$

and

$$D = dN_t$$

where b and d are the per capita birth and death rates in the population. In other words, b is the number of births per individual organism and d is the individual probability of dying in the population in a given time period. If we substitute these values in the above expression, then the change in population numbers becomes

$$N_{t+1} - N_t = (b - d)N_t.$$

Obviously, if the birth rate is higher than the death rate, the population size will increase, and if the death rate is higher, then the population size will decrease. For example, if $b = 0.1$ per year and $d = 0.05$ per year and the population size is 1,000 at time t, then

$$N_{t+1} - N_t = (0.1 - 0.05)1000 = 50$$

and the population will increase by 50 individuals. In Chapter 12, we will discuss in more detail how changes in the birth and death rates may contribute to the overall regulation of population numbers.

I. GEOMETRIC GROWTH

The simplest models of population growth assume that the population size changes at a constant per capita (per individual) rate, independent of density. To begin with, let us examine such a population growth model, called the geometric growth model, in which reproduction takes place once a generation and in which the reproductive individuals die after reproduction. These characteristics occur in such organisms as annual plants and univoltine insects—that is, organisms with annual generations. Let us assume that each individual on the average produces R_0 progeny where R_0 is defined as the *net replacement rate* per generation. Therefore, R_0 is

$$R_0 = \frac{N_{t+1}}{N_t}$$

or the ratio of the number of individuals in generation $t + 1$ over the number in generation t. Rearranging this expression, then

$$N_{t+1} = R_0 N_t.$$

As a result, we can say that the number in generation 1 is

$$N_1 = R_0 N_0$$

and that the number in generation 2 is

$$N_2 = R_0 N_1.$$

By substitution, then the number in generation 2 is

$$N_2 = R_0^2 N_0$$

and, in general, the number of individuals in generation t is

$$N_t = R_0^t N_0.$$

If R_0 is greater than unity, then the population size increases over time. For example, if $R_0 = 2$, and $t = 4$, then $N_4 = 16N_0$. On the other hand, if R_0 is smaller than unity, the population size will decrease. When R_0 is exactly 1.0, then there is no change in population numbers. Figure 11.1 gives the numbers in a population over time for four different values of R_0, two greater than unity, unity, and one smaller than unity. For the largest values of R_0, the curve

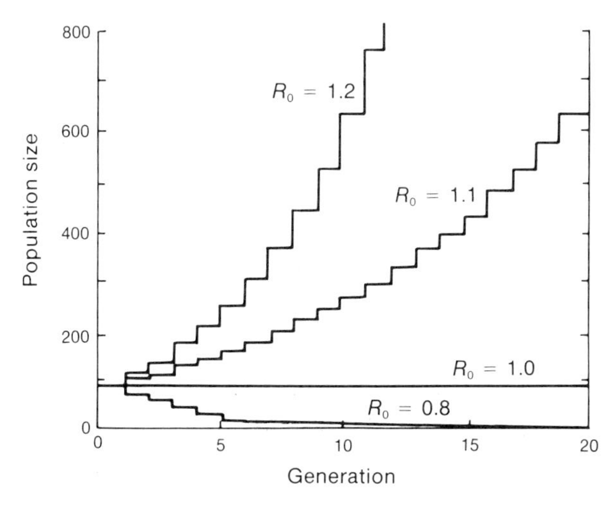

Figure 11.1. The growth of a population when the initial size is 100 for four different values of R_0.

of population growth is J-shaped—that is, the increase in numbers begins slowly and then increases very quickly at higher population numbers. Notice that the population numbers change in discrete jumps for each generation.

Many organisms reproduce multiple times and, for example, many large

---

Example 11.1. Both in laboratory experiments and in natural populations, many populations, after either an initial colonization or a bottleneck (see Example 1.2), initially exhibit population growth patterns that appear similar to geometric growth. For example, pheasants were introduced to an island off the coast of Washington in the 1930s; Figure 11.2 gives the numbers monitored in subsequent years (from Hutchinson, 1978). The population increased slowly over the first few generations and then increased very quickly. In the mid-1940s, the birds were used for food by soldiers, and the rate of increase was therefore somewhat limited. However, the pattern of increase before this time can reasonably be approximated by using a geometric growth model with a finite rate of increase of $\lambda = 1.46$. This theoretical curve is given as a broken line in Figure 11.2, beginning with 50 individuals in late 1937. Notice that overwintering mortality reduces the observed population number in the spring of each year from that in the previous fall, resulting in a zigzag in the curve. This suggests that a more detailed model that includes alternating growth and mortality would more precisely predict the changes in population size.

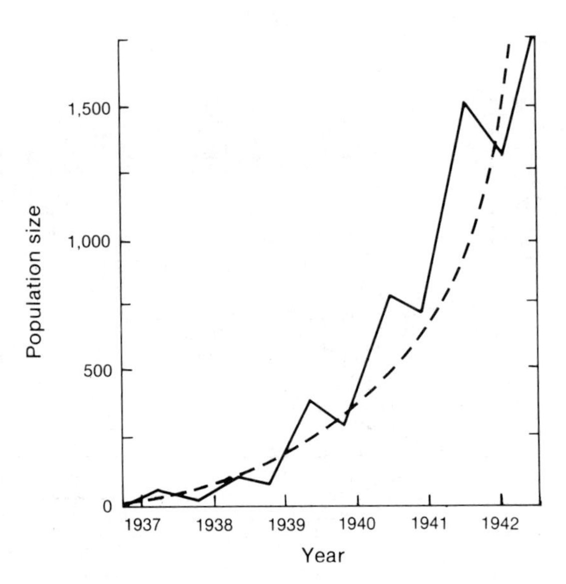

Figure 11.2. The growth of a pheasant population (solid line) and that predicted if $\lambda = 1.46$ (after Hutchinson, 1978). For clarity, the theoretical curve is given as a smooth line.

vertebrates may reproduce once each year (see Example 11.1). In this case, we can designate the rate of growth for such a time period, whether it be a year, month, or other time interval, as the *finite rate of increase*, λ. Therefore, the numbers at time $t + 1$ can be given as

$$N_{t+1} = \lambda N_t$$

where t indicates time units. As when there are discrete generations, this expression may be generalized over a number of time units as

$$N_t = \lambda^t N_0.$$

II. EXPONENTIAL GROWTH

In other organisms, there is nearly continuous reproduction and there are no particular intervals in which reproduction takes place. In this case, the change in population size may be indicated by the differential equation

$$\frac{dN}{dt} = (b - d)N$$

where dN/dt indicates the change in population numbers over a very short time interval and b and d are per capita birth and death rates in this same short time interval. The birth and death rates can be combined into one value so that

$$r = b - d$$

where the value r is defined as the *instantaneous* or *intrinsic rate of increase*. By substitution, then, the rate of change in population numbers is

$$\frac{dN}{dt} = rN.$$

Obviously, if r is greater than zero, there will be an increase in population number. If r is less than zero, the population number will decline, and if r is exactly zero there will be no change.

This equation may be used in several ways. First, if we divide both sides of this equation by N, then we can calculate the growth rate per individual, or per capita, which becomes

$$\frac{1}{N}\frac{dN}{dt} = r.$$

In other words, r is the per capita rate of increase when population growth is exponential. Notice that for this model the per capita growth rate is

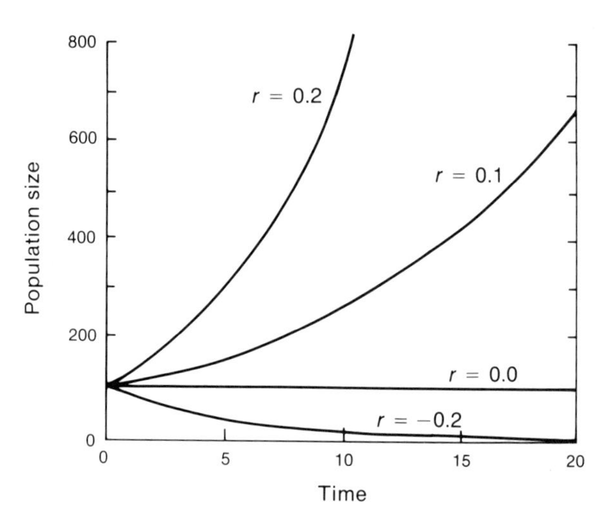

Figure 11.3. The growth of a population when the initial size is 100 for four different values of r.

independent of population numbers. Second, we can solve this expression (using some elementary calculus) so that the number of individuals at time t is

$$N_t = N_0 e^{rt}$$

where e is the base of the natural logarithms. Using this expression, we can calculate the future population size. For example, Figure 11.3 gives the population number over time for four different values of r, two greater than zero, zero, and one smaller than zero. Notice that the population numbers change continuously, and, as a result, the curves are smooth. For larger values of r the curve is J-shaped (or exponential), as was the growth from the geometric growth model. This expression is similar to that given when there is a finite rate of increase, except λ is replaced by e^r. In other words,

$$\lambda = e^r$$

an expression that may be solved to give r as a function of λ, or

$$r = \ln \lambda.$$

III. LOGISTIC GROWTH

The geometric and exponential growth models predict that the size of a population will always continue to increase, eventually to infinity, if R_0 or λ are greater than unity or if r is positive. In general, populations appear to be limited in their numbers by the amount of food, space, and other resources available in the environment (or possibly by other species, as discussed in

Chapters 13 and 14). In other words, the per capita birth rate, death rate, or both change with population density, due to increased intraspecific competition for resources with higher densities. Thus, these numbers are equal for some density (see the general discussion concerning population regulation in Chapter 12). This limit is determined by the resources available in a particular environment and is called the *carrying capacity*, the number of individuals that can be supported by a given environment. In some simple experimental cases, the size of the carrying capacity appears to be directly related to the food supply. For example, the number of *Daphnia* that can exist in a container increases linearly with the amount of food available. The size of the carrying capacity for natural populations, is also determined in large part by the level of resources available in a given environment.

The carrying capacity, symbolized by K, may be used in a growth equation so that, as the population number increases, the growth rate decreases. When the population size equals the carrying capacity, population growth ceases and there is no change in population numbers (see Example 11.2). To accomplish this, the exponential growth equation can be expanded with a term that includes K so that

$$\frac{dN}{dt} = rN\left(\frac{K - N}{K}\right)$$

where the term in the parentheses is known as the logistic term. This term becomes negative when N is greater than K, is positive when N is smaller than K, and is zero when $N = K$. This term also has a braking effect on the change in population numbers; thus, the population numbers asymptotically approach the carrying capacity, forming a flat S-shaped (or sigmoid) curve as shown in Figure 11.4, in contrast to the J-shaped curve for the exponential

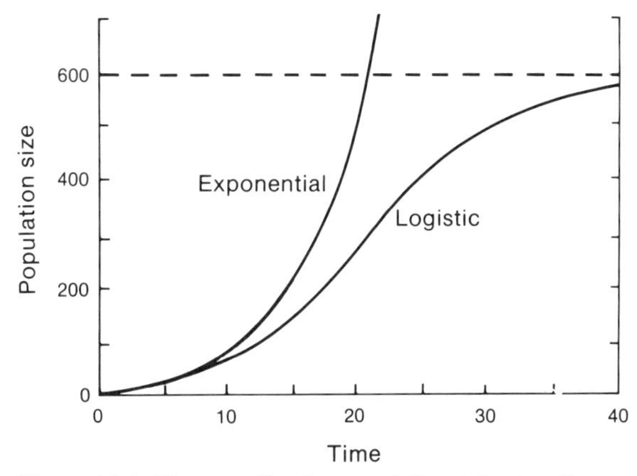

Figure 11.4. The growth of a population following the exponential and logistic curves when $N_0 = 10$, and $r = 0.2$ and $K = 600$ (as indicated by the broken line) for the logistic.

model. In addition, because the population size decreases from values above K and increases from values below K, K is a stable equilibrium for the population number in this environment (see Figure 11.5).

There are two ways in which the growth rate for the logistic model may be analyzed. First, the per capita growth rate is

$$\frac{1}{N}\frac{dN}{dt} = r\left(\frac{K - N}{K}\right)$$

for the logistic growth equation. When the population density is low, then

$$\frac{1}{N}\frac{dN}{dt} \cong r$$

Example 11.2. To examine the growth of a population in nature, one would like to follow population numbers over time after colonization. The growth of populations that are limited by space, such as barnacle populations, are often consistent with the predictions of the logistic model. For barnacles, the limit to population size is space on a rock, and once all the available space is taken, then further increases in density cannot occur. As an example, Figure 11.6 gives the density of barnacles over time for two freshly exposed rocks (from Connell, 1961). Both the carrying capacity and the pattern of density increase are similar for the two rocks.

Once a population size nears its carrying capacity for a given environment, one would expect that if the density becomes greater than K, the density would decline, and when the density is smaller than K, it

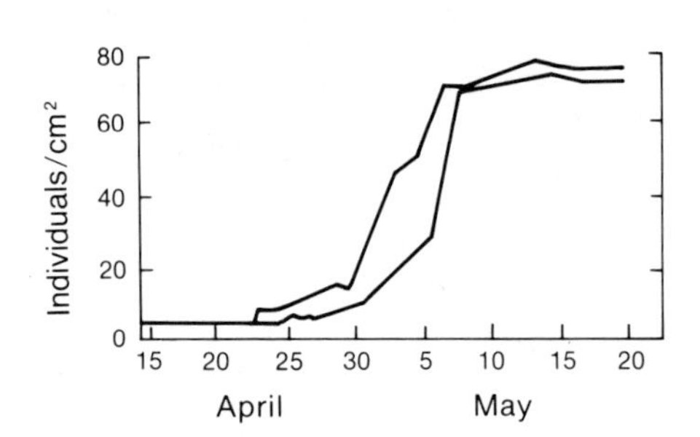

Figure 11.6. The density of barnacles on two rocks over time (from Connell, 1961).

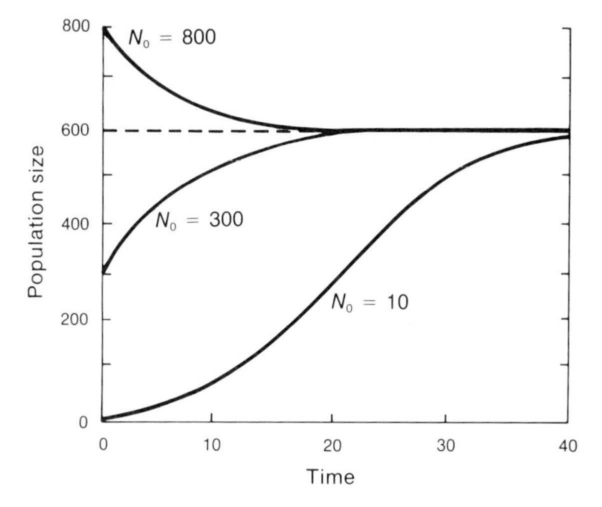

Figure 11.5. The population numbers over time when a population is started with 10, 300, or 800 individuals.

would increase. In a study of a population of ovenbirds, *Seiurus auroea-pillus*, this expected negative association of density and change in density in a woods was observed (Figure 11.7, after MacArthur, 1960). In this case, the density increased when it was low (smaller than 15) and decreased when it was high (greater than 20). When the density was between 15 and 20, the density increased part of the time and decreased part of the time. From these data, it appears that the carrying capacity in this area is between 15 and 20 birds.

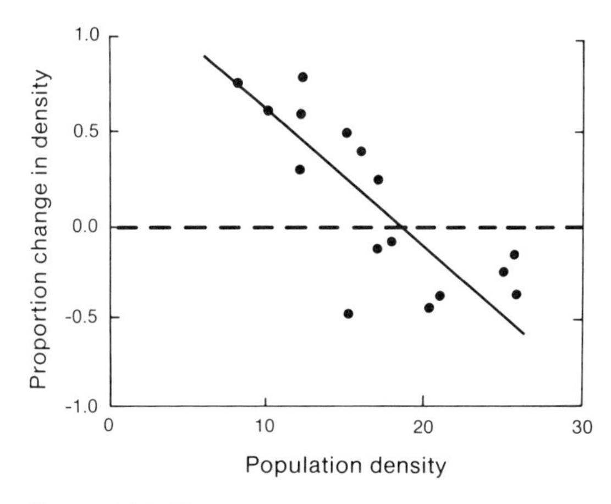

Figure 11.7. The proportion of change in population density in a population of ovenbirds at different densities (after MacArthur, 1960).

as for exponential growth (see Figure 11.8). However, as N increases, then the per capita rate of growth declines to zero linearly as K approaches N. Because the per capita growth rate is zero at K, there is no population change.

On the other hand, the change in population numbers, dN/dt, the population growth rate, has a different pattern, as shown by calculating dN/dt for different values of N (Figure 11.9). Obviously, dN/dt is at a minimum when $N = 0$ or $N = K$ and reaches a maximum value at an intermediate density.

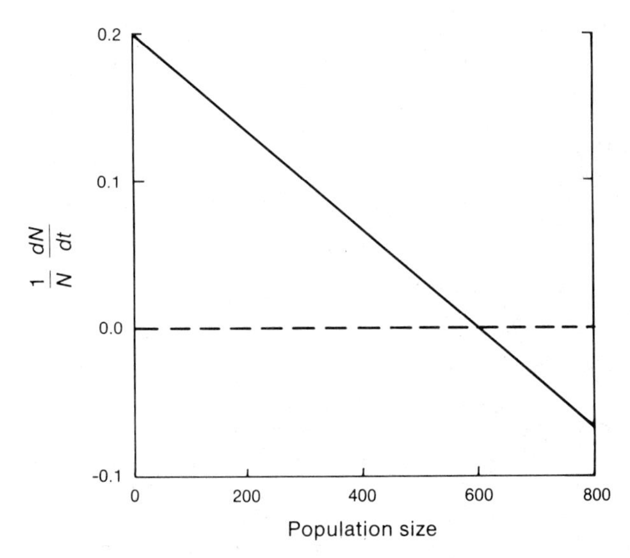

Figure 11.8. The per capita rate of growth for different population densities with $K = 600$ and $r = 0.2$.

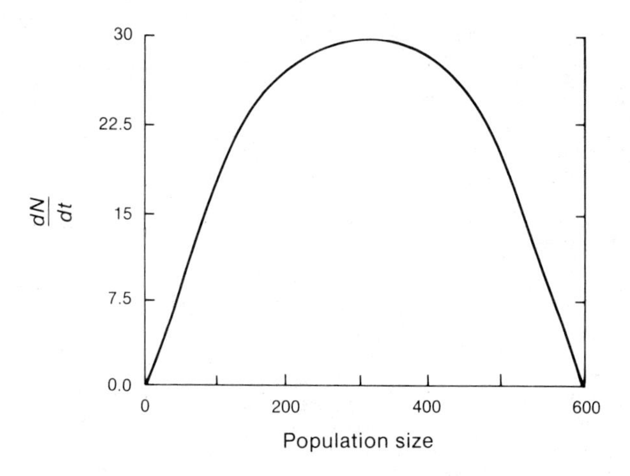

Figure 11.9. The change in population numbers for different population sizes with $K = 600$ and $r = 0.2$.

We can calculate the value of this maximum by taking the derivative of dN/dt, setting it equal to zero, and solving for N:

$$\frac{d}{dN}\left(\frac{dN}{dt}\right) = \frac{d}{dN}\left[rN\left(\frac{K-N}{K}\right)\right]$$

$$0 = \frac{d}{dN}\left(rN - \frac{rN^2}{K}\right)$$

$$0 = r - \frac{2rN}{K}$$

$$N = \frac{K}{2}.$$

Therefore, the maximum population growth rate occurs when $N = K/2$. This value is also the population density at which the population growth curve changes from concave to convex (see Figure 11.4).

We can use the logistic equation to predict the population growth rate at different densities by solving the equation to give the actual population number at any given time throughout the sigmoid growth phase (see Example 11.3). In other words, the population number at a given time is

$$N_t = \frac{K}{1 + \left(\frac{K}{N_0} - 1\right)e^{-rt}}$$

where N_0 is the initial population size. For example, if $K = 1{,}000$, $N_0 = 100$, $r = 0.1$, and $t = 5$, then

$$N_t = \frac{1000}{1 + \left(\frac{1000}{100} - 1\right)e^{-(0.1)5}}$$

$$= \frac{1000}{1 + 9e^{-0.5}}$$

$$= 183.2.$$

IV. MODIFICATIONS OF THE GROWTH MODELS

A number of assumptions are built into these population growth curve expressions, assumptions that can be relaxed to make the model more reflective of a real biological situation. For example, one of the assumptions of the logistic model is that the per capita growth rate declines linearly as density increases. An experiment in *Daphnia* (Smith, 1963) demonstrated that this

Example 11.3. Experiments with a number of laboratory organisms, such as *Paramecium*, *Drosophila*, and yeast, have demonstrated that population growth generally follows a simple sigmoid pattern under controlled laboratory conditions. Figure 11.10 gives the growth of a yeast population that is consistent with logistic predictions (from Pearl, 1927). In this case, the carrying capacity is approximately 665 and the value of r per hour is approximately 0.53. Therefore, the equation to predict the population numbers is

$$N_t = \frac{665}{1 + \left(\frac{665}{N_0} - 1\right)e^{-0.53t}}.$$

Let us assume that $N_0 = K/2 = 332.5$ and calculate the expected population size three hours later, and, as a result

$$N_t = \frac{665}{1 + \left(\frac{665}{332.5} - 1\right)e^{-0.53(3)}}$$

$$= \frac{665}{1 + 0.20} = 554.$$

Other values can be calculated in a similar manner, and the theoretical curve generated in this manner is given in Figure 11.10. Notice that the theoretical curve closely approximates the observed change in population numbers, suggesting that the logistic model is a good predictor of population growth in this case.

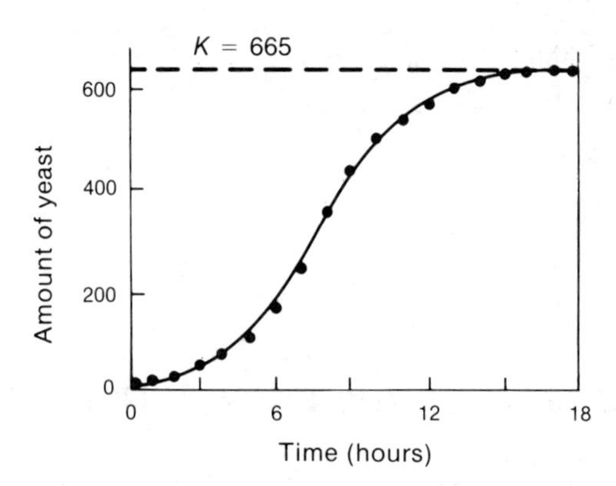

Figure 11.10. The growth of a yeast population (closed circles) and a logistic growth curve fit to these data (after Pearl, 1927).

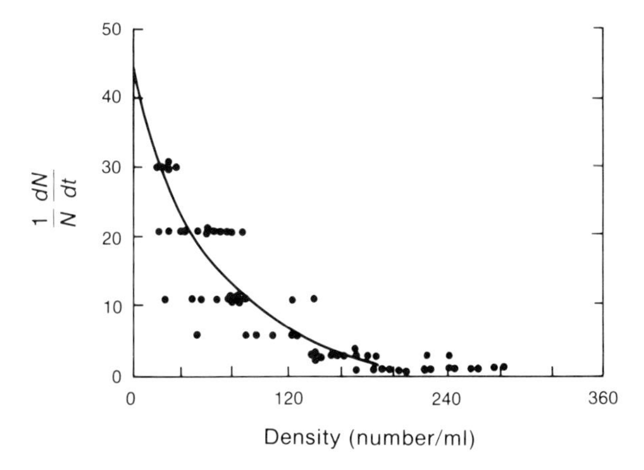

Figure 11.11. The per capita growth rate of a *Daphnia* population for different densities (after Smith, 1963).

relationship may be concave, with the per capita rate lower at intermediate densities than a linear relationship would predict (Figure 11.11).

Another assumption is that the effect of density on the population growth rate is instantaneous and not the result of past densities. However, in some cases, some time may pass before the density affects reproduction or mortality. For example, the number of eggs in a clutch may be determined to some extent by the food available when the eggs are formed, not the food available exactly at the time of hatching. The logistic equation can be modified to allow nonlinear changes in per capita growth rate or to allow time lags that can account for past densities (see Example 11.4). Below we briefly consider two other simple modifications to illustrate how these models can be expanded.

a. Variation in Finite Growth Rate

In our discussion until now, we have assumed that the values determining growth rates, such as R_0, λ, r, and K, are constant over time. Obviously, if the environment changes over time, then birth and death rates will vary and consequently cause variation in these values. To investigate the effect of this variation on population growth, let us examine the geometric growth equation in which λ is not constant.

Assume that λ varies in different years, so that λ_0 and λ_1 are the finite rates of increase in years 0 and 1, respectively—that is,

$$N_1 = \lambda_0 N_0$$

and

$$N_2 = \lambda_1 N_1.$$

By substitution as before, then,

$$N_2 = \lambda_0 \lambda_1 N_0.$$

In general, the population number in generation t is

$$N_t = \lambda_0 \lambda_1 \lambda_2 \ldots \lambda_{t-1} N_0 = \prod_{i=0}^{t-1} \lambda_i N_0$$

where Π is used to indicate the product of a series of numbers in the same way Σ is used to indicate the sum of a series of numbers. For example, if $N_0 = 100$ and $\lambda_0 = 1.2$ and $\lambda_1 = 1.6$, then

$$N_2 = (1.2)(1.6)(100) = 192$$

or there would be 192 individuals after two years.

Example 11.4. Although many laboratory experiments exhibit growth in accord with the prediction of the logistic expression, some experiments have led to large-scale oscillations in population numbers. For example, populations of the sheep-blowfly, *Lucilia cuprina*, showed oscillations approximately every 40 days in which the population size changed more than tenfold (Figure 11.12). This pattern of population change can be adequately explained by modifying the logistic expression with a time lag of nine days (broken line, Figure 11.12, after May, 1975). In this modification, the effect of density on population growth is delayed for nine days, approximately the time necessary for a larva to mature to an adult. Although one could construct a more detailed model, this simple modification produces a theoretical prediction that is close to the observed patterns.

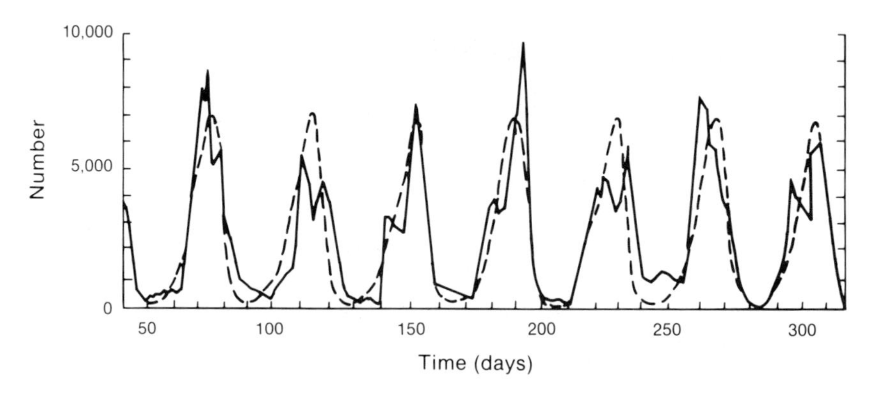

Figure 11.12. The number of individuals over time in a blowfly population (solid line, from Nicholson, 1958) and a similar pattern (broken line) using a logistic model with a time lag (May, 1975).

The growth of the population in which λ varies can be compared to the growth in a population with a constant λ that gives the same total growth. For example, if we take the square root of the product of the λ values used in the numerical example above

$$[(1.2)(1.6)]^{1/2} = 1.386$$

then the population size after two years is also

$$N_2 = (1.386)(1.386)(100) = 192$$

the same as when λ was variable. In fact, this can be generalized for any number of years, so that

$$N_t = (\lambda')^t N_0$$

where the constant λ' is

$$\lambda' = \left(\prod_{i=0}^{t-1} \lambda_i\right)^{1/t}$$

or the tth root of the product of the λ values. The measure λ' is an example of a geometric mean of a series of numbers, as contrasted to our usual arithmetic mean.

Let us consider an example in which λ is always greater than unity and in ten consecutive years is 1.2, 1.4, 1.4, 1.3, 1.1, 1.3, 1.2, 1.2, 1.3, and 1.1. Given that a population begins with 100 individuals, then the population grows as indicated by the solid line in Figure 11.13, reaching a size of 900 individuals

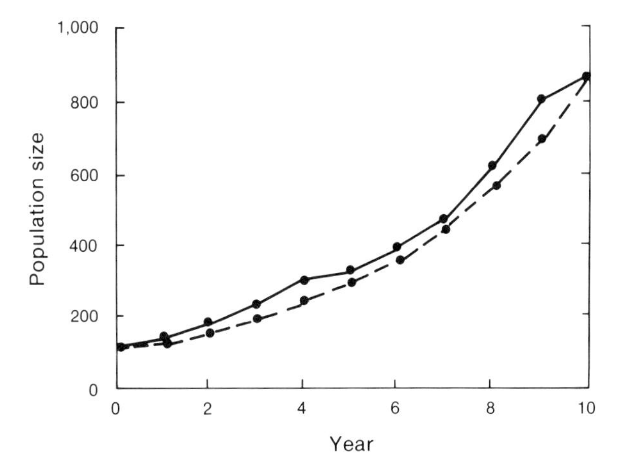

Figure 11.13. The population size over time when λ is variable (solid line) or constant (broken line).

after ten years (for clarity, continual growth throughout the year is shown in Figure 11.13, rather than the annual steps predicted by the geometric model). If we take the geometric mean of the product of the λ values

$$\lambda' = (9.0)^{1/10} = 1.245$$

then we have a value that gives the same total growth at a constant rate per year. Such a population is indicated as the broken line in Figure 11.13.

Example 11.5. Several bird species appear to have become extinct because human hunting reduced their numbers to such low levels that even when protected they could not reestablish themselves (see the discussion of extinction in Chapter 14). For example, the highly social passenger pigeon, which was once the most numerous bird species in North America, was reduced primarily by hunting to several thousand individuals by 1880. Although these few birds were no longer hunted, they could not reproductively sustain themselves and the species became extinct in 1914.

The final years leading to the extinction of the American heath hen was followed in some detail. The number of this species was reduced to less than 100, primarily because of recreational hunting and habitat destruction in the early part of the century (see Figure 11.14). At this point, strong protective measures were instituted and the population number rebounded. However, the population crashed a few years later after a fire and the species became extinct in 1932. The exact cause of this final collapse is not clear, but the species did form large mating aggregations, and, perhaps once this social mating pattern was disrupted, the species could not maintain itself.

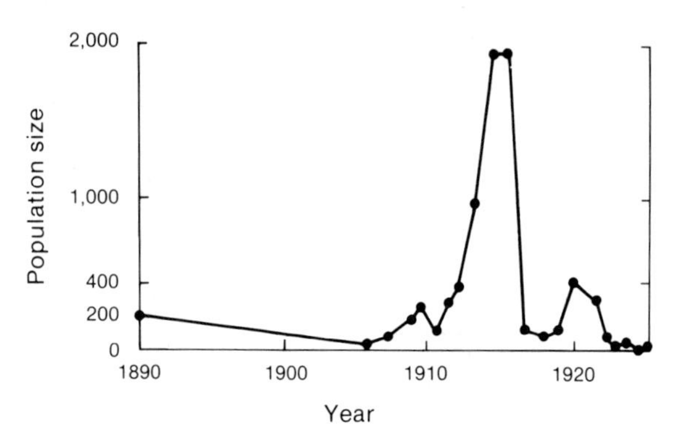

Figure 11.14. The numbers of the heath hen in the years just prior to its extinction (after Ziswiler, 1967).

---

b. Minimum Critical Population Density

The growth equations assume that the maximum per capita growth occurs at the lowest population number. In fact, at very low population numbers some species may decline in size and become extinct because the per capita growth rate is reduced (this is sometimes called the *Allee effect*). The reason for this may be that some minimal population size may be necessary for the organisms to find mates efficiently or to avoid predators. Such factors are probably most important in organisms with social structures that increase overall mating success or survival (see Example 11.5).

To alter the growth model so that the minimum critical density is considered, we can add another term to the logistic growth equation. In this case, we will introduce M, the minimal population size for population growth. Thus, the equation becomes

$$\frac{dN}{dt} = rN\left(\frac{K - N}{K}\right)\left(\frac{N - M}{N}\right).$$

Notice that this new term, $(N - M)/N$, is negative when N is less than M and positive when N is greater than M. Figure 11.15 shows how the per capita rate of increase varies with N with this expanded expression. There are now two points at which the per capita growth rate is zero and at which there are

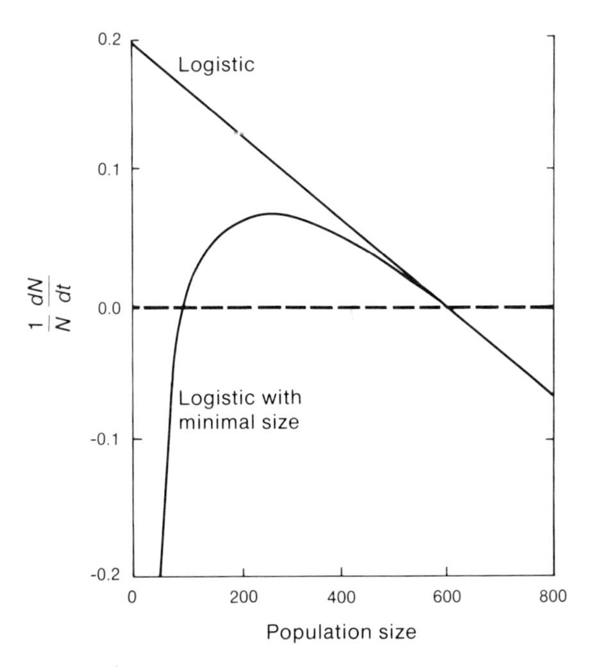

Figure 11.15. The per capita growth rate for the logistic with a minimal size constraint of $M = 100$ when $K = 600$ and $r = 0.2$.

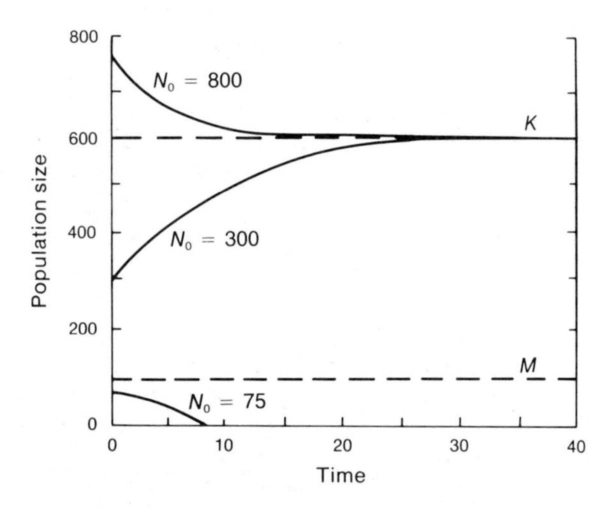

Figure 11.16. The change in population size over time for three initial population sizes when $K = 600$ and $M = 100$ (as indicated by broken lines).

consequently equilibria. The equilibrium at M is unstable because below M the per capita growth rate is negative and just above M, the growth rate is positive. In other words, the population number moves away from M in both directions. The equilibrium at K is stable as before. The population numbers over time for three different starting population numbers are shown in Figure 11.16. Notice that if N is initially smaller than M, then the population will become extinct, and if it is greater than M, the population will go to the stable equilibrium at K.

V. HUMAN POPULATION GROWTH

Archaeological and historical evidence obtained from various sources permit us to estimate changes in human population size. Obviously, the human population has increased in historical times but even more striking is that the human population has increased at an ever-growing rate—that is, estimates of r or other population growth parameters appear to be getting larger (Table 11.1). Several major factors in particular appear to be responsible for this historical increase. First, the spread of agriculture, beginning around 10,000 years ago, allowed a higher population density to exist and probably greatly increased the population size from preagricultural times. Second, the industrial revolution increased the productivity of groups of individuals and changed population patterns so that greater numbers of people could live together in high-density situations. Last, the improvement of medical care, the eradication or reduction of many infectious diseases, such as smallpox, diptheria, and plague, and the improvement of public health conditions greatly reduced death rates.

What will eventually stop the growth of the human population?

Table 11.1
The growth of the human population and the estimated time for population doubling (from Ehrlich and Ehrlich, 1972).

Date	*Estimated population*	*Doubling time (years)*
8000 B.C.	5,000,000	1,500
1650 A.D.	500,000,000	200
1850 A.D.	1,000,000,000	80
1930 A.D.	2,000,000,000	45
1975 A.D.	4,000,000,000	35

Table 11.2
Factors that may lead to an increased death rate or a decreased birth rate.

Increased death rate	*Decreased birth rate*
Famine and malnutrition	Sterilization
Disease	Abortions
War	Other forms of birth control
Energy shortages	
Shortages of other resources	
Infanticide	

Obviously, there is some limit to the number of humans that the earth can support, but the factors that will result in a cessation of growth are not clear. In the eighteenth century, Malthus first discussed in some detail factors that potentially could limit population growth. A number of factors are listed in Table 11.2 and are categorized according to whether they would primarily increase death rates or reduce birth rates. Of course, some factors, such as famine and malnutrition, would result both in an increased death rate and a decreased birth rate. Considering that some or many of these factors will probably be influential in limiting population numbers, these grim prospects should be of great concern.

There are cases in which a rise in death rates has already resulted from such factors, limiting population numbers in particular areas. For example, famine and malnutrition in sub-Saharan Africa have led to extremely high death rates in the past decade, and the plague in Europe during the Middle Ages killed a large proportion of the population. War and its resultant political upheavals resulted in very high death rates both in Europe during World War II and in southeast Asia during the 1960s and 1970s (see Example 11.6).

Overall, the most acceptable means of limiting population growth is reducing birth rates. A number of voluntary means of birth control have been employed in different societies. For example, birth control pills and birth control devices such as diaphragms and condoms are widely used in the

United States. Voluntary abortions in eastern Europe and Japan are the basis of the low birth rates in these countries, where the number of abortions may sometimes be larger than the number of live births.

It appears that an effective incentive for birth control is economic opportunity or development. In Japan and western Europe, economic conditions appear to have been responsible in large part for voluntary reductions in birth rates. However, it may be impractical to wait for economic development to occur in many of the developing countries. For example, in China birth rates have been greatly reduced by a social revolution, which has resulted in effective birth control through the raising of the marriage age and the limiting of family size. However, in other areas, such as much of Central and South America, tropical Africa, and Pakistan, the birth rate and growth rate continue to be high (see Example 11.7).

Let us assume that some factor or combination of factors will begin to slow down the worldwide population growth rate. As an illustration, assume that the net replacement rate R_0 either reaches 1.0 (*zero population growth*) in the period 1980–1985 or it drops gradually to 1.0 by the interval 2000–2005. The projections of population size for these two cases are given in Figure 11.17

Example 11.6. It is obvious to most people that undesirable factors, such as war, famine, and disease, will be important in limiting human populations in the future. Malthus identified these as major factors that would limit population numbers, but he did not discuss in any detail other factors of his time already affecting population numbers. Two checks that appear to have had an important impact in the period around 1800 in Europe are celibacy and infanticide (Langer, 1972).

First, the average marriage age in this period was in the middle or late twenties for both men and women. However, a large number of men made careers in the army or clergy and many women became domestics, occupations that greatly reduced the likelihood of marriage. In addition, many countries placed legal restrictions on marriage, making potential husbands present proof that they could support families. It is estimated that because of these and other factors, 30 to 40 percent of the population in a number of European countries was celibate.

Infanticide was also a major factor in reducing population numbers. Unwanted children were actively killed by starvation, smothering, or dosing with narcotics or alcohol. Many children were also abandoned at the doorstep of a church or a house where they often died of exposure. In England, those who survived were generally placed in "workhouses," where the mortality rate in the eighteenth century was more than 60 percent. In France, 138,000 babies were abandoned in 1822, with approximately 80 percent dying in their first year.

(after Frejka, 1973). Notice that even when $R_0 = 1.0$, in the 1980–1985 period, the population is still growing because of the large number of reproductive individuals. In this case, the population size levels off at 6.4 billion. If R_0 reaches 1.0 by 2000–2005, the population continues to grow, reaching a value of 8.4 billion. Of course, if R_0 does not decline to unity by the 2000–2005 interval, and it does not appear to be doing so, then the world population will be larger than either of these predictions.

Example 11.7. The population of India, the second largest of any country in the world, is predicted to reach approximately one billion by the year 2000. In the early 1950s, India was the first country to embark on a national family planning program. As a result, the use of birth control, primarily in the form of sterilization and intrauterine devices, has increased in the past two decades (Table 11.3). A sizable proportion of the population now employs birth control, although this usage is primarily limited to five states containing only 33 percent of the population. One state has passed a law calling for the compulsory sterilization of all men with three living children. The areas having most success in reducing birth rates are urban areas with relatively high educational and economic development. In rural areas, resistance to birth control has been greatest and is influenced by a number of cultural factors. For example, a young village wife generally receives favored treatment when she is pregnant. In addition, there is an economic advantage to farmers in having large families, and the Hindu religion requires a son to perform certain functions of a father's funeral. Because both economic and educational development are occurring slowly in India, compulsory sterilization and other strong measures to curb birth rates are advocated by the Indian government.

Table 11.3
Birth control usage and percent of couples using birth control in India (Gulhati, 1977).

Period	*Sterilizations*	*Intrauterine devices*	*Other contraceptives*	*Percent couples protected*
1955–1959	152,677	—	—	0.2
1960–1965	1,373,166	812,713	582,141	3.0
1966–1969	4,391,996	2,057,436	960,896	8.7
1970–1974	9,003,626	2,149,160	3,009,995	16.4
1974–1979*	18,500,000	5,700,000	10,000,000	35.9

* Goals for this period

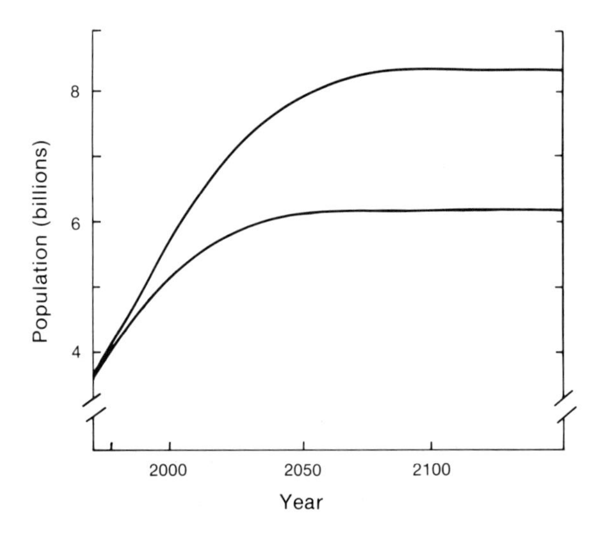

Figure 11.17. Two projections of world population size, where $R_0 = 1$ in 1980–1985 (lower line) or 2000–2005 (upper line) (after Frejka, 1973).

SUMMARY

Unlimited population growth can be described by either the geometric or exponential models. Assuming that population growth is limited by the carrying capacity of the environment, the logistic population growth model may be used. These simple models can be modified to include such factors as time lags, temporal variation in growth rates, and a mininum critical population density. The growth of the human population and factors affecting it are discussed in terms of these principles.

PROBLEMS

1. Given that $R_0 = 3$ and $N_0 = 10$, what is N_4? Given that $r = 0.1$ and $N_0 = 10$, what is N_4 using the exponential growth curve?
2. What is the per capita growth rate? Calculate it for $r = 0.1$ and $K = 400$ when $N = 100, 200$, and 300. What is dN/dt for $N = 100, 200, 300$? Contrast the changes in dN/dt and the per capita growth rate for different population sizes.
3. Given that $K = 400$ and $r = 0.1$, what is N_3, if $N_0 = 200$? If $\lambda_0 = 1.5$, and $\lambda_1 = 6.0$ what is N_2, given that $N_0 = 100$? What value of λ' would give the same amount of population growth?
4. Given that $K = 400$ and $M = 50$, what would be the eventual population size if the initial population size was $N_0 = 500, 100$, or 25?
5. What factors do you think will eventually limit the human population? Support your answer.

CHAPTER 12

Demography

In developing his culture and precipitating the technological, industrial, and scientific revolutions, man has profoundly altered the rhythm of his own reproduction. He has destroyed the equilibrium between the birth rate and the death rate which existed for most of the millennia he has been on this globe.

PHILIP HAUSER

The growth of a closed population depends upon the increase in the population due to births and the decrease due to deaths. *Demography* is the study of birth and death rates in a population and their impact on population growth and age structure. Birth and death rates may vary in a number of important ways in a population, and this variation may influence population growth. For example, birth and death rates may vary with the age of an organism with the maximum birth rate generally at some intermediate age and the maximum death rate at some very young age or at an old age. In addition, the density of a population may affect birth and death rates, with mortality increasing and birth rates decreasing as density grows. By accounting for age-dependent or density-dependent birth and death rates, we can understand more fully the dynamics of population growth. In this chapter we will discuss how such specific measures of birth and death rates may be utilized to explain the age structure of a population and the regulation of population size. The evolution of demographic characteristics will be discussed in Chapter 19.

I. FECUNDITY AND SURVIVAL SCHEDULES

Neither the exponential nor the logistic growth equations take into account the fact that birth and death rates are not uniform over the whole population. Both birth and death rates are age-specific, with the birth rate of juveniles zero or very low and the death rate of older individuals usually higher than that of younger individuals. Age-specific birth and death rates can be summarized in a *fecundity schedule*, which gives the average number of births per individual of a given age for individuals of all possible ages, and a *survival schedule*, which gives the probability of surviving from birth to a

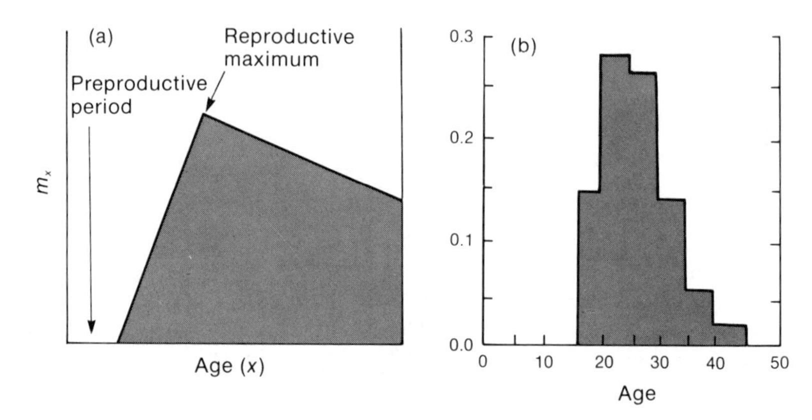

Figure 12.1. (a) A general fecundity schedule with an early high fecundity declining with age and (b) the fecundity schedule for United States women in 1973 (from Krebs, 1978).

particular age. The information in a survival schedule is often referred to as life table data.

A fecundity schedule is composed of m_x values, which are the average number of births per individual of age x (such data are also often called fertility schedules). A general fecundity schedule in an organism that reproduces multiple times can be graphed as in Figure 12.1a. Here there is a prereproductive stage in which no reproduction occurs, a reproductive stage with initially high reproduction, and then a decline in reproduction with increasing age. Most organisms, except for humans and some large mammals, do not have a postreproductive stage. In some organisms, such as many trees, the m_x values may continue to increase with age. The fecundity schedules of many human populations have been calculated, and Figure 12.1b gives the fecundity schedule for the white population in the United States. Notice that the m_x values here refer to the number of female births per female of a given age, a convention used in many organisms because it is often difficult to identify male parents. The schedule is divided into five-year intervals so that the average fecundity of individuals in the 15–19 year age class is 0.142, the number of female offspring for females in this age class. The maximum fecundity is in the age class 20–24, after which it declines to virtually zero in the 45–49 age class.

A survival schedule is somewhat different in that it is a cumulative value over age. A survival schedule gives the probability of survival from birth to age x, which is signified by l_x. Generally, the vertical axes in graphs of survival schedules are plotted on a logarithmic scale; thus, if the per capita death rate is constant for different ages, then the l_x values form a straight line. For example, if the probability of surviving each time interval is 0.10, then $l_0 = 1.0$ (all individuals are alive at age 0), $l_1 = 0.1$, $l_2 = 0.01$, and $l_3 = 0.001$. Figure 12.2a gives this situation, which is labeled a type II curve. Two other basic types of survival curves are given here. Type I indicates a survival curve with relatively

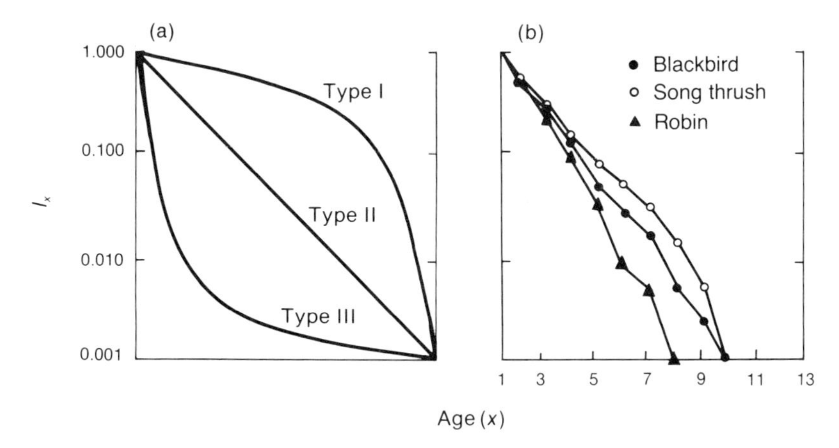

Figure 12.2. (a) Three basic types of survival schedules as a function of age and (b) the survival schedules of three species of birds in adults (Hutchinson, 1978).

low juvenile mortality, minimal accidental death in the middle age classes, and eventual death due to senescence. Type III indicates the survival schedule for an organism with high juvenile mortality and evenly spread mortality in the adult age classes. Figure 12.2b gives examples of data from three bird species that are similar to the type II curve (after Hutchinson, 1978).

Type I curves are generally found in vertebrates with extensive maternal care. Type II curves often occur in smaller organisms, such as bacteria, protozoa, and other organisms that do not change greatly in size with age. Type III curves are found in organisms with high fecundity and subsequent high juvenile mortality, such as many mollusks and plants. Of course, some organisms may have elements of more than one type of mortality schedule, such as some trees that have high juvenile mortality and low mortality in the intermediate age classes and that eventually die of senescence.

There are two basic ways to construct a fecundity or survival schedule, the cohort and the static approaches. The cohort approach relies on the monitoring of a *cohort*—a group of individuals born at the same time—throughout its lifetime (see Example 12.1). In other words, one records the proportion that dies or gives birth as the cohort ages. Obviously, such a study must last as long as individuals in the cohort are still alive. On the other hand, the static approach is carried out for only one time interval, and the survival and birth rates of individuals of different ages are estimated from the individuals alive in the time interval observed. For example, the death rate of individuals aged 50–54, 55–59, and 60–64 may be observed for a five-year time period, say, 1975–1979. These approaches may give similar fecundity and survival schedules if no change occurs in environmental conditions over the time period considered. However, if, as in contemporary human populations in developed countries, environmental changes over time increase survival, then a cohort-derived survival schedule would differ from one based on the static approach.

II. DEMOGRAPHIC MEASURES

Fecundity and survival schedules can be used to calculate a number of measures that are useful in interpreting changes in population numbers. However, because fecundity and survival schedules may include many age classes and consequently many l_x and m_x values, it is often easier to understand changes in population numbers by using measures that combine l_x and m_x values. Below we introduce several such measures, demonstrate how they are calculated, and illustrate their application.

a. The Net Replacement Rate

As discussed in Chapter 11, the ratio of the numbers in generation $t + 1$ over the numbers in generation t is equal to R_0, or

$$R_0 = \frac{N_{t+1}}{N_t}.$$

This measure of population change is generally applied to populations with nonoverlapping generations, such as annual plants or univoltine insects. For organisms with overlapping generations, such as humans and trees, the l_x and m_x values can be used to estimate R_0. In this case, the value of R_0 is the sum of the product of the fecundity and survival value for each age class, or

$$R_0 = \sum_{x=0}^{n} l_x m_x$$

where n is the number of age classes in the population. This value gives the average number of offspring produced per individual over its entire lifetime.

To illustrate the calculation of R_0, let us use the data given in Table 12.1. In this case, the probability of surviving from one age class to the next is 0.5. As a result, the proportion of individuals remaining alive decreases by one-half in

Table 12.1
The l_x and m_x values that are used to calculated several demographic measures.

Age (years)	l_x	m_x	$l_x m_x$
0	1.0	0	0
1	0.5	0	0
2	0.25	5	1.25
3	0.125	5	0.625
4	0.0	0	0.0
			1.875

each age class until the beginning of age class 4, by which time it is assumed that all individuals have died. The m_x values reflect no reproduction in the two juvenile age classes and five offspring per individual in the adult age classes. Therefore,

$$R_0 = \sum_{x=0}^{4} l_x m_m$$

$$\begin{aligned} R_0 &= l_0 m_0 + l_1 m_1 + l_2 m_2 + l_3 m_3 + l_4 m_4 \\ &= (1.0)(0) + (0.5)(0) + (0.25)(5) + (0.125)(5) + (0.0)(0) \\ &= 1.875. \end{aligned}$$

In keeping with our earlier interpretation of R_0, this value of R_0 indicates that the population is growing and that each individual produces on the average 1.875 offspring. Remember that if R_0 is less than 1.0, then the population would be expected to decline, and if it is exactly 1.0, then the population should remain constant in size.

b. Generation Length

When generations are overlapping, the length of a generation is not clear. One way to visualize the generation length in this case is as the average age of an individual when offspring are born. This can be calculated approximately as

$$T \cong \frac{\sum_{x=0}^{n} x l_x m_x}{\sum_{x=0}^{n} l_x m_x} \cong \frac{\sum_{x=0}^{n} x l_x m_x}{R_0}.$$

where T is the generation length. Using the data in Table 12.1 and ignoring the products that are zero, then

$$\begin{aligned} T &\cong \frac{(2)(0.25)(5) + (3)(0.125)(5)}{1.875} \\ &= 2.33 \text{ years.} \end{aligned}$$

c. The Instantaneous Rate of Increase

The instantaneous rate of increase can be estimated two different ways using fecundity and survival data. First, the instantaneous rate of increase is approximately

$$r \cong \frac{\ln R_0}{T}$$

where R_0 and T are as defined above. Using the data in Table 12.1, then

$$r \cong \frac{\ln(1.875)}{2.33} = 0.27.$$

The instantaneous rate of increase can also be calculated using a formula (called the Euler equation) directly relating r to the l_x and m_x values, which is

$$\sum_{x=0}^{n} l_x m_x e^{-rx} = 1.$$

Here, if the sum of the product on the left side of the equation is 1.0, then the value of r used in the exponent is the instantaneous rate of increase corresponding to the given fecundity and survival data. If the l_x and m_x values are known, then r is the only unknown in the equation, and it can then be calculated.

To illustrate the calculation of r using this approach, let us again use the data in Table 12.1. Because the product $l_x m_x$ is 0.0 for age classes 0, 1, and 4, we need only to consider age classes 2 and 3. In other words, we need to find a value of r such that

$$l_2 m_2 e^{-2r} + l_3 m_3 e^{-3r} = 1$$

$$1.25e^{-2r} + 0.625e^{-3r} = 1.$$

This value can be found by trial and error—that is, by substituting in values of r and checking to determine if the sum is equal to 1.0. To begin with, let $r = 0.27$, the value obtained by the first approach above. Then

$$1.25e^{-0.54} + 0.625e^{-0.81} = 1.006.$$

This sum is slightly greater than unity, so we need to try a larger value of r, say 0.28 (r is in the denominator, so a larger r value makes the sum smaller). In this case, the sum is 0.984, somewhat too small. After trying several more values, we find $r = 0.273$ to be the closest estimate (see Example 12.1).

d. Life Expectancy

One demographic measure of particular importance in human populations is the *life expectancy*—that is, the expected duration of life from a given age. For

Example 12.1. Determining the survival and fecundity schedules in vertebrate populations is often difficult because it is generally hard to monitor a complete population. However, a long-term study of ground squirrels in Utah using live traps and observation towers documented birth and death rates over a seven-year period (Slade and Balph, 1974). These squirrels hibernated approximately seven months each year and were restricted by physiographic features to the study area. During the first four years of the study, the population density was undisturbed and averaged slightly more than 200 animals. The population was then artificially reduced to a lower level, approximately 100 animals. The l_x and m_x schedules for these two densities and the instantaneous rates of increase estimated using both approaches above are given in Table 12.2. Notice that the values of r at the two densities are quite different. At the high density, the natural density before human intervention, r is very close to zero, reflecting the fairly stable population number observed in this population. On the other hand, at the low density r is much greater than zero and indicates the increase in population number observed after the artificial reduction.

Table 12.2
The l_x and m_x schedule observed for a ground squirrel population at two different densities and the demographic measures calculated from these data (from Slade and Balph, 1974).

x (*age in years*)	*High density*		*Low density*	
	l_x	m_x	l_x	m_x
0	1.000	0.00	1.000	0.00
0.25	0.662	0.00	0.783	0.00
0.75	0.332	1.29	0.398	1.71
1.25	0.251	0.00	0.288	0.00
1.75	0.142	2.08	0.211	2.24
2.25	0.100	0.00	0.167	0.00
2.75	0.061	2.08	0.115	2.24
3.75	0.026	2.08	0.060	2.24
4.75	0.011	2.08	0.034	2.24
5.75	0.000	—	0.019	2.24
6.75	—	—	0.010	2.24
R_0	0.927		1.686	
T	1.616		1.961	
$r = \ln R_0/T$	−0.047		0.266	
r (from Euler equation)	−0.046		0.306	

example, e_0 is the life expectancy of an individual at birth, which in many human populations of developed countries is about 70 to 75 years (see Example 12.2). Insurance companies are particularly interested in life expectancy so they can charge enough money before an individual dies to make a profit. In natural populations of other species, life expectancy values may be used to determine the expected longevity of individuals of a known age. The calculation of life expectancy is relatively straightforward, although it involves several steps.

First, using an l_x schedule, let us calculate the average proportion of a cohort alive in any time interval. Remember that l_x is the proportion alive at the beginning of the time interval and that l_{x+1} is the proportion alive at the end of the time interval (or the beginning of the next). Therefore, the average proportion alive is

$$L_x = \frac{l_x + l_{x+1}}{2}.$$

Second, we need to calculate the average amount of time to be lived by the individuals surviving to a given age. This value is the sum of the L_x values from the age of interest to the last age class. For example, for the zero age class, this value is

$$T_0 = \sum_{x=0}^{n} L_x$$

(note that T_x is different from the generation length T). Finally, because in all age classes except the zero class some individuals have already died, the life expectancy must be standardized by the proportion of individuals alive at the beginning of the time period, so that in general

$$e_x = \frac{T_x}{l_x}.$$

Let us calculate the life expectancy for a simple numerical example and use the survival schedule in Table 12.1. For example, with these data

$$L_0 = \frac{l_0 + l_1}{2} = \frac{1.0 + 0.5}{2} = 0.75.$$

After calculating the other L_x values, then T_0 is

$$\begin{aligned} T_0 &= L_0 + L_1 + L_2 + L_3 \\ &= 0.75 + 0.375 + 0.1875 + 0.0625 \\ &= 1.375. \end{aligned}$$

Table 12.3
A survival schedule and the values used to calculate life expectancy.

Age (years)	l_x	L_x	T_x	e_x
0	1.0	0.75	1.375	1.375
1	0.5	0.375	0.625	1.25
2	0.25	0.1875	0.25	1.0
3	0.125	0.0625	0.0625	0.5
4	0.0	—	—	—

The life expectancy at age zero is then

$$e_0 = \frac{T_0}{l_0} = \frac{1.375}{1.0} = 1.375 \text{ years.}$$

The life expectancy for individuals of older age classes is smaller, and, for example, the life expectancy of an individual at the beginning of the third age class is

$$e_3 = \frac{T_3}{l_3} = \frac{0.0625}{0.125} = 0.5 \text{ years.}$$

The values used in these calculations and those for other age classes are given in Table 12.3. Example 12.2 illustrates the calculation of life expectancy in a barnacle population and discusses life expectancy in humans.

Example 12.2. Sessile organisms, such as plants and some mollusks, can be studied demographically because individuals can be followed over time. For example, the number of barnacles in a cohort that settled on a rock within a two-month period were observed for mortality over a nearly ten-year period (Connell, 1970). Initially there were 142 barnacles, and this number declined to 62 after a year, to 34 after two years, etc. (see Table 12.4). From these data, a survival schedule can be calculated (third column of Table 12.4). For example,

$$l_2 = \frac{34}{142} = 0.24.$$

Using this information, then the L_x, T_x, and e_x values can be calculated as demonstrated above. When examining the life expectancy values, notice that the life expectancy is at maximum for e_3. In other words, individuals entering the third age class have the highest life expectancy, higher than for individuals when they first settle on the rock. This delayed maximum is the result of quite a high mortality in the first two years.

The average human life expectancy has increased substantially in recent decades in societies with modern medical care. In the United States, the life expectancy from birth has increased from 47 to 73 years in this century (Figure 12.3; Fries, 1980). Notice, however, that the life expectancy for the older ages—for example, at age 75—has changed very little. In addition, the proportion of extremely longlived individuals has not increased in recent years, with only approximately one in 10,000 individuals living beyond 100 years in developed countries.

The cause of death in the earlier age classes in the early part of the century was infectious diseases, such as smallpox, diphtheria, tuberculosis, polio, and pneumonia. At present these diseases are less than 2 percent of the health problems they once were. The decline in these diseases can be primarily attributed to immunization, antibiotics, improved nutrition, and better public health. As a result, there has been a rise in the proportion of deaths and disability caused by chronic diseases, such as arteriosclerosis, cancer, diabetes, arthritis, and pulmonary diseases. These diseases often symptomatically begin in early life and develop gradually over a lifetime until they become debilitating. Death from such chronic diseases can often be delayed and individuals can live until a relatively old age with diseases such as diabetes or arteriosclerosis.

It has been suggested that there is a limit to how long death due to chronic disease or its related effects can be postponed. As a result of this postponement and of what appears to be a biological limit to life, death incidence is becoming compressed into a shorter age span. Figure 12.4

Table 12.4
The number of a cohort of barnacles alive on a rock over time and the values used to calculate life expectancy.

Age (years)	*Number alive*	l_x	L_x	T_x	e_x
0	142	1.0	0.72	1.575	1.58
1	62	0.44	0.34	0.855	1.94
2	34	0.24	0.19	0.515	2.15
3	20	0.14	0.12	0.325	2.32
4	15.5*	0.11	0.10	0.205	1.86
5	11	0.08	0.06	0.105	1.31
6	6.5*	0.05	0.03	0.045	0.90
7	2	0.01	0.01	0.015	1.50
8	2	0.01	0.005	0.005	0.50
9	0	0.0	—	—	—

* Estimated number.

gives the survival schedule for 1900, 1980, and a theoretical curve for the future, with death compressed into a short time interval (Fries, 1980; the vertical axis here is not logarithmic, as it was before). Notice that the elimination of premature deaths resulted in a rise in the curve from 1900 to 1980 and that the additional increase predicted in the future will further compress mortality from natural causes into the seventies and eighties.

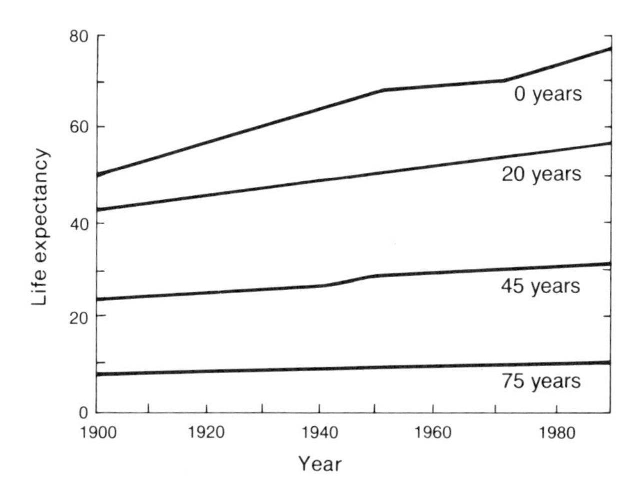

Figure 12.3. The life expectancy for different ages over recent years (Fries, 1980).

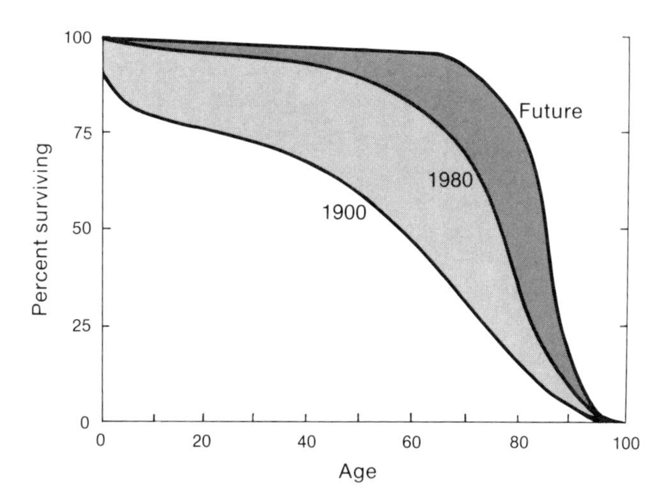

Figure 12.4. The proportion surviving to different ages for three different times in the United States population (Fries, 1980).

e. Reproductive Value

As the life expectancy of an individual varies with age, the expected contribution of individuals in terms of progeny also varies with age. One measure of this contribution is the *absolute reproductive value* v_x, which is the expected number of future offspring for an individual of a given age. The absolute reproductive value for age class 0 is

$$v_0 = \frac{1}{l_0} \sum_{y=0}^{n} l_y m_y$$

where l_y and m_y are the same as l_x and m_x before. This expression simplifies to

$$v_0 = R_0$$

Example 12.3. In order to calculate reproductive values in a population, we need both survival and fecundity schedules, data that are uncommon in many species. First, let us go through the complete calculation of reproductive values for some data on a white-crowned sparrow population in California and then present the results for the ground squirrel data given previously.

The survival and fecundity schedules of the white-crowned sparrow population, combined for two cohorts, are given in Table 12.5 (from Baker et al., 1981). Notice that there is a particularly high loss, more than 80 percent, in the first year, and that the fecundity values are relatively constant for the adult age classes. The v_x and v_x' values are quite low for the zero age class because most of these individuals die before they reproduce. The general trend is then a decline in the reproductive values

Table 12.5
The survival and fecundity schedules and reproductive values for a population of white-crowned sparrows (Baker et al., 1981).

Age (years)	l_y	m_y	$l_y m_y$	$\sum_{y=x}^{n} l_y m_y$	v_x	v_x'
0	1.000	0.0	0.0	1.033	1.033	1.000
1	0.174	3.142	0.547	1.033	5.937	5.747
2	0.084	3.333	0.280	0.486	5.786	5.603
3	0.044	3.556	0.156	0.206	4.682	4.532
4	0.009	3.750	0.034	0.050	5.556	5.379
5	0.004	4.00	0.016	0.016	4.000	3.872

because $l_0 = 1.0$. The absolute reproductive value at other ages is the sum from that age through the rest of the age classes or in general

$$v_x = \frac{1}{l_x} \sum_{y=x}^{n} l_y m_y.$$

For example, using the data in Table 12.1,

$$v_1 = \frac{1}{l_1}(l_1 m_1 + l_2 m_2 + l_3 m_3 + l_4 m_4)$$

$$= \frac{1}{0.5}[(0.5)(0.0) + (0.25)(5) + (0.125)(5) + (0.0)(0)]$$

$$= 3.75$$

---

with age. The v_x and v_x' values are similar because the population seems to be nearly stable with $R_0 = 1.033$.

The relative reproductive values for the ground squirrel data of Slade and Balph (1974; see Example 12.1) are given in Figure 12.5. Notice that the patterns are quite similar for the two densities, with a maximum in age classes 1–3. The oscillations result from overwintering periods in which there is no fecundity and some mortality.

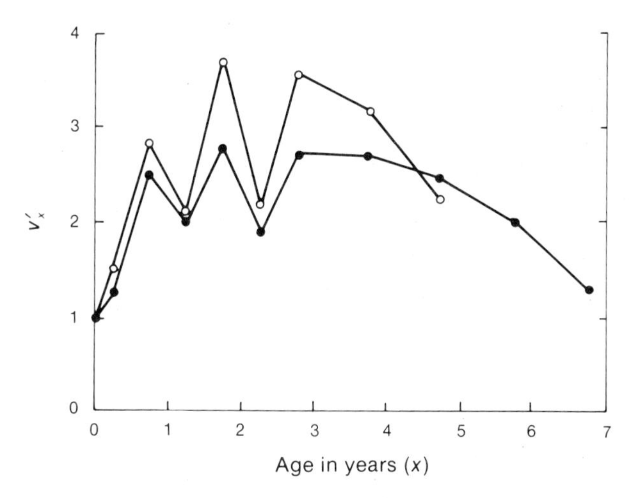

Figure 12.5. The relative reproductive values for a ground squirrel population at a high density (open circles) and a low density (closed circles) (data from Slade and Balph, 1974).

or, on the average, an individual entering age class 1 will have 3.75 progeny. Notice that this value is higher than v_0, which is 1.875, because individuals entering age class 1 have already survived a period of high mortality. In general, v_x is at a maximum at the beginning of the reproductive period, and as an individual ages until first reproduction at age x, v_x increases. After that time, v_x declines, reaching a relatively low value at the older ages.

Generally, reproductive values are given relative to v_0 (remember this is equal to R_0), so that populations that have different growth rates can be compared. The *relative reproductive value* is, therefore,

$$v'_x = \frac{v_x}{v_0}$$

As an example, the reproductive values are calculated in Example 12.3 for some data from two species (see Example 12.3).

Individuals for which the reproductive value is at a maximum will contribute the most to future population growth. On the other hand, if the reproductive value is low, as in the zero age class or the older age classes in the examples above, then these individuals will contribute less to future population growth. As a result, removal of individuals with low reproductive values, either by human harvesting or by predators, should have the smallest effect on population growth. As an extreme example, the taking of salmon after they have spawned (they only do this once in a lifetime and then die) removes individuals with a zero reproductive value and does not affect future population numbers.

III. PROJECTING POPULATION SIZE

One of the important attributes of demographic values is that they can be used to predict the future population size and the distribution of the population into different age classes. Such predictions can be used in understanding population growth in any organism for which good demographic information exists, and they have been widely used to predict human population trends.

In predicting population size, one can use the fecundity schedule as given above, calculating the survival probability for each time period, instead of using the cumulative survival schedule as given in the l_x values. For example, the survival in the age interval from x to $x + 1$ is

$$p_x = \frac{l_{x+1}}{l_x}$$

or the proportion surviving at the end of the time period divided by the proportion alive at the beginning of the time period. For example, again using the data in Table 12.1,

$$p_1 = \frac{l_2}{l_1} = \frac{0.25}{0.125} = 0.5$$

or half the individuals entering the age class survived it. Often, we calculate the mortality for specific age classes. The age-specific mortality is the complement of age-specific survival and is

$$q_x = 1 - p_x.$$

An example of age-specific mortality in Dall sheep and domestic sheep is given in Figure 12.6 (Caughly, 1966). Notice that there is relatively high mortality in the first year and then high mortality in the older age classes.

The number of individuals in all the age classes, except for the zero age class, is determined by the number in the next younger age class in the previous time period and the survival probability for that age class. In other words, the number in age class $x + 1$ at time $t + 1$ is

$$N_{x+1 \cdot t+1} = N_{x \cdot t} p_x.$$

For example, given that 100 individuals are in age class 1 at time t, $N_{1 \cdot t} = 100$, and assuming $p_1 = 0.5$, then

$$N_{2 \cdot t+1} = N_{1 \cdot t} p_1 = (100)(0.5) = 50$$

or there should be 50 individuals in age class 2 in the next time period.

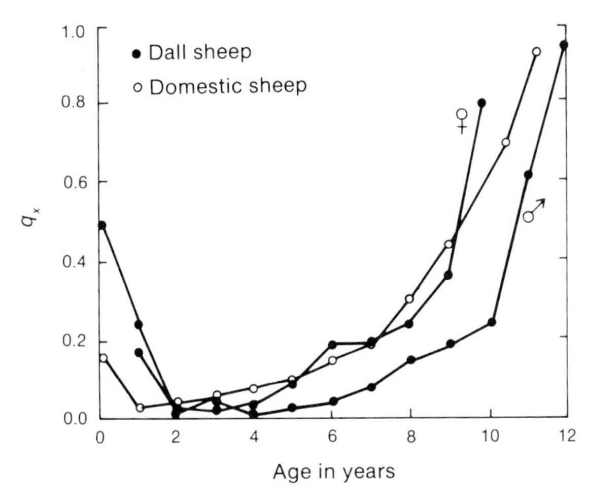

Figure 12.6. Age-specific mortality in Dall and domestic sheep (from Caughly, 1966).

Calculating the number in the zero age class is more complicated, because individuals enter this age class as births from individuals of several different ages. For example, if there are 100 individuals in age class 2 and $m_2 = 5$, then it would be expected that 500 individuals would be produced that enter the zero age class. In general then,

$$N_{0 \cdot t+1} = \sum_{x=1}^{n} N_{x \cdot t+1} m_x$$

where the number of newborns is the sum of the product of the m_x values and the numbers in the various age classes. Notice that the number of adults used to determine the number of newborns is the number at time $t + 1$. In other words, the number of individuals in the higher age classes must be calculated first and then the number of newborn individuals calculated.

Let us consider a numerical example to illustrate the prediction of population size and the age-class distribution over time. Assume that the l_x and m_x values are as in Table 12.6. In this case, $R_0 = 1.0$ and, using the techniques described before, $r = 0.0$ and $\lambda = 1.0$. The survival values for each of the age classes is given in the last column. Assume that a population begins with 20 individuals in each of the age classes at time 0 (see Table 12.7). The

Table 12.6
The survival schedule, age-specific mortality, and fecundity schedule used in the population projection in Table 12.7.

Age	l_x	m_x	p_x
0	1.0	0	0.2
1	0.2	3	0.5
2	0.1	4	0.0
3	0.0	0	—

Table 12.7
The projection of population numbers over time, using the survival and fecundity schedules in Table 12.6.

Age	Time						
	0	1	2	3	4	5	6
0	20	4(3) + 10(4) = 52	38	44	43	47	47
1	20	20(0.2) = 4	10	8	9	9	9
2	20	20(0.5) = 10	2	5	4	5	5
	60	66	50	57	56	61	61

number in age class 1 at time 1 is then

$$N_{1\cdot 1} = N_{0\cdot 0}p_0 = (20)(0.2) = 4.$$

Likewise, the number in age class 2 at time 1 is

$$N_{2\cdot 1} = N_{1\cdot 0}p_1 = (20)(0.5) = 10.$$

The number in age class 0 at time 1 is the total derived from births to individuals in age classes 2 and 3. Therefore,

$$N_{0\cdot 1} = N_{1\cdot 1}m_1 + N_{2\cdot 1}m_2 = (4)(3) + (10)(4) = 52.$$

Overall, then, there are 66 individuals at time 1, a slight increase from time 0. Such an increase would not have occurred in a population without age structure and only resulted here because initially there were a relatively large number of individuals in the reproductive-age classes 2 and 3.

The population numbers can be predicted for future time intervals as given in Table 12.7. Notice that at time interval 5, the population ceases to grow and that there is a constant number in each age class from this time on. When there is a constant proportion in each age class over time, then this is termed a *stable age distribution*. In this example, there are 0.770, 0.148, and 0.082 of the population in the three age classes where a stable age distribution exists. A population with other l_x or m_x values could be growing or declining in numbers but still maintaining constant proportions in all age classes (see the discussion of age pyramids later in this chapter).

Where a stable age distribution exists, the population grows at a rate identical to that determined by λ, the finite rate of increase. Obviously, in this case, there is no change in population number, so that, for example

$$\lambda = \frac{N_6}{N_5} = \frac{61}{61} = 1.$$

If we calculate r using the approach described above, then

$$l_1m_1e^{-r} + l_2m_2e^{-2r} = 1$$
$$0.6e^{-r} + 0.4e^{-2r} = 1.$$

In this case, if $r = 0.0$, then this equation is equal to unity. Because $\lambda = e^r$, then $\lambda = 1$ as observed from the ratios of population numbers from one time interval to the next or from the ratios of each age class from one time period to the next.

Once the stable age distribution has been approached, the population will change in a geometric manner as determined by the value of λ. To make the model more realistic, l_x and m_x values may be made a function of density, indicating increased intraspecific competition at higher densities, so that the population size will approach some equilibrium density. One way to accomplish this is to use the logistic term, so that

$$l'_x = l_x\left(\frac{K - N}{K}\right)$$

$$m'_x = m_x\left(\frac{K - N}{K}\right)$$

with the restriction that l'_x and m'_x are positive. Notice that the equilibrium density here is not K but the density at which the l'_x and m'_x values result in a λ value of unity.

IV. POPULATION REGULATION

The principal factors determining the density of a population are the subject of a longtime controversy in population ecology (for a discussion, see Tamarin, 1978). Expressed as a simple dichotomy, some ecologists feel that intraspecific competition for resources, competition which is more intense at higher densitites, regulates population numbers while others feel that control of population numbers is independent of density and that extrinsic factors, such as the weather, are of primary importance. Obviously, the relative importance of these factors, which can be simply called density-dependent and density-independent factors, depends upon the organism involved (see below and Chapter 19). The effects of other interacting species, such as competitors and predators, on population density are discussed in Chapters 13 and 14.

In Chapter 11, we considered the three basic descriptions of population growth as given by the geometric, exponential, and logistic equations. The geometric and exponential growth models assume that the per capita growth rate is constant for all densities—that is, that the growth rate is density independent. On the other hand, the logistic model assumes that the per capita growth rate decreases as population density increases towards the carrying capacity. As we stated earlier, the per capita growth rate is a function of the combined birth and death rates and, as a result, a reduction in growth rate may result from either a decrease in birth rate or an increase in death rate with increasing density or both.

Density-dependent and density-independent effects on birth and death rates can be compared in a graphical manner, as in Figure 12.7. Here the horizontal lines indicate the density-independent rates, birth or death rates that do not change with population density. On the other hand, when there are density-dependent effects, the birth rate decreases and the death rate increases with increasing N. Assuming that both birth and death rates are density dependent, then the two density-dependent curves can be combined in one graph (Figure 12.8a). At the intersection of the two lines, the per capita birth and death rates are equal; thus, the per capita growth rate is zero. The density at which the per capita growth rate is zero is the carrying capacity for the population in a particular environmental situation.

In reality, the curves for birth and death rates may have a variety of slopes and shapes. For example, the curves may be density dependent but not linear at very low densities, resulting in an increase in the per capita death rate and a decrease in the per capita birth rate as numbers decrease. The death rate may have such a pattern in social animals in which social organization at higher densities enhances survival. Likewise, the birth rate may be low at lower densities because mates are difficult to find. As a result, there may be another intersection of the two curves at low density, indicating a minimal critical density of the kind discussed in Chapter 11. Such an example is indicated in Figure 12.8b, where the per capita birth and death rates are equal at a low density, M, as well as at a high density, K.

Density-independent factors are factors whose influence on the per capita birth or death rates is independent of density. For example, the death rate resulting from volcanic ash covering an area would probably be independent of the density of the organism. Climatic factors, noninfectious diseases, and pollution are generally density-independent in their effects. On the other hand,

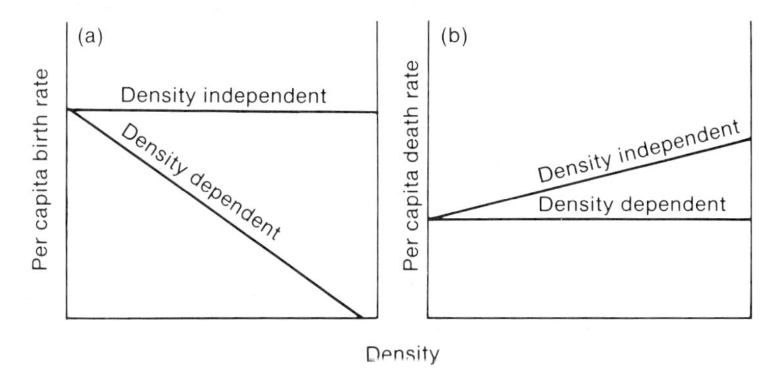

Figure 12.7. A schematic representation of density-dependent and density-independent per capita (a) birth and (b) death rates.

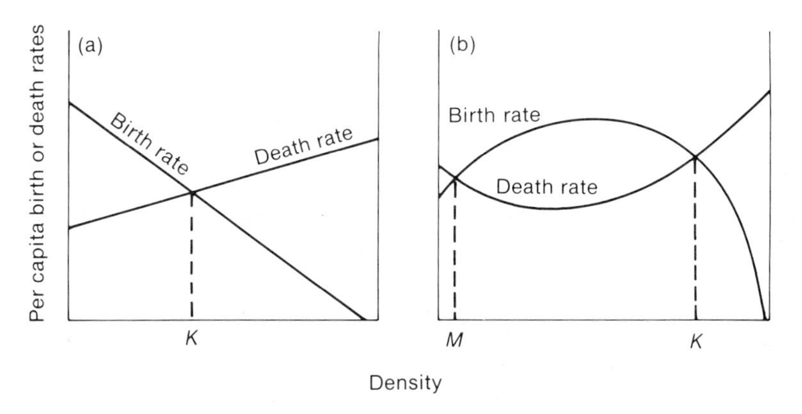

Figure 12.8. Examples of density-dependent birth and death rates when (a) they are linear and (b) curvilinear.

density-dependent factors are factors whose influence on the per capita birth or death rates is dependent upon the density. For example, infectious diseases and intraspecific competition for food have their greatest effect on the death rate at high density. Some organisms, such as many insects and protozoans,

---

Example 12.4. In natural populations, to ascertain the effect of density on birth and death rates, observations are usually made over a period of time. As a result, variation in density is often accompanied by changes in other abiotic or biotic factors, and it is sometimes difficult to attribute trends to density because of the variation in birth or death rates due to other factors. Such an example is given in Figure 12.9a for mortality in the east African buffalo at different densities (Sinclair, 1974). The mortality in adults does appear to increase with density, nearly doubling over the range of density examined. Juvenile mortality, although higher at all densities than adult mortality, does not show a clear trend with density. It appears that juvenile death, which mainly appears to be the result of malnutrition and disease, is density-independent or that the effects of other factors besides density on mortality are large enough to obscure density-dependent effects.

A second example of density dependence is the average clutch size of the great tit in a woods over a period of years in which the breeding density varied (Figure 12.9b, from Perrins, 1965). In the years when the number of breeding pairs was high, the clutch size was low, while in the years of low density, the clutch size was quite high. This inverse relationship is consistent with that expected with density-dependent birth rates.

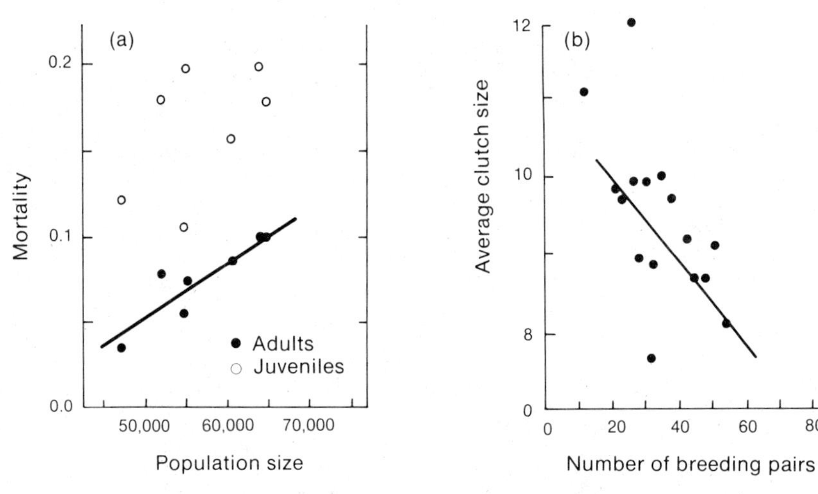

Figure 12.9. (a) The mortality of buffalos at different densities (Sinclair, 1974), and (b) the clutch size of the great tit at different densities (Perrins, 1965).

---

appear to be influenced primarily by density-independent factors, while other organisms, such as many large vertebrates, appear to be primarily affected by density-dependent factors (see Example 12.4).

Density regulation in vertebrates (and perhaps some invertebrates) may be partly the result of dispersal (and probably subsequent increased mortality) of some individuals to less favorable habitats. The dispersal event itself may be precipitated by intense intraspecific competition for resources through territorial behavior (see Wynne-Edwards, 1962).

To develop an approach to detect density dependence, let us use the exponential expression

$$N_{t+1} = N_t e^r$$

where we are considering the numbers of individuals from time t to time $t + 1$. If we take natural logarithms of this expression, then

$$\ln N_{t+1} = \ln N_t + r.$$

If r is constant—that is, if there is no density dependence—then $\ln N_{t+1}$ and $\ln N_t$ should be linearly related. Figure 12.10a gives the theoretical relationship between these values when r is constant, either equal to zero or to a value greater than zero (solid lines). However, when there is density dependence, the observed slope is not unity (a 45° angle). For example, the broken line in Figure 12.10a gives a theoretical situation in which the slope is less than unity, $\ln N_t < \ln N_{t+1}$ for low density and $\ln N_t > \ln N_{t+1}$ for high density. The point where this line crosses the line for $r = 0$, where $\ln N_t = \ln N_{t+1}$, indicates the carrying capacity for this example.

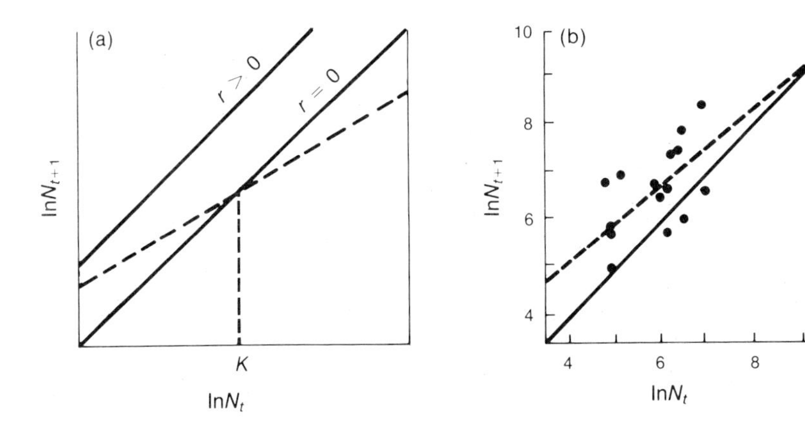

Figure 12.10. (a) The relationship between the natural logarithm of population density at two different times where there is density independence (solid lines) and density dependence (broken line). (b) The relationship between population density in successive years in the European spruce sawfly (after Neilson and Morris, 1964). The solid line indicates a constant $r = 0$ value, and the broken line is fitted to the observed points.

This approach has been used to demonstrate density dependence in a number of organisms (see Example 12.5). However, this general approach must be expanded to account for the particular factors important in causing the density dependence or the life stages at which they are important. One such approach is the key-factor method, which allows identification of the life stages or ages at which mortality is most important in regulating population density (see, for example, Begon and Mortimer, 1981).

In Chapter 11, we discussed some of the factors that may limit the growth of the human population. In fact, the present-day population growth rates (actually the annual finite rate of increase here) among countries and different continents vary considerably (Table 12.8). In general, the more developed areas, such as Europe and Japan, have lower growth rates, while the developing areas, such as Central and South America, have high growth rates. A number of factors, such as industrial development, medical care, tradition, and religion, appear to contribute to these differences. Although the basis of these growth rates varies from country to country, it is possible to describe in a general way the patterns of change observed historically in low-growth countries.

Many countries with low population growth have gone through a sequential change in death and birth rates in the past century. For example, in many parts of Europe at the end of the nineteenth century, both birth and death rates were high, with birth rates slightly higher, resulting in a slight net increase in population growth. With the introduction of modern advances in medicine and public health and other social changes, the death rate dropped dramatically to approximately one-third of its previous value, and, as a result, the growth rate increased substantially. After several decades, the birth rate began to drop, eventually reaching a level at which the net population growth

---

Example 12.5. Knowledge of the factors determining the density of a pest organism is crucial in devising an effective pest-control program. For example, the European spruce sawfly, *Diprion hyrcyniae*, which in high densities can cause extensive defoliation in spruce forests, is the object of extensive study in Canada. The size of sawfly populations varies greatly, with densities differing by ten- or even a hundredfold in a few years. When the logarithms of the densities from year to year are plotted, the slope is 0.82, somewhat less than unity (Figure 12.10b, Neilson and Morris, 1964), suggesting that some stabilizing factors were affecting density. Further examination indicated that parasites and disease organisms contributed some of this density-dependent effect and that high precipitation also tended to increase sawfly density but not in a density-dependent manner. However, not all of the deviation from a slope of unity could be accounted for, indicating that some other factor(s) must be influencing sawfly numbers in a density-dependent manner.

Table 12.8
The population growth rate for different continents, the population size in 1970 in millions, and the predicted population size in 2000.

Continents	*Growth rate*	*Population size* 1970	*Population size* 2000
North America	1.1	225	375
Central and South America	2.9	276	750
Europe	0.8	456	550
Africa	2.4	344	550
Asia	2.0	1,990	4,500

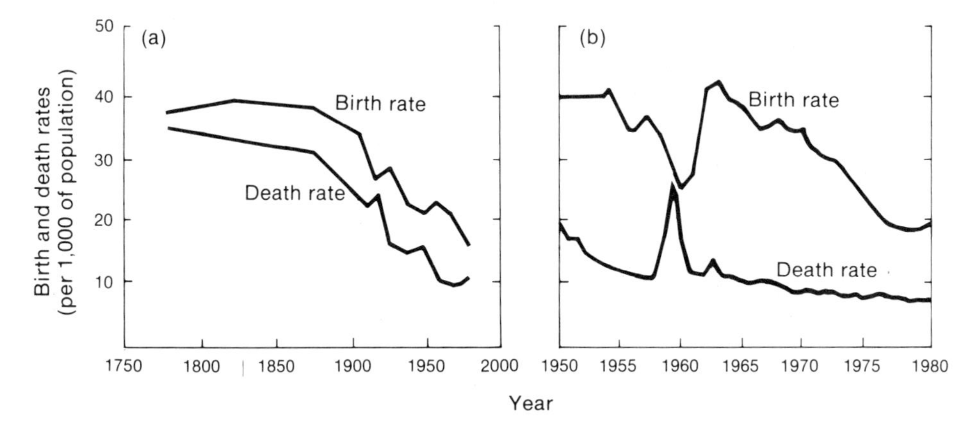

Figure 12.11. (a) The birth and death rates in developed countries over the past two centuries (from Westoff, 1978) and (b) birth and death rates in China over the past three decades (from Coale, 1983).

rate was again low. This change from a combination of a high birth rate–high death rate to one of high birth rate–low death rate and eventually to low birth rate–low death rate is termed the *demographic transition*. Figure 12.11 illustrates this change (birth and death rates are given as the number of births or deaths per thousand individuals).

It appears that many developing countries are still in the intermediate part of this transition, with relatively high birth rates and low death rates resulting in an extremely high population growth rate. China, which, since the establishment of the People's Republic after World War II, has undergone dramatic political and social change, has shown a decline in both birth and death rates in the past three decades (see Figure 12.11b). Notice here the sharp increase in the death rate and decrease in the birth rate during and after the Great Leap Forward between 1959 and 1963. Although some underreporting of births and deaths occurred during this period, it is obvious that this political turmoil resulted in large demographic changes.

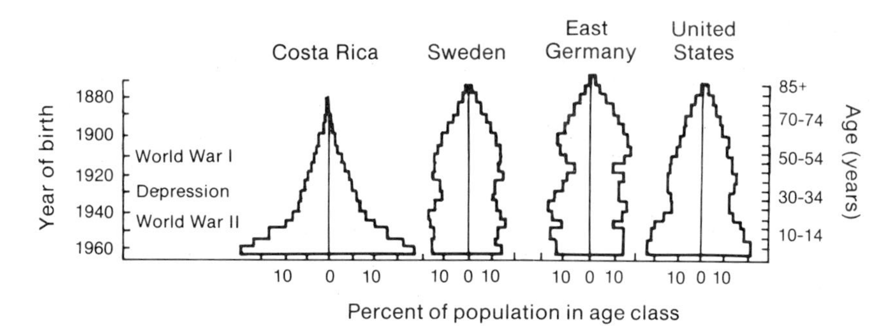

Figure 12.12. Age pyramids in four different countries. The left axis gives the year of birth and the right axis gives age at the time of the survey (after Ricklefs, 1973).

One way to examine more closely the trends in populations over time is to look at *age pyramids*—that is, graphical breakdowns of populations into age and sex categories. We have not considered sexual demographic differences, but important differences do occur; for instance, women have a greater life expectancy in most societies. The ratio of the number of males to females in a population is known as the *sex ratio*. The sex ratio is age dependent and as a result is measured at different ages. The primary and secondary sex ratios are measured at conception and birth, respectively, and for both, the ratio is greater than unity. However, the sex ratio after birth begins to shift, because of generally higher mortality in males, so that in cohorts of older individuals, the sex ratio may be much smaller than unity.

Figure 12.12 gives four age pyramids, one from the growing population of Costa Rica, two from the relatively stable populations of Sweden and East Germany, and one from the United States (from Ricklefs, 1973). The left side of these pyramids is the percent for males and the right side the percent for females. First, notice in the Costa Rican pyramid that there are many very young individuals because of the high birth rate and that high death rates result in a sharp dropoff in population numbers. In the pyramid of Sweden, the numbers in the young age classes are low because of low birth rates, and there are similar proportions in the adult age classes because of the low death rates. The Swedish age pyramid may indicate the stable age distribution in that country because survival and fecundity rates have been relatively constant for several generations. On the other hand, the Costa Rican age pyramid is not a good indicator of the stable age distribution because of recent rising birth rates and dropping death rates.

The East German and the United States pyramids illustrate the effect of sociopolitical events on the age structure and consequent population growth. The small numbers of individuals in the 45–49 age class of the East German population reflect both low fecundity after World War I and high mortality in World War II. Notice that the sex ratios in this age class and in those following are distorted with many more females because of higher male mortality in

World War II. The compression of the United States pyramid in the 24–34 age class occurred because of the low birth rates during the depression of the 1930s. The bulge in the 0–14 age class represents the "baby boom" in the prosperity after the end of World War II. This bulge has led to an increase in overall birth rates in recent years because of the larger number of individuals now in the reproductive age classes. However, the increase has not been as great as originally predicted because it appears that many women either are postponing marriage and childbirth or are having no children, thereby lowering fecundity rates for these women (Westoff, 1978). We should note that delayed fecundity will reduce the rate of increase of a population. In other words, if on the average women have two children in their thirties instead of two in their twenties, the measures of population growth that we discussed earlier, such as R and r, are substantially reduced.

Postponement of childbirth may lead to a smaller total number of births for these individuals because they may finally decide to have no children at all, or they may be infertile by the age they decide to have children. A decline in female fertility was shown in a study of 2,193 French women whose husbands were sterile and who were impregnated by artificial insemination (Schwartz and Mayaux, 1982). In this survey, the proportion of successful impregnations was approximately 0.75 for women in their twenties, nearly the overall population average. However, for women between 31 and 35, the proportion of successful impregnations dropped to 0.62, and, for women over 35, the rate dropped to 0.54. It appears that fertility may drop rather quickly with increasing female age and that delaying childbirth may lead to a permanent reduction in the number of offspring.

SUMMARY

Birth and death rates are often age and density dependent. Age-dependent birth and death rates are given in the form of survival and fecundity schedules. These data can be used to estimate a number of demographic measures, including the instantaneous rate of growth, generation length, life expectancy, and reproductive value. In addition, changes in both the population size and its age structure may be predicted. The regulation of population numbers is discussed in terms of density–dependent and density–independent factors.

PROBLEMS

1. Given a type II survival schedule, what would the l_x values be if 20 percent of the individuals survived each age class and there were five age classes?
2. Given the following survival and fecundity schedule what is R_0? What is T? What is r calculated both ways? What is e_x for all age classes? What is v'_x for all age classes?

Age (x)	l_x	m_x
0	1.0	0
1	0.2	0
2	0.1	10
3	0.05	5
4	0.0	0

3. Given that initially there are 10 individuals in each of the first four age classes, using the $l_x m_x$ schedule above, what would be the numbers (total and in each age class) after 1, 2, and 3 time units? Interpret these results.
4. How is population growth with age classes different from that without age classes?
5. What do you think is the relative importance of density–dependent and density–independent factors?

CHAPTER 13

Interspecific Competition

If turf which has long been mown, and the case would be the same with turf closely browsed by quadrupeds, be let to grow, the more vigorous plants gradually kill the less vigorous, though fully grown plants; thus out of twenty species growing on a little plot of mown turf (three feet by four) nine species perished, from the other species being allowed to grow up freely.

CHARLES DARWIN

A particular species may affect the density and distribution of other species in a number of different patterns. Some of the potential interspecific interactions for a pair of species in terms of their effect on density are given in Table 13.1. The first, *interspecific competition*, in which each species has a negative effect on the density of the other, is the subject of this chapter (the evolution of interspecific competitive ability is discussed in Chapter 20). The second and third patterns, *predator–prey* and *parasite–host interactions* are discussed in Chapter 14. In these two cases, the predator and the parasite, indicated as species 1, have negative effects on the density of the prey and the host, indicated as species 2, respectively. On the other hand, the density of the predator and the parasite are increased by the presence of the prey and host, respectively. However, the actual effects of these types of interactions are more complicated than this description suggests.

Table 13.1
Different types of interactions between species and the effect on density, where −, 0, and + indicate a reduction, no change, and an increase, respectively, in the density of a particular species due to the presence of the other species.

Interactions	*Species* 1	*Species* 2
Competition	−	−
Predator–prey	+ (predator)	− (prey)
Parasite–host	+ (parasite)	− (host)
Commensalism	+	0 (host)
Mutualism	+	+
Protocooperation	+	+
Amensalism	−	0

In addition to these interactions, four other types of interactions are given in Table 13.1 and we will only mention them here. *Commensalism* occurs when one species benefits from the presence of another but the density of the second species does not appear to be affected by the first. Commensalism often occurs between a sessile host, such as a plant or a mollusk, and a motile organism. For example, birds may live in trees and crabs may live in mollusk shells, thereby benefiting from these hosts while the hosts appear to experience little or no benefit from the presence of the birds or crabs.

Example 13.1. Behaviorally aggressive interspecific encounters have been observed among a number of bird species that appear to be competing for space or food. One example of such interference competition is between two species of blackbirds, the red-winged (*Agelaius phoeniceus*) and the yellow-headed (*Xanthocephalus xanthocephalus*), which frequently breed and are common in the same marshes in western North America. The male red-winged generally establishes territories (see Chapter 18) on the marsh about a month before the arrival of the yellow-heads. Figure 13.1a indicates the 13 red-winged territories established early in the spring on a marsh in Washington (from Orians and Willson, 1964). However, when the larger yellow-head males arrive they actively evict the red-wings from the parts of the marsh with deeper water and less vegetation. After this period, the red-wing territories are generally on the periphery of the marsh, with the yellow-head territories in the center (Figure 13.1b).

Interspecific competition, generally of an exploitative nature, may occur between diverse groups of organisms that utilize the same resource. For example, ants, rodents, and birds all use seeds as a food resource in a number of environmental situations. In such a habitat, the

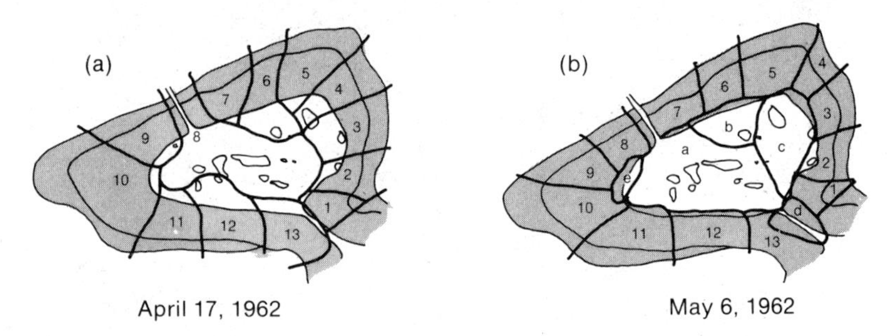

Figure 13.1. The territories of (a) the redwinged blackbird before the arrival of the yellowhead blackbird and (b) the territories of the two species together. The letters indicate yellowhead territories (after Orians and Willson, 1964).

When the interaction between the two species is beneficial and necessary for both species, the relationship is termed *mutualism* (see the discussion of a model of mutualism in Chapter 20). Often, mutualist species have quite different living requirements. For example, some ants provide a good habitat for fungi that decompose litter, making the litter digestible for the ants. In this case, the ants and the fungus are mutualists, each species obligated to and benefiting from the presence of the other. Similarly, nitrogen-fixing bacteria and legumes are mutualists, with the legumes deriving nitrogen from the

---

Arizona desert, Brown and Davidson (1977) examined seed utilization and competition by ants and rodents. They found that when the ants and rodents coexist, the seed size utilized by the two groups overlaps but that ants tend to take smaller seeds. However, there is considerable overlap in these seed-utilization curves, suggesting the importance of interspecific competition.

To examine more critically interactions between ants and rodents and their effect on seed density, Brown and Davidson set up several types of areas: a control area in which both ants and rodents were present, a rodent-removal area in which rodents were trapped out and then excluded by fencing, and an ant-removal area in which ants were eliminated by insecticide application. Table 13.2 gives the observed ant colony and rodent density for the control and these two treatment areas. When the rodents are removed, the number of ant colonies is 543, a large increase from the 318 colonies in the control. When ants are removed, then the rodent number is 144, also an increase from the 122 in the control. Therefore, it appears that the presence of rodents reduces ant density and the presence of ants reduces rodent numbers. In addition, the seed density for both these treatments and the control was nearly identical, indicating that the remaining organisms use the seeds formerly taken by the removed species. These experimental results are consistent with the hypothesis that exploitative competition for seeds occurs between these species.

Table 13.2
The number of ant colonies and rodents in control areas and two types of treatment areas in which rodents and ants were removed (Brown and Davidson, 1977).

	Control	*Rodent removal*	*Ant removal*
Ant colonies	318	543	—
Rodent number	122	—	144

bacteria and the bacteria obtaining nutrients from the plants. The last two types of interaction in Table 13.1 are *protocooperation*, which is similar to mutualism except that the interaction is not obligatory, and *amensalism*, in which one species is inhibited and the other is unaffected.

Species inhabiting the same area may reduce the density of each other because they compete for the same resources. Resources include a variety of factors, such as light, water, space, and nutrients for plants and food, space, and water for animals. Competition from other species may result in either a decreased birth rate or an increased mortality. For example, the number of overwintering birds that survive in colder climates depends to a large extent on their winter food supply. In this case, the presence of other bird species competing for the same food supply may result in increased winter mortality or perhaps in a lowered birth rate the following spring. It should be noted that there may be competition for resources both between individuals of the same species—that is, intraspecific competition (as we discussed in Chapter 12)—and between individuals of different species, interspecific competition.

Sometimes competitive interactions are quite apparent, as when one species harms another, or when one species physically or behaviorally excludes another from a resource. For example, barnacles may crowd out and overgrow other species, and one plant species may block the sun from another species. In other cases, competition is more subtle, as when plants compete for water via their root systems or secrete allelopathic chemicals (see Chapter 9), or when foraging competitors are separated temporally as nocturnal and diurnal organisms that feed on the same food resource.

In a general way, we can distinguish two types of competition—*interference competition*, in which, for example, one species may behaviorally exclude another from a resource, and *exploitative competition*, in which one species utilizes a resource also important for the second species while having no direct contact with that species. (Interference and exploitative competition are similar to the notions of contest and scramble competition, respectively.) Some clear examples of interference competition occur between rodents or between birds, in which particular species are aggressively prevented from using a resource (see Example 13.1). Good examples of exploitative competition are between protozoa, bacteria, or algae, in which one species utilizes nutrients before another species (also illustrated in Example 13.1). In a number of cases, competition appears to be some combination of these two classifications of competitive interactions.

I. COMPETITIVE EXCLUSION

What happens if populations of two species compete for the same limited resource in a constant environment? A series of pioneering laboratory experiments by the Russian ecologist G.F. Gause (1934) demonstrated that in a simple environmental situation one species will always eliminate the other. This observation was developed into the *principle of competitive exclusion*, which states that one competing species will eventually eliminate the other species (or, as it is sometimes stated, no two species can occupy the same niche simultaneously).

The early laboratory competition experiments using species of yeast, *Paramecium*, or flour beetles all resulted in the elimination of one of the two species—that is, competitive exclusion (see Examples 13.2 and 13.3). Because many similar species appear to coexist in nature, these results were initially a paradox to many ecologists. However, the basis for this discrepancy appears to be that the quite simple environment (both temporally and spatially) used in the laboratory situations contains essentially only one niche, while the complex and heterogeneous environment in nature allows partitioning into a number of different niches. When the environment in a laboratory experiment is made more complex, then two species can coexist (see Example 13.2), supporting the notion that competitive exclusion occurred in the early experiments because of environmental homogeneity.

One way to visualize competitive exclusion and coexistence is to utilize the niche concept discussed in Chapter 9. For example, two species of birds may compete for insects as a food resource, with that food being a limited resource, given the densities of the two species. Let us assume that the insects vary in size and that the feeding areas of the birds are at different heights in trees. For example, the relationship of the feeding niches of the two species of birds might be that in Figure 13.3a, which indicates a great deal of overlap. In this case, one would expect one species to be excluded from the area because of the similarity in resource use or the large overlap in niches. The size of insects and the area in which they are present would be important in determining which species would be eliminated (see below). On the other hand, the relationship of the feeding niches of the two birds might be that in Figure 13.3b, in which there is little niche overlap. In this situation, one could expect the two species to coexist, since they would be utilizing different food resources. However, this prediction depends upon the resources available; if only the environmental combination utilized primarily by one species is present (which would mean that only one niche was present), the other species might be eliminated. For example, if most insects are large and high in trees, then species 1 would be at an advantage and species 2 might be eliminated.

II. THEORETICAL MODELS

In Chapter 11, we discussed the regulation of population growth in a density-dependent manner using the logistic growth model. In other words, the population growth rate of a species—let us call it species 1—using the logistic growth model is

$$\frac{dN_1}{dt} = r_1 N_1 \left(\frac{K_1 - N_1}{K_1} \right)$$

where the subscripts are used to indicate values for species 1. Let us assume

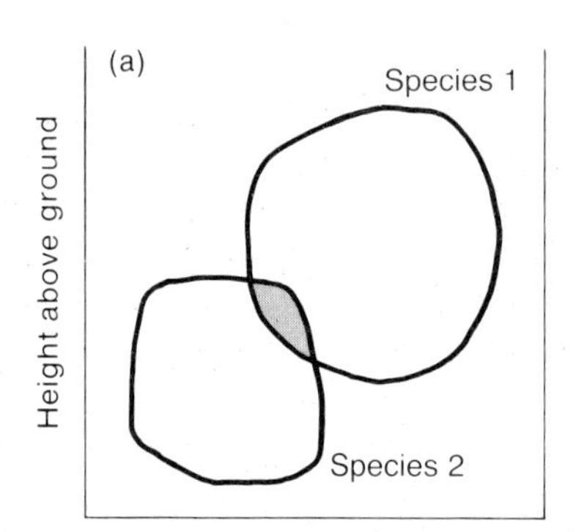

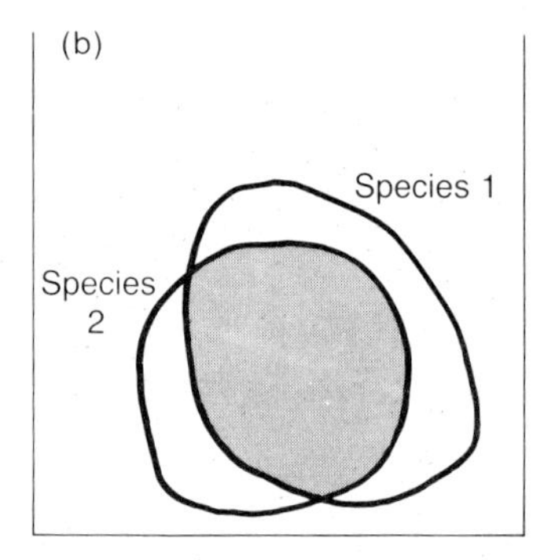

Prey size

Figure 13.3. Hypothetical niches of two bird species for insect size and height above ground in which the insects are found (a) where there is large niche overlap and (b) where there is little niche overlap.

Example 13.2. Different species of flour beetles have been used to investigate a number of phenomena in population ecology. One of the advantages of this organism is that all life stages—eggs, larvae, pupae, and adults—live in flour and can be separated and counted. When two species of flour beetles are placed in a container with only flour, then over a period of time one species will eliminate the other, consistent with the predictions of competitive exclusion. Which of the species eliminates the other is primarily dependent upon the environmental conditions. For example, Table 13.3 gives the competitive outcomes for two species of *Tribolium* under six different environmental conditions (Park, 1954). *T. castaneum* was the winner—that is, it eliminated *T. confusum*—more frequently under moist conditions, while *T. confusum* generally was the winner under cold conditions.

Table 13.3
The proportion of experiments that the two species of flour beetles, *Tribolium castaneum* and *T. confusum*, win under different environmental conditions (Park, 1954).

Environment (temperature, humidity)	*T. castaneum*	*T. confusum*
Hot, moist (34°C, 70%)	1.0	0.0
Hot, dry (34°C, 30%)	0.1	0.9
Temperate, moist (29°C, 70%)	0.86	0.14
Temperate, dry (29°C, 30%)	0.13	0.87
Cold, moist (24°C, 70%)	0.31	0.69
Cold, dry (24°C, 30%)	0.0	1.0

that a second species competes with species 1 for some resource, such as food or space, and that each individual of species 2 uses an amount α of this resource. The value of α indicates the relative per capita effect of species 2 on species 1 and is often between zero and unity for species of similar size. For example, if there are 100 individuals of species 2 and each uses 0.5 as much of the resources necessary for species 1 as an individual of species 1 ($\alpha = 0.5$), then the resource available for species 1 is reduced to the amount it would be if 50 more individuals of species 1 were present.

The effect of the competitor on the population growth rate of species 1 can be incorporated into the logistic equation, so that

$$\frac{dN_1}{dt} = r_1 N_1 \left(\frac{K_1 - N_1 - \alpha N_2}{K_1} \right)$$

where αN_2 is the product of the per capita effect of species 2 on species 1 and

---

When *T. castaneum* competes with another species of flour beetle, *Oryzaephilus suranimensis*, and the two species are placed in the same container with flour as food, then *Oryzaephilus* is always eliminated (see Figure 13.2a; Crombie, 1945). However, when fine glass tubing is added to the flour, then the two species can coexist (Figure 13.2b). The tubing provides a hiding place for *Oryzaephilus*, which is smaller than *Tribolium*, so that it is not eliminated. In other words, the tubing adds another niche, allowing the coexistence of the two species. It appears from this experiment and others that the basis for the coexistence of similar species in a particular environment is the fact that the environment is heterogeneous and allows partitioning into a number of different niches.

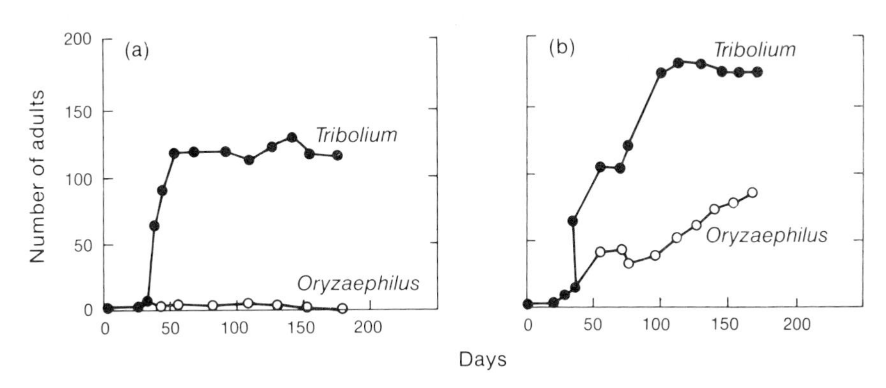

Figure 13.2. The population numbers of two species of flour beetles when (a) they are in containers with only flour and (b) when fine glass tubing is added (after Crombie, 1945).

the number of species 2, N_2. In essence, the remaining carrying capacity available for species 1 is a function of both the number of species 1 present and the resources used by individuals of species 2. If $\alpha = 0$, then the second species does not affect the growth of species 1, and if $\alpha = 1$, then an individual of species 2 is equivalent in its resource usage to an individual of species 1. If $\alpha > 1$, then this may indicate that each individual of species 2 consumes resources much faster than each individual of species 1 or that there is direct aggression between the two species and that species 2 is much larger or a more successful aggressor.

Similarly, the rate of population growth of species 2 is affected by the numbers and resource utilization of species 1, so that

$$\frac{dN_2}{dt} = r_2 N_2 \left(\frac{K_2 - N_2 - \beta N_1}{K_2} \right)$$

where β indicates the effect of species 1 on the resources for species 2. The values α and β are called the *competition coefficients* and indicate the relative per capita effect of species 2 on species 1 and the relative per capita effect of species 1 on species 2, respectively. Note that α and β are not necessarily equal but that in many cases they are probably of similar magnitude, particularly when the competing species are related (see the discussion in the next section on a method of estimating competition coefficients). These equations, giving the population growth rates of two competing species, are often called the Lotka-Volterra competition equations after the two ecologists, A. J. Lotka and V. Volterra, who first used them.

The Lotka-Volterra equations can be used to predict the eventual outcome of interspecific competition. Overall, there are four possible outcomes: species 1 wins, species 2 wins, either species 1 or species 2 wins depending upon the starting composition, or the two species coexist. Here we will concentrate on the first and last of these cases—that is, the conditions under which we could expect species 1 to win and under which we could expect a stable coexistence of the two species.

Let us begin by determining the conditions for the stable coexistence of two species. As before, we know that if $dN_1/dt = 0$, then there is an equilibrium for species 1. The only nontrivial way (as earlier, $r_1 = 0$ or $N_1 = 0$ are trivial situations) in which this can happen is if the numerator

$$K_1 - N_1 - \alpha N_2 = 0.$$

This equation can also be written as

$$N_1 = K_1 - \alpha N_2$$

indicating that there is no change in species 1 numbers for the values of N_1 that satisfy this equation. The solution to this equation is a straight line, with $N_1 = K_1$ when $N_2 = 0$ and $N_2 = K_1/\alpha$ when $N_1 = 0$. This line for species 1

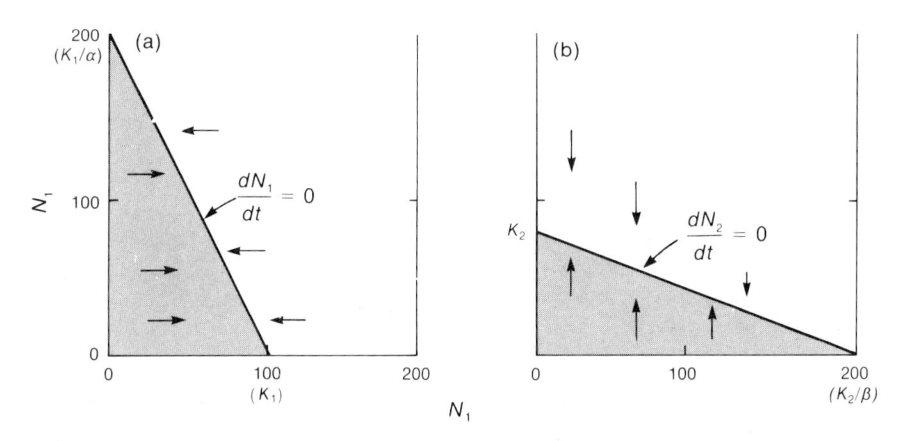

Figure 13.4. The change in numbers of (a) species 1 and (b) species 2 when there is interspecific competition.

(the values for which $dN_1/dt = 0$) and the analogous one for species 2 are called *zero isoclines*.

To clarify, let us give the constants K_1 and α particular values and calculate the line of points that satisfies the above equation. If we assume that $K_1 = 100$ and $\alpha = 0.5$, then

$$N_1 = 100 - 0.5N_2.$$

If $N_2 = 0$, then $N_1 = 100$, the carrying capacity for species 1, and if $N_1 = 0$, then $N_2 = 100/0.5 = 200$. This isocline is drawn in Figure 13.4a, where the number of species 1 is on the horizontal axis and the number of species 2 is on the vertical axis. This isocline also divides the area into two regions in which dN_1/dt is either positive (the shaded region below the line) or negative (the region above the line). For example, if the numbers of N_1 and N_2 are at some combination below the line, then dN_1/dt is positive and the numbers of N_1 will increase. This occurs because if we look at the numerator in the population growth equation,

$$N_1 + \alpha N_2 < K_1$$

making the numerator and the equation positive. On the other hand, if

$$N_1 + \alpha N_2 > K_1$$

then the numerator is negative, and the number of species 1 declines as indicated by the arrows above the line.

However, we have only given the equilibrium number of species 1. In order for the numbers of both species to remain constant, then dN_2/dt must also equal zero. Analogously to species 1, this occurs when

$$K_2 - N_2 - \beta N_1 = 0$$

or

$$N_2 = K_2 - \beta N_1.$$

Again, the solution to this equation is a straight line, in which $N_2 = K_2$ when $N_1 = 0$ and $N_1 = K_2/\beta$ when $N_2 = 0$. Let us assume that $K_2 = 80$ and $\beta = 0.4$, so that

$$N_2 = 80 - 0.4N_1.$$

If $N_1 = 0$, then $N_2 = 80$, the carrying capacity for species 2, and if $N_2 = 0$, then $N_1 = 80/0.4 = 200$. This isocline is drawn in Figure 13.4b, and it divides the graph into regions in which $dN_2/dt < 0$, above the line, and $dN_2/dt > 0$, the shaded region below the line.

In order for there to be no change in the number of either species, then both dN_1/dt and dN_2/dt must equal zero. In the numerical case discussed above, this is the point where the two isoclines cross. If we superimpose the two parts of Figure 13.4, then, as given in Figure 13.5, the two lines cross and divide the total area into four parts. The change in numbers for both species together is given by the arrows (or vectors). In the unshaded region both species decline in numbers, and in the darkly shaded region both species increase. The numbers of the two species change so that from any point in the total area they

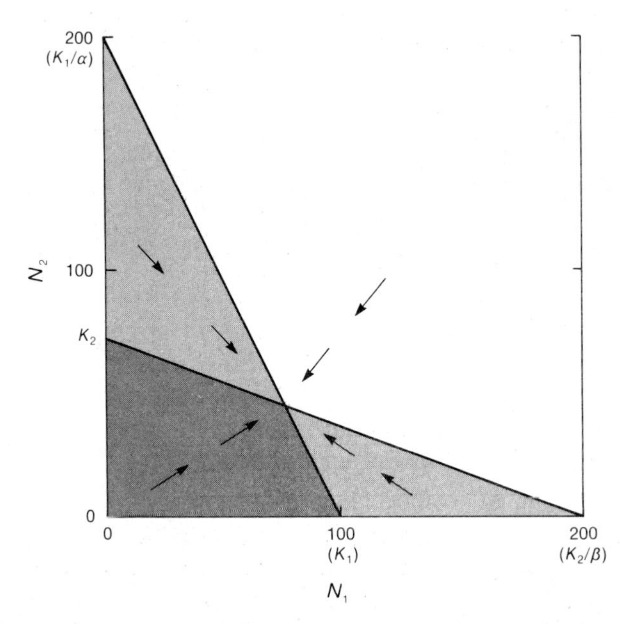

Figure 13.5. The change in numbers of two competing species when there is stable coexistence.

converge on a single value, the stable equilibrium. For example, if we let $r_1 = 0.1$ and $r_2 = 0.1$ and assume that initially we have 30 individuals of species 1 and 120 of species 2 (still assuming $K_1 = 100$ and $K_2 = 80$), then we can calculate the expected change in population numbers for both species. This combination of N_1 and N_2 is in the upper shaded region of Figure 13.5, so we would expect N_1 to increase and N_2 to decline. For species 1, the change is

$$\frac{dN_1}{dt} = (0.1)(30)\left[\frac{100 - 30 - (0.5)(120)}{100}\right]$$
$$= 0.3.$$

For species 2, the change is

$$\frac{dN_2}{dt} = (0.1)(120)\left[\frac{80 - 120 - (0.4)(30)}{80}\right]$$
$$= -7.8.$$

Therefore, the numbers are expected to change from 30 and 120 of species 1 and 2 to 30.3 and 112.2, respectively, after one time unit.

What are the numbers of the two species at the stable equilibrium? These numbers can be obtained by solving the two equations that were set equal to zero above. In other words, the values of the two unknowns, N_1 and N_2, that satisfy equations

$$N_1 = K_1 - \alpha N_2$$
$$N_2 = K_2 - \beta N_1$$

are the equilibrium numbers. If we substitute the values used above for the carrying capacities and competition coefficients, we can solve these equations. Inserting $K_1 = 100$, $K_2 = 80$, $\alpha = 0.5$, and $\beta = 0.4$, then

$$N_1 = 100 - 0.5\ N_2$$
$$N_2 = 80 - 0.4\ N_1.$$

Let us substitute the value of N_1 from the first equation into the second so that

$$N_2 = 80 - (0.4)(100 - 0.5N_2)$$
$$N_2 = 50.$$

This value can be inserted in the first equation so that

$$N_1 = 100 - (0.5)(50) = 75.$$

Therefore, the equilibrium numbers of species 1 and 2 in this case are 75 and

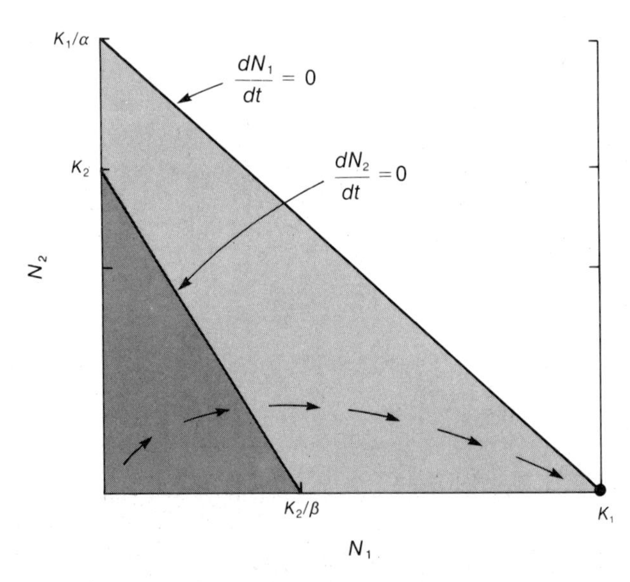

Figure 13.6. The change in numbers of two competing species when species 1 eliminates species 2.

50, respectively. This is the point where the two isoclines cross, indicated by a closed circle in Figure 13.5. Notice that the total number of individuals of both species at equilibrium, 125, is higher than the carrying capacities of either of the species when they are present by themselves.

Let us briefly consider the case in which one of the species eliminates the other. From these equations it is clear that elimination occurs if the isocline for one species completely encloses the isocline for the other species. Figure 13.6 gives such an example, in which the zero isocline for species 1 is always above that for species 2. As a result of this condition, there is an area in which species 1 will still be increasing while species 2 is decreasing, indicated by the lightly shaded area in Figure 13.6. The path of the change in numbers of two species over time for a particular starting composition is given. In this example, the numbers of both species were initially low and both increased (in the darkly shaded region). Then, after the numbers exceeded that for the species 2 zero isocline, N_2 begins to decline, while N_1 continues to increase. Eventually, species 2 is eliminated, $N_2 = 0$, and species 1 approaches its carrying capacity, K_1 (see Example 13.3).

Using the Lotka-Volterra competition equations, we can identify the conditions that lead to the four possible theoretical outcomes (Table 13.4). For stable coexistence, the intersections of the isoclines with the axes must be as in Figure 13.5—that is, $K_1/\alpha > K_2$ and $K_2/\beta > K_1$. These conditions can be written so that for stable coexistence

$$\alpha < \frac{K_1}{K_2} \qquad \text{and} \qquad \beta < \frac{K_2}{K_1}.$$

If we assume that $K_1 = K_2$, then these conditions imply that both competition coefficients must be smaller than unity. In other words, the effect of species 2 on species 1 is smaller than the effect on itself, and the effect of species 1 on species 2 is smaller than the effect on itself. The conditions for the other three cases are also given in Table 13.4. For example, for species 1 to eliminate species 2, then species 2 should have a relatively small effect on species 1, while species 1 has a relatively large effect on species 2.

Notice that these conditions suggest that information on single species, such as the r and K values, may not be sufficient to predict the outcome of

---

Example 13.3. As we discussed in Example 13.2, when two species of flour beetles are placed in a simple laboratory environment, one species eliminates the other. Figure 13.7 gives another example of such competitive exclusion, this time in two species of *Paramecium*, a protozoan (after Gause, 1934). The culture was initiated with low numbers of the two species, *P. aurelia* and *P. caudatum*, and initially both species increased. After only a few days, *P. caudatum* began to decrease, eventually to be eliminated. On the other hand, *P. aurelia* increased to its carrying capacity, a density slightly over 100, in the same time period. However, the rate at which *P. aurelia* reached its carrying capacity was slowed by the presence of *P. caudatum*, suggesting the existence of interspecific competition. The broken line gives the density over time for *P. aurelia* when it is in a monoculture and reaches the carrying capacity much more quickly. The difference between these patterns illustrates that, even though *P. caudatum* was eliminated, it had a temporary effect on the density of *P. aurelia*. The dynamics of change in this *Paramecium* example are consistent with changes predicted by the model given in Figure 13.6, assuming that *P. aurelia* and *P. caudatum* are species 1 and 2, respectively.

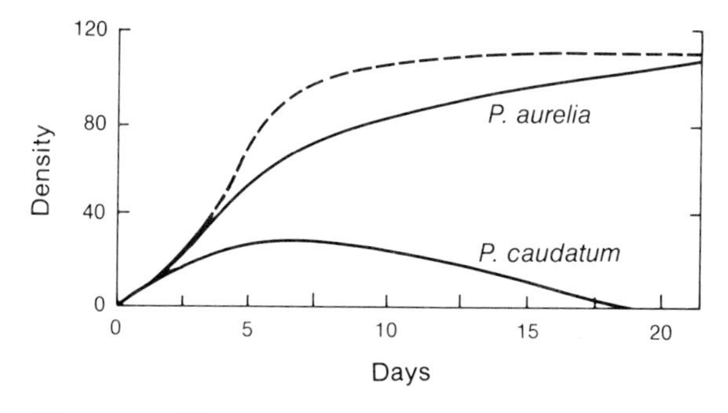

Figure 13.7. The numbers of two species over time when *P. aurelia* eliminates *P. caudatum*. The broken line indicates the change in *P. aurelia* when alone (after Gause, 1934).

Table 13.4
The conditions that lead to the four possible outcomes of interspecific competition.

Outcome	Conditions	
Coexistence, stable equilibrium	$\alpha < \frac{K_1}{K_2}$	$\beta < \frac{K_2}{K_1}$
Species 1 wins	$\alpha < \frac{K_1}{K_2}$	$\beta > \frac{K_2}{K_1}$
Species 2 wins	$\alpha > \frac{K_1}{K_2}$	$\beta < \frac{K_2}{K_1}$
Unstable equilibrium	$\alpha > \frac{K_1}{K_2}$	$\beta > \frac{K_2}{K_1}$

interspecific competition. For example, given that $r_1 > r_2$ and $K_1 > K_2$, species 2 may still be the winner in interspecific competition because α may be greater than K_1/K_2 (remember that, as suggested earlier, α may be greater than unity).

We have only considered competition between two species, but in fact several species may be competing for the same resources. When a species competes simultaneously with several other species, this can be called *diffuse competition*, and we can use an extension of the Lotka-Volterra model to examine population growth rates or coexistence in such a multispecies group. For example, the growth rate of a particular species may be written as a function of its r and K values and as a term that includes the competition coefficients and carrying capacities of all other competing species. Of course, the conditions for coexistence in this situation become more complicated, since they involve the competition coefficients and carrying capacities of all the species.

Let us briefly discuss another approach, developed by the Dutch botanist C. T. de Wit, to examine interspecific competition in plants. This approach also utilizes changes in population numbers from one time to another but presents it in a somewhat different manner, using the ratios of species numbers. For a given overall density, the ratio of the number of two competing species may vary from 0 to ∞. For example, when the total density is 100 and there are 80 of species 1 and 20 of species 2, the ratio of the numbers of species 1 to species 2 is $N_1/N_2 = 80/20 = 4$. On the other hand, when there are 20 of species 1 and 80 of species 2, then the ratio is $20/80 = 0.25$. Generally, to make these ratios symmetrical around one, natural logarithms of the ratios are used. For example, $\ln(4) = 1.39$ and $\ln(0.25) = -1.39$, values equidistant from zero, the logarithm of one.

If the ratio of the numbers of the two species in the parents or seeds (the input ratio) is the same as that in the progeny (the output ratio), then there is no change in relative species numbers. When the ratio increases or decreases from the input to the output, then this indicates an increase or a decrease in the relative numbers of species 1, respectively. As with the Lotka-Volterra

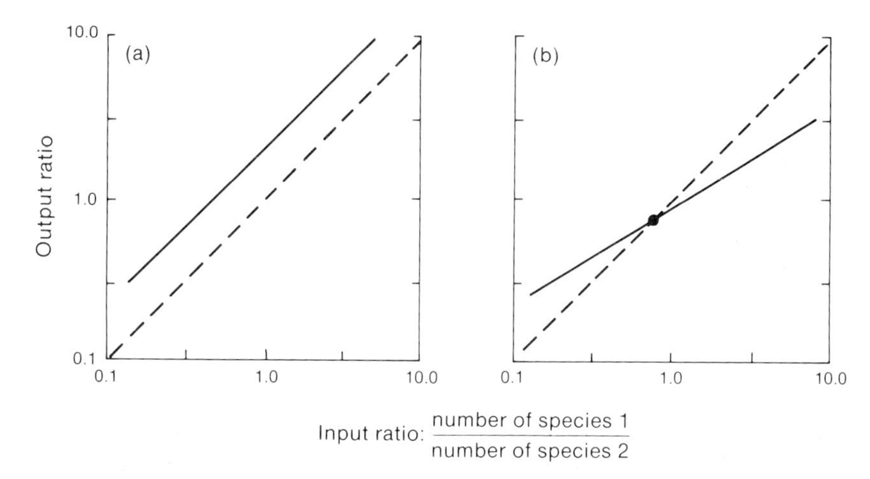

Figure 13.8. Input–output ratio diagrams when (a) species 1 is always favored and (b) when there is coexistence.

equations, there are four basic possible outcomes. Figure 13.8 gives the two examples in which species 1 wins and in which there is coexistence (notice that both axes are logarithmic scales). The broken line in both graphs indicates equal input and output ratios—that is, no change in the species composition. When the observed values (solid line) are always above this diagonal, as in Figure 13.8a, then the relative number of species 1 increases for all input proportions. As a result, over time species 1 should eventually eliminate species 2. In Figure 13.8b, the output ratio is also higher than the input ratio for the lower input ratios (when species 1 is in low frequency). However, the output ratio is lower than the input ratio for high input ratios. This suggests that the frequency of species 1 will increase when low and will decrease when high, reaching an intermediate equilibrium proportion indicated by the closed circle at which this line crosses the diagonal line. Note, however, that these conclusions are for one overall density. To understand the total competitive picture, similar experiments at various densities are necessary (see Example 13.4).

III. COMPETITION IN NATURAL POPULATIONS

Intuitively, one might expect that interspecific competition is a widespread phenomenon and that species densities and distributions are affected to a large extent by competition. For example, different species of birds foraging for similar food, or different species of trees in a forest should have important effects on one another. Although interspecific competition may be important in many situations, other factors may also play an important role in determining species densities. For example, the distribution of trees in tropical forests appears consistent with a model that assumes their number is primarily regulated by dispersal and density-independent mortality (Hubbell, 1979), with all tree species being competitively equivalent (all competition coefficients equal to unity). On the other hand, some species pairs of birds initially thought to have similar food resources have been found, upon closer study, to have somewhat different diets (Krebs, 1978).

In this latter case and others, the fundamental niches of related species may not be as different as the observed or realized niches. In other words, an observation that the resource utilization of two species differs may be the result either of inherent differences between the species or of competitive

---

Example 13.4. Two species of wild oats, *Avena barbata* and *A. fatua*, occur together in many grassland areas of California. Marshall and Jain (1969), in an effort to understand the competitive interactions between these species, grew them together in a greenhouse in various initial proportions and densities. The output was measured in the number of spikelets produced by each species, assuming that there was a similar number of seeds per spikelet. Figure 13.9 gives the results for a low and a high density, 32 and 256 seeds per pot, respectively. For both densities, the relative output of *A. fatua* increases at the low frequency. On the other hand, at the high initial frequency the relative output of *A. fatua* decreases at the low density, while at the high density the input and output ratios are nearly equivalent. Therefore, from these experiments one could predict that the two species would coexist and that, as the density increases, there would be a higher proportion of *A. fatua*.

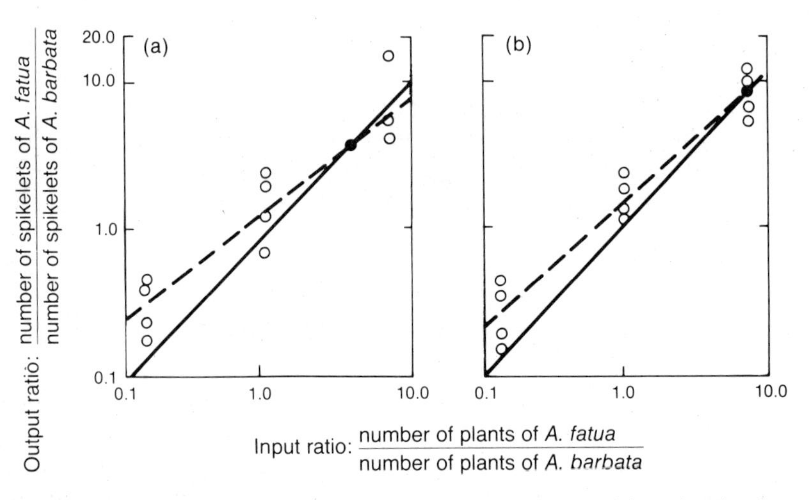

Figure 13.9. The input–output ratios for two species of wild oats, (a) at a low density, 32 seeds per pot, and (b) at a high density, 256 seeds per pot (Marshall and Jain, 1969).

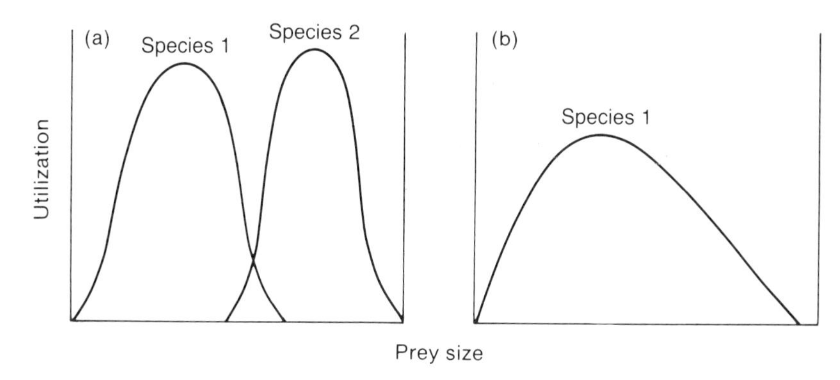

Figure 13.10. The utilization of different prey sizes when (a) two species are present and (b) species 2 is absent.

interactions. One way to differentiate between these causes is to examine situations in which one species is absent to determine if there is *competitive release*—that is, changes in resource utilization, population numbers, or geographic distribution of one species as the result of the absence of a competitor. For example, Figure 13.10a gives the theoretical size of prey utilized by two bird species, two nearly nonoverlapping curves. If interspecific competition is responsible in part for determining the resource utilization of species 1, then the resource utilization of species 1 should increase when species 2 is absent, as shown in Figure 13.10b, assuming that the resource utilization curve is behaviorally labile (see also the latter part of Example 13.1). Furthermore, in the continued absence of species 2, the density of species 1 should increase. The combination of niche expansion and the increase in density resulting from the absence of a competitor is good circumstantial support for the importance of interspecific competition. However, examination of the mechanisms of competition, either from experimental manipulations or detailed ecological analysis, is essential as support for such correlative patterns (see Example 13.5).

When are two competitors able to coexist? As suggested from the ant–rodent experiment in Example 13.1, coexistence may occur when competing species utilize somewhat different although similar resources. In this example, the ants utilized smaller seeds than the rodents when the two groups coexisted. Such a division of resources is termed *resource partitioning* and may occur for such factors as food or habitat. For coexistence, the amount of resource partitioning must be sufficient for each of the competing species to sustain itself in the environment. In other words, the utilization of resources should be different enough that the realized niches of the species do not greatly overlap.

How much resource partitioning (niche differentiation) is necessary for the coexistence of two competitors? To understand this problem, let us first examine the resource utilization curves for two competing species. If we assume that prey size is an important aspect of their niches and that each has different optimum prey sizes, then for narrow-niched and broad-niched

Example 13.5. A series of islands or an island and the nearby mainland are good locations in which to examine the predictions of competitive release. For example, the meadow vole, *Microtus pennsylvanicus*, and the red-backed vole, *Clethrionomys gapperi*, exist in close proximity to each other in grassland and woodland areas, respectively, in northeastern North America. However, on Newfoundland and a number of smaller islands where the red-backed vole is absent, the meadow vole occurs commonly in both woodlands and grasslands (Cameron, 1964). This niche expansion suggests that the meadow vole is competitively excluded from woodlands areas by the red-backed vole. To examine the hypothesis, Grant (1972) established experimental enclosures that were half grassland and half woodland. When only one species was present, it occupied both types of habitat. However, when both species were introduced, the meadow vole was restricted to the grassland area and the red-backed vole to the woodland area, indicating that each species behaviorally excluded the other species from its favored habitat.

Diamond (1975) examined the species distributions for three similar species of ground doves in New Guinea and nearby islands. On the island of New Guinea, all three species are present, with one species present on the coast, one in the light forest, and the third in the rain forest (Table 13.5). However, on other smaller islands, at least one of the species is missing, permitting a consequent niche expansion of another species. For example, on Bagabag the rain forest species is missing and the light forest species has expanded into this region. On Espiritu Santo, both the inland species are missing and the coastal species has expanded its distribution into these areas. Although there is no direct evidence of interspecific competition among these birds, these geographic patterns suggest that there has been competitive exclusion on New Guinea and competitive release on the other islands (however, see Simberloff, 1978).

Table 13.5
The presence of three species of ground doves; (a) *Chalcophaps indica*, (b) *C. stephani*, and (c) *Gallicolumba rufigula* in New Guinea and nearby islands (after Diamond, 1975).

	Coastal scrub		*Light forest*		*Inland rain forest*
New Guinea	a		b		c
Bagabag	a		b	→	b
New Britain	b	←	b	→	b
Espiritu Santo	a	→	a	→	a

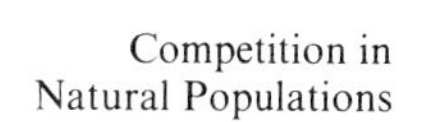

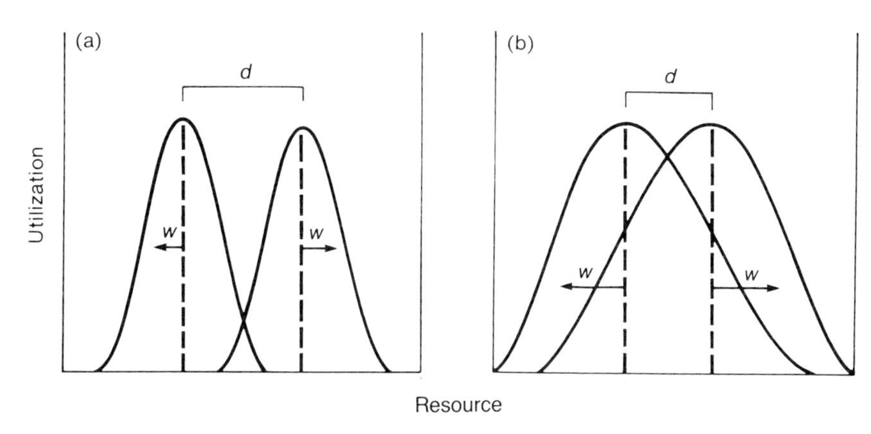

Figure 13.11. The resource utilization of (a) two narrow-niched and (b) two broad-niched species. The standard deviation of the curves, an estimate of niche width, is indicated by w, and the mean niche difference between the two species is measured by d.

species, theoretical resource utilization curves are given in Figures 13.11a and 13.11b. The niche width can be estimated by the standard deviation of these resource utilization curves and is indicated by w. For a narrow-niched species, w is smaller than for a broad-niched species. The difference in the niches between the two species (d) can be measured as the difference in optimal prey size for the two species. The extent of resource partitioning (or niche overlap) is then a function of the ratio d/w. Thus, when this ratio is large, there is little niche overlap (as in Figure 13.11a) and when it is small, there is substantial niche overlap (as in Figure 13.11b).

Considerable niche overlap implies that there will be intense interspecific competition for limiting resources. Unless the resource utilization curves separate, one species should be competitively eliminated. On the other hand, if there is very little niche overlap, intraspecific competition will become more important and there should be an expansion of resource utilization to include underutilized food items. In other words, the conflicting effects of interspecific and intraspecific competition should result in a limit to the similarity of the niches of competing species. Theoretical consideration of such a limiting-similarity model suggests that the ratio d/w should be equal to or slightly greater than unity (May, 1975).

The amount of similarity of competing species can be quantitatively measured by comparing the observed utilization of resources by the species. One way to evaluate niche overlap is to calculate the observed competition coefficients between species 1 and 2 using the formula

$$\alpha = \frac{\sum_{i=1}^{n} p_{1i} p_{2i}}{\sum_{i=1}^{n} p_{1i}^2}$$

$$\beta = \frac{\sum_{i=1}^{n} p_{1i}p_{2i}}{\sum_{i=1}^{n} p_{2i}^2}$$

where p_{1i} and p_{2i} are the proportions of resource of type i utilized by species 1 and 2, respectively, of the n resource types present. If both species use the same resources in identical proportions, then $\alpha = \beta = 1$, and, if they use completely different resources, $\alpha = \beta = 0$. The theory of limiting similarity predicts in general that if $d/w = 1$, then competition coefficients for coexisting species calculated in this manner would be less than 0.8. If for species 1, $p_{11} = 0.2$ and $p_{12} = 0.8$ and for species 2, $p_{21} = 0.6$ and $p_{22} = 0.4$, then

---

Example 13.6. To evaluate resource utilization among species, one needs to examine species for which a limiting resource can be defined and its utilization documented. Several species of sunfishes of approximately the same size exist together in small ponds in central North America. Werner and Hall (1976) examined the stomach contents of these fish to determine their prey utilization when raised together in experimental ponds. Table 13.6 gives the proportions of four prey categories used by the pumpkinseed sunfish (*Lepomis gibbosus*) and the green sunfish (*L. cyanellus*). Obviously, these species utilize prey differently: the pumpkinseed eats more benthic epifauna and the green more vegetation dwellers. Using these data, $\alpha = 0.72$ and $\beta = 0.82$, values consistent with our predictions from the limiting-similarity model. Furthermore, when the pumpkinseed and the green were raised alone in separate ponds, the prey utilization of the two species was virtually identical (see values in parentheses in Table 13.6), suggesting that interspecific competition resulted in a divergence of prey utilization.

Table 13.6
The proportion of four prey types in the diet of two coexisting sunfish, the pumpkinseed and the green (from Werner and Hall, 1976). The values in parentheses indicate diets when the species is alone in a pond.

	Vegetation dwellers	*Benthic epifauna*	*Zooplankton*	*Other*
Pumpkinseed	0.05 (0.41)	0.34 (0.12)	0.06 (0.01)	0.55 (0.47)
Green	0.40 (0.43)	0.12 (0.23)	0.04 (0.00)	0.44 (0.33)

$$\alpha = [(0.2)(0.6) + (0.8)(0.4)]/[(0.2)^2 + (0.8)^2] = 0.65.$$

Example 13.6 gives some results that are consistent with this expectation, but it should be noted that, although this approach has intuitive instructive value, there are limited supporting data and a number of the assumptions have been questioned (see Schoener, 1982, for a review).

Direct evidence of the effect of interspecific competition is most easily gathered when interference competition takes place that leads either to behavioral interaction between the species or other physical contact (see Example 13.1). Another classic example of interference competition is the

---

Example 13.7. The red scale for many years was a major pest on orange trees in southern California. A species of wasp, *Aphytis chrysomphali*, was accidently introduced from Europe into this area and became widespread as a parasite on the red scale (Figure 13.12a; after DeBach and Sundby, 1963). In 1948, a related species of wasp, *A. lingnanensis*, was introduced from China. Within a decade, the second species had essentially eliminated the first species from the areas it had occupied (Figure 13.12b). The competitive displacement took place over a large area in a relatively short period of time. Laboratory experiments demonstrated that the two species could not coexist. However, no actual behavioral mechanism of interference between the two species could be identified, and food did not appear to be a limiting resource.

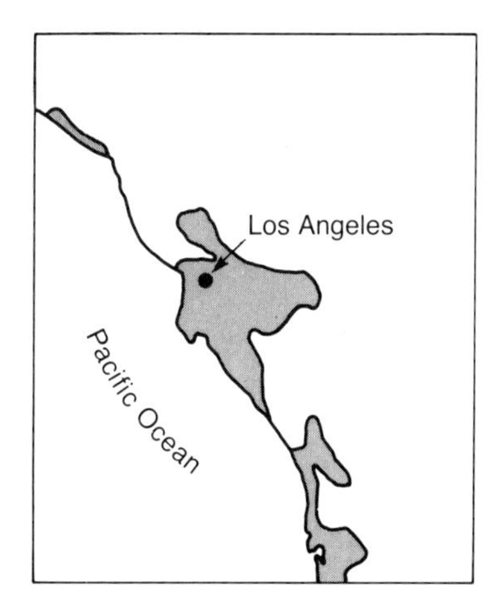

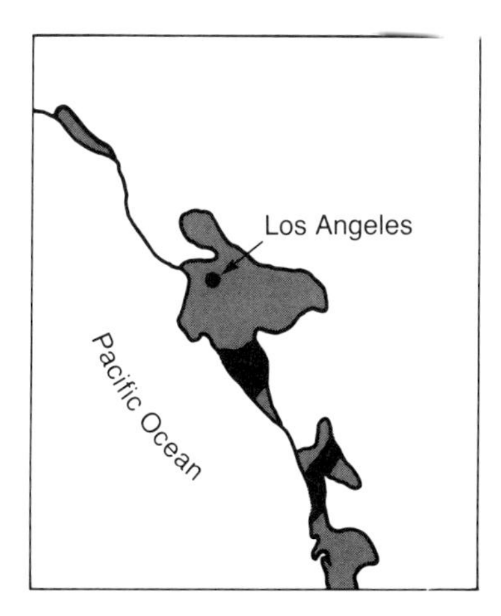

Figure 13.12. The distribution of two wasp species in (a) 1948 and (b) in 1959 (DeBach and Sundby, 1963). The light shading is for *Aphytis chrysomphali*, the intermediate shading *A. lignanensis*, and the dark shading both species.

exclusion of a barnacle species from a particular area of the intertidal zone by another barnacle species (Connell, 1961). In this case, the excluded species settles as larvae in this area, but the adults are displaced as the other species overgrows them, actually causing the species' physical removal.

Some good examples of competitive exclusion occur when one species is eliminated by the introduction or spread of a second species (Example 13.7). Introductions onto islands have often resulted in the reduction of the native fauna and flora, in part owing to interspecific competition. However, in many cases such introductions may simultaneously change other factors. For example, they may bring new diseases to which the native organisms are not resistant, such as the bird malaria that appears to have been important in the elimination of many species of native Hawaiian birds. Competition in this case does not appear to be the only cause of the change in numbers, but the introduced birds have filled many of the niches formerly occupied by endemic species.

The extent of interspecific competition between species presumably varies both over space and time. For example, if resources are not limiting in a particular environment, then one might expect other factors, such as predation or dispersal ability, to be more important in the structure of a community. In migratory birds, overwinter survival in other areas could be of greater significance than interspecific interactions during the breeding season. In nonmigratory organisms, periodic bad resource years could force interspecific competition, while in good years resources may not be limiting.

Schoener (1982) has suggested that the importance of interspecific competition or resource usage may be evaluated by comparing niche overlap in good and bad times. In good years or seasons, one would expect more niche overlap, because resources would not be limiting and related species would tend to use the most abundant (or most profitable) resources. However, in bad years or seasons, the limitation of resources would result in interspecific competition, divergent use of resources by competitors, and consequent low niche overlap. Consistent with these predictions, in 27 of 30 examples in a variety of competitors, Schoener (1982) found the least niche overlap in bad times. For example, competing species of tropical frogs had the least niche overlap in the kind and size of food in the season of resource shortage, the dry season (Toft, 1980). In other words, even though interspecific competition may be periodic or even occasional, it may still have important effects on species that use similar resources.

SUMMARY

Interactions between different species may be categorized according to their effects on the densities of the involved species. Interspecific competition results in a reduction in density of competing species or competitive exclusion of one species if the environment is homogeneous. The logistic growth model can be modified to predict both changes in density of two competitors and coexistence of two species. Evidence for the importance of competition in natural populations comes from changes in resource utilization when one species is missing or in times of low resource availability.

PROBLEMS

1. Discuss and differentiate among the various species interactions. Discuss and differentiate between interference and exploitative competition.
2. What is competitive exclusion and why does it not apply to many natural populations? What experiments would you carry out to demonstrate that competition is the cause of a species being excluded from an area?
3. If $\alpha = \beta = 0.6$, $K_1 = K_2 = 100$, what are the equilibrium population numbers? Given that $r_1 = r_2 = 0.1$, $N_1 = 150$, and $N_2 = 50$, what are dN_1/dt and dN_2/dt?
4. In a study using the input–output approach of competition between two plant species, the input proportions of species 1 were 0.1, 0.5, and 0.9, and the output proportions were 0.2, 0.4, and 0.7, respectively. What are the input and output ratios and what would be the eventual predicted outcome of competition (assuming similar results for other densities)?
5. The diets of two species of hawks living in the same area were composed of the following types of prey:

Species	*Rodents*	*Birds*	*Other*
1	0.2	0.6	0.2
2	0.7	0.2	0.1

Calculate values of α and β from these data. Are these values consistent with the model of limiting similarity? If species 1 were absent, can you predict the change in resource utilization for species 2?

CHAPTER 14

Predator–Prey Interactions

> Predation should be regarded as being of primary importance, whether directly determining species composition or in preventing competitive exclusion, except where the effect of predation is reduced for some reason.
>
> JOSEPH CONNELL

Where interspecific competition occurs, the numbers of each of the competing species may be reduced by the presence of the other species. On the other hand, when one species is a predator and another is prey, then the effect of predation is to reduce the numbers of the prey and increase the number of predators. Here we will define *predation* in a very broad sense. First, predation describes the interaction between carnivores, such as lions or hawks, and their prey, such as antelope or rodents. This is the classical concept of a predator–prey interaction and includes all situations in which one species attacks, kills, and consumes another species. Second, the interaction of a parasite or parasitoid and its host can be considered in the same context as a predator–prey interaction. Parasites are mainly animals that live in an obligatory association with their hosts most of their lives, and parasitoids are mainly hymenoptera that are free living as adults but lay their eggs on other insects. In many ways, parasites and parasitoids are analogous to predators in that they cause a reduction in the numbers of another species. For example, large numbers of a parasitoid that kills its host will result in a reduction in host numbers, with an effect much like that of predators on their prey. Last, the interaction of herbivores and the plants they feed on can be considered in the same general way. This parallel is particularly apparent when a herbivore feeds on seeds that are potentially whole organisms of another species. Grazers and browsers seldom kill the plants they feed on, but the plants constitute a resource population that is reduced by herbivory and that provides a potential for increase in the herbivore numbers, a relationship essentially the same as a classical predator–prey interaction.

In this chapter, we will discuss together these three types of interactions: predator–prey, parasite- or parasitoid–host, and herbivore–plant. As a result, we will consider the population of one species in the interaction—the predator, herbivore, parasite, or parasitoid—as the *predator population*, and the population of the other species—the prey, plant, or host—as the *resource population*. The predator in general has a negative effect on resource numbers, and the resource population may have a positive effect on predator numbers. We will discuss evolution of a predator–resource system in Chapter 20.

We may divide predators into two arbitrary classes, specialists and generalists. *Specialists* are predators that are restricted to a particular resource species—for example, parasites or parasitoids that are host specific, such as the cicada killer wasp, or animals that are monophagous (feeding on only one

---

Example 14.1. The diet of a generalist predator may vary over time or space if the numbers in the resource populations also vary. Predators exhibiting this behavior are opportunistic feeders, switching from one resource type to another. An example of this type of feeding behavior was given by Kephart and Arnold (1982) in the garter snake, *Thamnophis elegans*. They studied the diet of this snake for seven consecutive years at a lake in California by forcing individual snakes to regurgitate and examining their stomach contents. During this time, the water level of the lake dropped nearly 2 meters, causing an important change in resource availability—that is, as the water level dropped, toads stopped breeding at the lake. This change was reflected in the diet of the snakes (Figure 14.1), which shifted from a high proportion of young toads when they were abundant to a low proportion when toads were scarce. The shift in diet is consistent with the observation that the toads are relatively easy for the snakes to catch. For example, the snakes could optimally capture about one toad per minute, while the optimal capture rate for fish was one per hour. As a result, these data appear to be generally consistent with the predictions from the optimal foraging strategy.

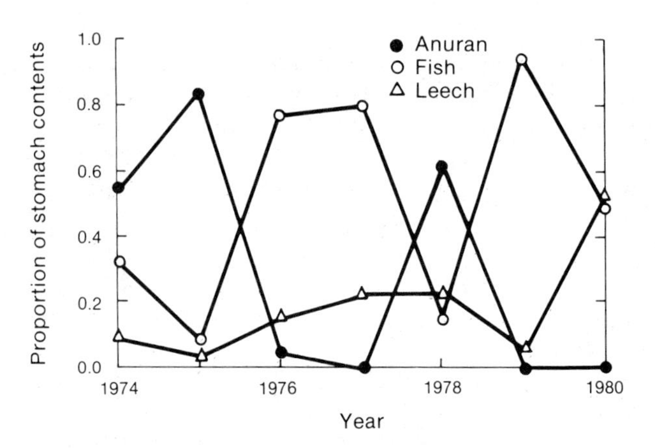

Figure 14.1. The composition of the diet of a population of garter snakes over seven years (after Kephart and Arnold, 1982).

---

resource species), such as the panda or the fly discussed in Example 16.9. For the most part, the discussion in the next section concerns specialist predators and their potential effect on the numbers in the resource population and the complementary effect of resource numbers on the numbers in the predator population.

Many species are *generalist* predators and can feed on a number of resource species. Generalists often have a preferred resource, on which they may feed exclusively when it is available (see Example 14.1). Some species will alternate among basic types of resources over time, as do many birds that feed on insects in the summer and seeds in the winter. When the favored resource species is available in limited numbers, a generalist may have a diet composed of a number of resource species. In theory, a generalist predator would select members of resource species that are easy to catch and eat and that are high in nutritive value. If these species are uncommon, then species that are next in order of catchability and nutritive value would be selected. This concept, called the *optimal foraging strategy*, is an attempt to explain the diet preference of generalist predators.

I. THEORETICAL MODELS

The first models to predict the changes in the population number of a predator population and its resource population were developed by A. J. Lotka and V. Volterra in the 1920s. In their equations, they accounted for the inverse effect of population numbers of one species on the other in a predator–resource system by assuming that the growth rates of a predator and its resource could be modeled using a modification of the exponential growth equation. The basic assumptions of this approach are that the birth rate of the predator is a function of the number of the resource population and that the death rate of the resource population is a function of the number of predators. We will designate the number in the predator population as P and the number in the resource population as R.

First, let us consider the growth rate of the resource population. When the predator is not present, then it is assumed that the resource increases exponentially, or

$$\frac{dR}{dt} = rR$$

where r is the intrinsic rate of increase for the resource population. However, when the predator is present, then the resource growth rate is reduced by the number of predators and their ability to obtain the resource. This reduction can be incorporated into the model as

$$\frac{dR}{dt} = (r - aP)R$$

where a is a constant that indicates the successful individual attack rate by the predators. If P were constant, then R would grow (or decline) exponentially.

Because the predator is dependent upon the resource population as a food resource, when the resource is absent the predator population declines exponentially. This rate of decrease can be symbolized as

$$\frac{dP}{dt} = -dP$$

where d is the death rate of the predator in the absence of the resource population. However, in the presence of the resource, the predator population will increase at a rate dependent upon the number of the resource population and the ability of the predator to utilize the resource population. The growth rate in the predator population is then

$$\frac{dP}{dt} = (-d + bR)P$$

where b is a constant that indicates the individual conversion rate by the predator of the resource population into new predators. If R were constant, then P would follow exponential growth.

Notice that the last terms in these two equations when they are multiplied out, $-aPR$ and bPR, involve the products of the numbers of the two species. In other words, either the reduction in the growth rate of the resource or the increase in the growth rate of the predator due to predation may change very quickly because the terms resulting in these changes are a function of the joint densities of the two species.

Let us solve these two equations for their equilibria. The equation for the growth rate of the resource population can only be zero if

$$r - aP = 0.$$

In other words, if

$$P = \frac{r}{a}$$

then $dR/dt = 0$, and there is no change in the numbers of the resource population. As we did for two competing species, let us plot this on a graph that gives the number of predators on one axis and the number in the resource population on the other axis. In general, the number in the predator population will be larger than the number in the resource population, so the scales on these two axes will be different. The horizontal line in Figure 14.2a indicates this equilibrium, the zero isocline for the resource population. Above this line, the number in the resource population decreases because of the large

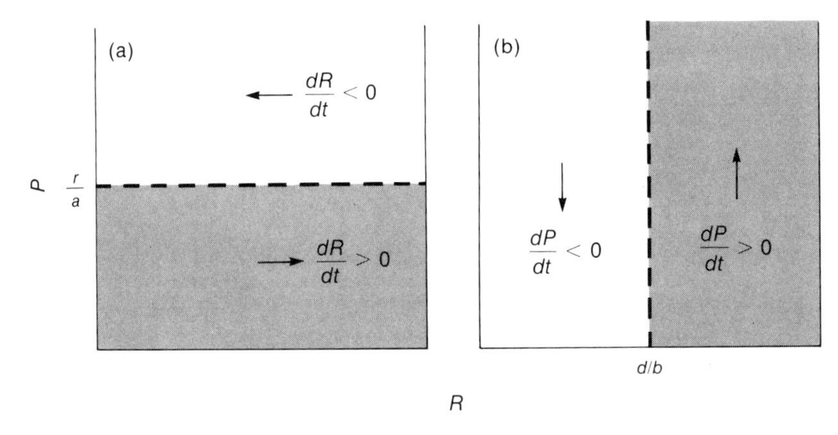

Figure 14.2. The change in population numbers for (a) the resource species and (b) the predator species.

number of predators, and below the line it increases because of the small number of predators.

Likewise, the equilibrium for the predator population occurs when

$$-d + bR = 0$$

because the growth rate of the predator population is zero. Therefore, if

$$R = \frac{d}{b}$$

then $dP/dt = 0$ and there is no change in the number of predators. This equilibrium is called the predator zero isocline and is plotted as a vertical line in Figure 14.2b. To the right of this line, the number of predators increases because of the excess of the resource, and to the left of it the number of predators decreases because there are few individuals in the resource population.

At first glance, it would seem that the point where these two lines cross—that is, where both $dR/dt = 0$ and $dP/dt = 0$—would be a stable equilibrium. However, if the two parts of Figure 14.2 are combined, as in Figure 14.3, then it is obvious that the numbers of the two species do not approach these equilibrium densities when they are initially different from these values. For example, in the upper right quadrant, the number of predators is increasing while the number of prey is decreasing. Likewise, in the lower left quadrant, the number of predators is decreasing and the number of prey increasing. As a result, the numbers never converge to the equilibrium but oscillate around it. The size of the oscillation is dependent upon the starting numbers and does not increase or decrease in size with time. However, random

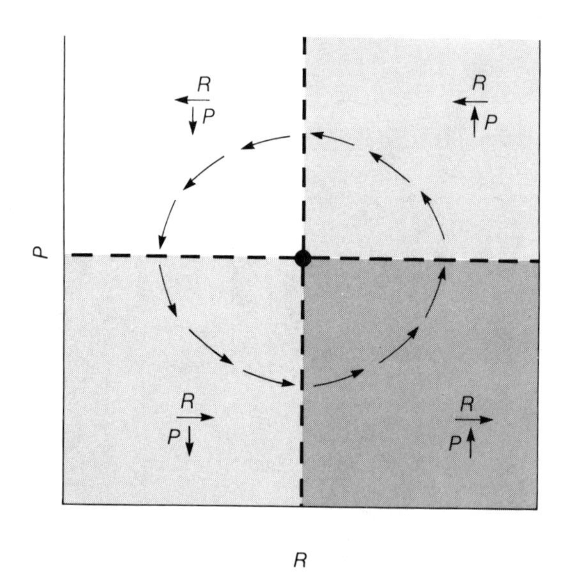

Figure 14.3. The joint change in numbers for the resource and predator populations.

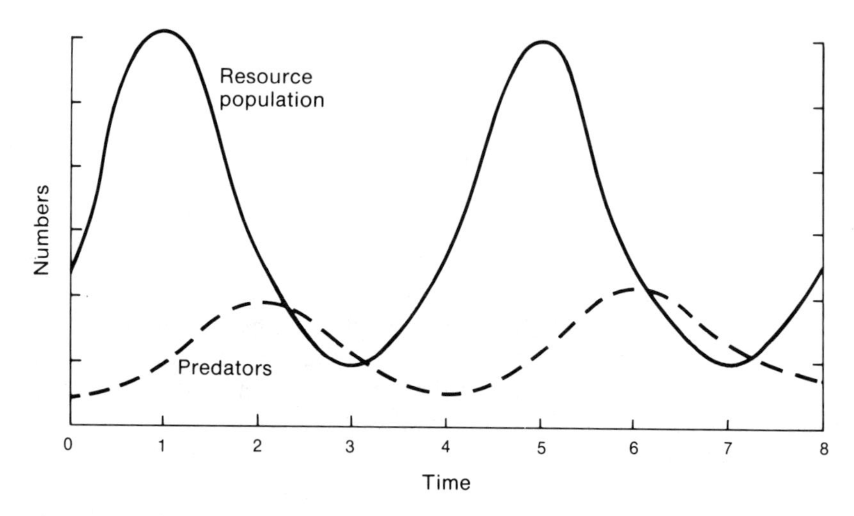

Figure 14.4. The change in population numbers over time for the resource and predator populations.

variation in the environment affecting the species numbers may move them to other predator–resource combinations. The numbers will then cycle from this point, indicating that the cycles do not converge to a single stable cycle. However, the equilibrium numbers do indicate the average expected number of each species over an extended period of time.

The numbers of the two species can also be plotted over time. Figure 14.4 gives the population numbers for both species for two cycles, beginning at a point in the middle of the lower section of Figure 14.3. Notice that the predator numbers follow the resource numbers, reaching maximums and minimums a quarter of a cycle later. The difference in the numbers of the two species results from a time lag effect of resource numbers on predator numbers. For example, in the lower left quadrant in Figure 14.3, resource numbers are increasing and predator numbers are still declining, and the predator numbers do not begin to increase until the lower right quadrant.

Let us illustrate these changes with a numerical example. Assume that $r = 0.2$ and $a = 0.01$, so that the number of predators at which $dR/dt = 0$, the resource zero isocline, is

$$P = \frac{r}{a} = \frac{0.2}{0.01} = 20.$$

Likewise, if $d = 0.2$ and $b = 0.001$, then the resource number at which $dP/dt = 0$, the predator zero isocline, is

$$R = \frac{d}{b} = \frac{0.2}{0.001} = 200.$$

If we let $P = 10$ and $R = 100$, a point in the lower left quadrant, then the change in resource numbers is

$$\frac{dR}{dt} = (r - aP)R = [0.2 - (0.01)(10)]100 = 10.$$

The change in predator numbers is

$$\frac{dP}{dt} = (-d + bR)P = [-0.2 + (0.001)(100)]10 = -1.$$

Therefore, the number in the resource population is increased to 110 and the number of the predator population decreased to 9. Similar changes can be calculated for points in the other three quadrants. One prediction of this simple model is that predator and prey numbers should oscillate. However, the first experiments conducted to determine whether this prediction was correct in a simple laboratory situation were unsuccessful and resulted in elimination of either the resource or the predator (see Example 14.2).

In addition, this model allows two predictions that may not be immediately obvious. First, if a, the successful attack rate, increases, then P decreases (remember, at equilibrium $P = r/a$). In other words, an evolutionary increase in the predator efficiency may lead to a decline in the numbers of the predator. Second, if a density-independent mortality factor affects both populations equally, then the equilibrium number of predators decreases and that of the resource increases. The increase in mortality for the predator reduces r, making the equilibrium r/a smaller; the increase in mortality for the resource increases d, so the equilibrium d/b is increased. This may explain why some applications of pesticides actually increase pest problems, assuming that the pest is a resource species.

Example 14.2. The predictions of oscillations by the Lotka-Volterra models were examined for a number of experimental, laboratory systems. For example, two species of mites, one species (the resource) that infests oranges and a predatory mite that feeds on the resource species, were studied by Huffaker (1958). When the predatory mite was introduced to an orange infested by the resource species, the resource species was quickly eliminated, with the elimination of predator species following because the predator had no food. Huffaker then made the environment more complex by introducing spatial heterogeneity. For example, when the feeding area (the oranges) was interspersed equally with areas without food, elimination of both species still resulted (see Figure 14.5a). However, when an even more complicated environment was used, 252 oranges with barriers between them, then oscillations resulted (Figure 14.5b). In this case, the predator numbers oscillated in a cycle slightly after the resource numbers, although the predators were eliminated after three oscillations.

Remember that the simple model predicted oscillations but that in such a laboratory situation the species were eliminated. Only in a more complex situation with a refuge did oscillations actually occur. In other words, even though the simple model is consistent with the observations in the complex situation, it is so for the wrong reasons.

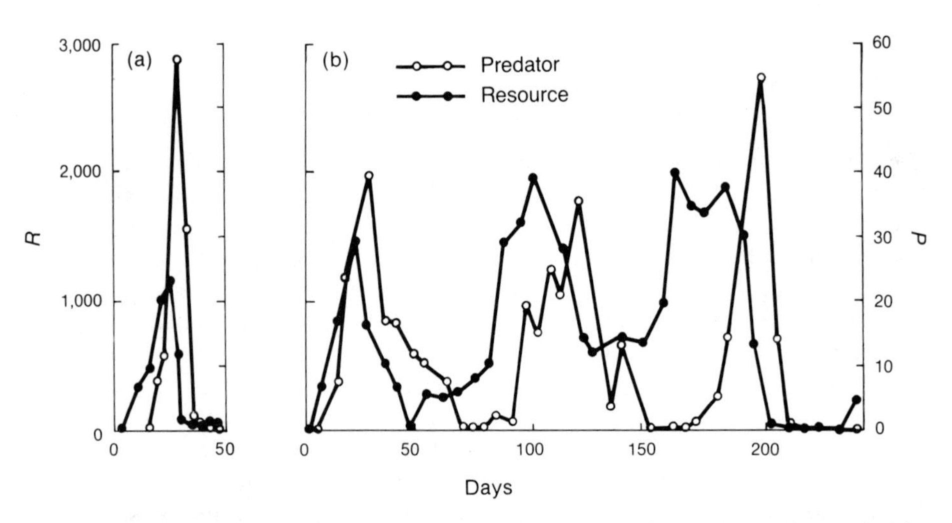

Figure 14.5. The change in population numbers over time for a predatory mite and its prey in (a) a relatively homogeneous environment and (b) a heterogeneous environment (Huffaker, 1958).

The Lotka-Volterra model for predator–resource interaction has been expanded in a number of ways to make the model more realistic (in Chapter 20 we will discuss genetic change that can dampen oscillations). An obvious extension is to make the rate of population growth for each species a function of their own population numbers—in other words, to include a logistic term. Let us modify the graph in Figure 14.3 to reflect this change for the resource species. Figure 14.6 gives such a modified resource zero isocline, in which the line bends and intersects the horizontal axis at the carrying capacity of the resource population. This figure also includes another extension, in that the resource zero isocline is allowed to increase where resource numbers are small. In other words, if resource numbers are small, then it is assumed that the predators will not attack them and that the resource population will increase. As a result, resource numbers will increase independently of predator numbers. Such situations may occur where a refuge exists in which a certain number of the resource population may hide or where the predators switch to a different resource when the resource numbers are low (this model appears to conform generally with the experimental situation illustrated in Figure 14.5b).

Given such a resource isocline, the change in population numbers over time depends upon the relative position of the predator isocline. For example, if the predator zero isocline intersects the prey zero isocline to the left of the hump, as in Figure 14.6, then the oscillations in population numbers will dampen and approach the equilibrium given by the intersection of the two lines. An example of such a trajectory is given in Figure 14.6. Alternatively, if the isoclines are somewhat different, the oscillations may approach a stable limit cycle—that is, a cycle of a given size that is reached from all starting combinations of the two species (see Example 14.3).

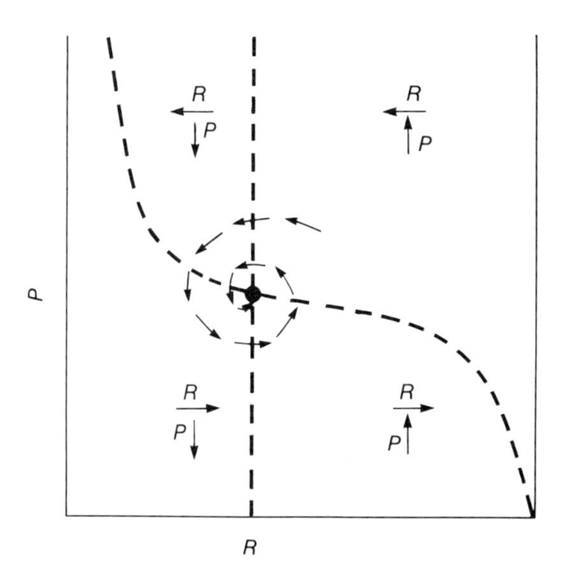

Figure 14.6. The change in population numbers when the resource isocline includes a refuge and density dependence.

Example 14.3. The persistence of a preferred resource species in the presence of intense predation may occur when a physical refuge is present. The importance of a refuge is most easily demonstrated when a predator is absent in one area and present in another. For example, the sea otter is a predator on both sea urchins and abalone. The sea otter was hunted to near extinction off the coast of California. Only about 50 animals were left in 1911, but with protection the number grew to 2,000 animals in 1978. In areas where the sea otter is present, sea urchins are absent, because sea urchins have no way of avoiding sea otter predation (see Example 14.7).

On the other hand, abalone can continue to exist in the presence of sea otters because they live in crevices in rocks, where the sea otters have difficulty removing them (Hines and Pearse, 1982). However, as a result, the average size of the abalone is much smaller in the presence of the sea otters. Figure 14.7 gives the size distribution of shells eaten by sea otters in an area just invaded, Point Santa Cruz, and an area in which sea otters have lived for a number of years, Hopkins Marine Life Refuge. The abalone size in the presence of the predator is about half that in the absence of the predator. Furthermore, virtually all abalone were in crevices where sea otters were present and were numerous outside crevices where sea otters were absent. The continued existence of abalone in the presence of the sea otter appears to depend upon the crevice refuge and a high productivity to counteract sea otter predation pressure.

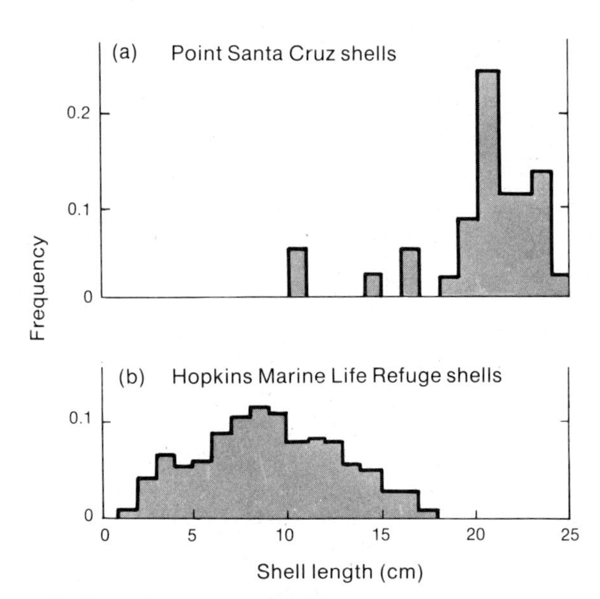

Figure 14.7. The shell length of abalone (a) where sea otters have just invaded and (b) where they have been resident for a number of years (after Hines and Pearse, 1982).

II. PREDATION IN NATURAL POPULATIONS

Intuitively, it would appear that the presence of predators would control the numbers of a resource species below the carrying capacity for the resource species. Examples in which predators have been removed by hunting that led to subsequent outbreaks of resource populations, such as deer or rabbits, support this contention, as do extinctions resulting from human predation (see discussion in Part 3). In some cases, however, predators may take primarily juvenile, injured, or older animals; thus, their effect on resource numbers is small and may even be beneficial by reducing intraspecific competition. Even when some reproductive individuals are removed by predators, the rate of recruitment in the resource population may consequently increase, so that the total consumption by the predator may remain nearly constant (see the discussion of harvesting in Chapter 21). It also appears intuitively that the number of a resource species would affect predator numbers. However, resource numbers may be secondary to other factors affecting predator numbers. For example, predator numbers may be limited by available territorial space, as in mountain lions, and by suitable unoccupied habitat, not by available prey. In other words, the numbers in a predator and a resource species may involve many other factors besides the numbers of the other species (see also Murdoch, 1966; Ehrlich and Birch, 1967) (see Example 14.4).

Some clearcut examples exist of disease-causing organisms, predators in the broad sense defined at the beginning of the chapter, affecting a resource species (see Example 14.5). In fact, the abundance and distribution of many organisms, both in plants and animals, may be primarily determined by disease. For example, plant rusts strongly affect wheat (see Example 20.5) and other grass populations, and plague has led to high mortality in human and other mammalian populations. In some cases, the effect of a disease may be difficult to detect, both because of its periodic nature and the fact that many animal and plant diseases are not well known.

The population density of some species, such as lemmings, voles, and snowshoe hares, varies somewhat regularly over time. Such variations were long thought to be the oscillations predicted by the Lotka-Volterra equations (see discussion in Hutchinson, 1978). In fact, the lynx, an important predator of the snowshoe hare, also had oscillations in density. However, the snowshoe hare has similar oscillations on an island in which the lynx is not present. In addition, the predators of voles, such as owls and hawks, do not vary in density in a pattern similar to the voles. It appears that other factors may be responsible in part for the regular variations in population numbers in some species (see Example I.1).

Predators may respond in two different ways to resource densities. First, the Lotka-Volterra equation for the change in predator numbers suggests that the increase in predator numbers due to predation is bRP. In other words, a *numerical response* occurs in the predator population owing to an increase in the resource population. In addition, predator numbers may also be increased by the immigration of predators feeding on a large resource population.

Second, predators may reduce the resource population by an amount $-aPR$, as suggested by the Lotka-Volterra equation for the change in resource numbers. In this case, as resource numbers increase, the number eaten by the predators increases at a linear rate; thus, doubling the number of the resource doubles the predation rate (the straight line in Figure 14.10a). The general relationship between the rate of predation and resource density is called a *functional response*. In natural populations, the rate of predation may

---

Example 14.4. The removal or introduction of a predator species offers reasonably direct evidence of predator–resource interactions. Introductions of a predator species into a new area can illustrate the effect of a predator species on the population number in a resource species. When the resource population is easily attacked and captured by the new predator species, a precipitous drop in the resource number may result. For example, the sea lamprey attaches to the sides of fish and sucks out their fluids. Lampreys entered the Great Lakes in the early 1920s via the canal built around Niagara Falls and by 1945 were found in Lake Superior, the westernmost lake. The lamprey spread resulted in a virtual elimination of the lake trout fishing industry in a matter of a few decades. Figure 14.8 gives the lake trout catch, a measure of relative density, in three of the Great Lakes over several decades (after Baldwin, 1964). The lamprey was first observed in Lake Michigan, Lake Huron, and Lake Superior in 1936, 1937, and 1945, respectively, and reached high numbers in only a few years. By the late 1940s, the lake trout catch had

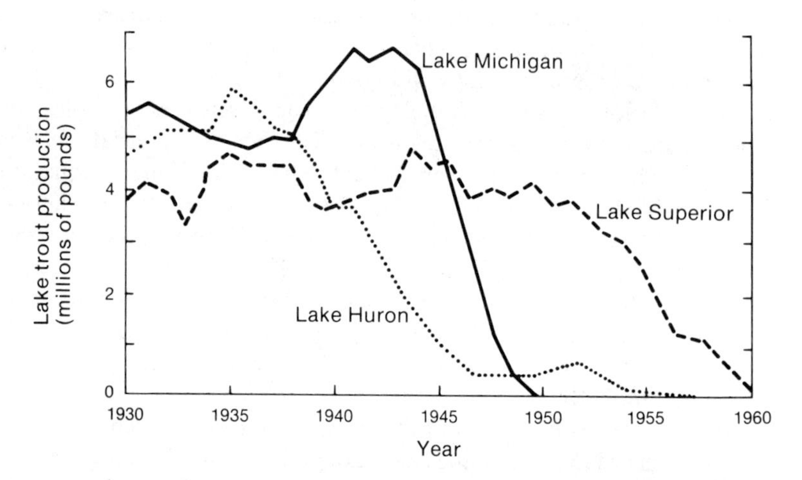

Figure 14.8. The production of lake trout in Lake Huron, Lake Michigan, and Lake Superior after the introduction of the sea lamprey (after Baldwin, 1964).

change in a variety of ways, two of which are given in Figure 14.10a. First, the intermediate curve illustrates the expected response for a predator that becomes satiated at high resource density. This curve appears typical of that for invertebrate predators, such as the praying mantis, which reduces its predatory effort as it becomes satiated, or for other predators for which handling time (pursuing and consuming) limits the number of the resource population eaten. The curve that has a somewhat sigmoid shape is for a

---

dropped to near zero in Lake Michigan and Lake Huron and by 1960 was approaching zero in Lake Superior.

On the other hand, removal of a predatory species should result in an increase in the resource population. For example, in the rocky intertidal shoreline of western North America a large starfish, *Pisaster ochraceous*, is the major predator on sessile mollusks. In particular, the starfish limits the lower distribution of the mussel *Mytilus californianus*. When the starfish is removed, the mussel spreads its distribution from its refuge in the upper tidal zone and extends lower into the intertidal zone. Figure 14.9 gives the change in distribution in one site when *Pisaster* was removed between 1963 and 1968 (Paine, 1974). The mussels extended their distribution by 0.85 m over a five-year period. When *Pisaster* was no longer removed, the distribution of the mussel returned to the level before the removal experiment. In nearby control areas in which *Pisaster* was not removed, the distribution did not change.

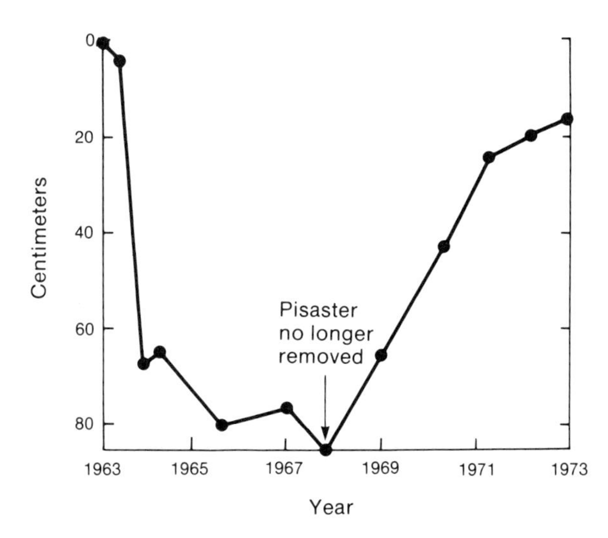

Figure 14.9. The distribution of a mussel when the starfish *Pisaster* was removed (Paine, 1974).

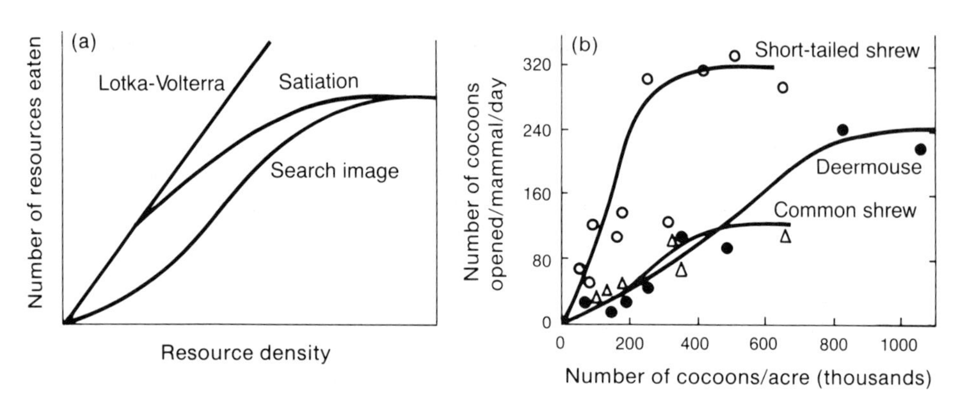

Figure 14.10. (a) Three types of theoretical functional responses by a predator to resource density and (b) the observed functional responses to sawfly cocoon density by three predators (from Holling, 1959).

predator that learns resource density and switches to the resource of high density. This type of response appears common for vertebrate predators, such as coyotes or jays, with a diet that may consist of several resource types. In this case, when a particular resource is at low density, the predator may switch to another resource species, so that its numbers are relatively unaffected. However, as the resource increases in density, the predator learns this by increased encounters and forms a search image on this now common resource. At high resource density, satiation occurs, resulting in an S-shaped functional response curve (see Example 14.6).

Example 14.5. Before 1800, there were no white-tailed deer in Nova Scotia, Canada, although archaeological findings suggest that they had once been present. There were, however, substantial populations of moose and caribou in Nova Scotia (Embree, 1979). As the deer recolonized this area in the nineteenth century, the caribou disappeared and were virtually extirpated by 1915. The moose population also began to decline, and it was thought that a combination of competition from the deer and increased hunting pressure were responsible for the decline in these populations. However, detailed studies of the moose population uncovered a nematode that was harbored by the deer and that was lethal to both the caribou and moose. Because the deer occupies the same general habitat as the caribou, the nematode infected and killed nearly all the caribou. Moose are more sedentary, often occupying areas that deer do not frequent, and therefore can avoid the nematode. In other words, what at first appeared to be a case of competitive exclusion was actually a change in species numbers apparently caused by differential susceptibility to a disease.

Until now we have considered only the populations of two species. However, most natural communities have more than one predator species and more than one resource species, although in some communities there may be only one predator of the highest trophic level (the *trophic level* is the position in the food chain, such as producer, herbivore, and carnivore) or one major resource species. Consideration of the complex interactions of all the species in a community is the province of community ecology and beyond the scope of our discussion here. However, let us briefly consider some aspects of systems with more than two species.

First, assume that a community has two predator species and one resource species. In this case, the two predators can be considered competitors for the resource, and one might expect from our discussion in Chapter 13 that one species would exclude the other. However, as suggested from the example of ant and rodent competitors on seeds (Example 13.1), predator species may coexist, given enough heterogeneity in the resource. In the ant–rodent case, the seeds consist of several species and are, as a result, quite variable in size, resulting in relatively little niche overlap.

Second, assume that there is one predator species and two resource species. In this case, the resource species are competitors, but the presence of the predator may have important effects on the numbers of the two competitors. For example, if the predator has no resource preference, it may reduce the density of both competitors. This may appear like competition between the resource species but actually may allow their coexistence because of their lowered densities. On the other hand, if the predator has a preference for one of the resource species, predation may result in a reversal of dominance among the competitors (see Example 14.7).

III. EXTINCTION

The cause(s) of recent extinctions vary with the organism, but in nearly every case human activity has been in large part responsible. In fact, if one examines the number of extinct birds and mammals from 1650 until the present, one finds a striking correlation with the exponential increase in the human population during this time period (Figure 14.14). In addition, if we examine the known causes of extinctions, we find that 91 percent and 59 percent of the extinctions of mammals and birds, respectively, were due to either human predation or predation by human-associated organisms, such as the rat, mongoose, cat, dog, and pig (see Table 14.2). For example, many large organisms, such as marine mammals, flightless birds, and sea turtles, were overhunted for their meat and fur. Some were hunted to extinction, while others, such as the northern elephant seal, have increased in numbers after nearly being hunted to extinction. Hunting for sport is still one of the major pressures on the populations of many large animals (see Example 14.8).

Because population sizes vary over time and environments change, there is a constant turnover in the number of species and the proportion of the total number of individuals in a community of a particular species. For a particular species, a probability exists that it will become extinct at any point in time because of variations in population number or because it cannot adapt to a changing environment. On a local scale, species that become extinct or are extirpated may be reintroduced by migration from other areas. On a global scale, an extinct species cannot be replaced exactly, although new species may be produced by speciation events (see Chapter 16). However, this natural process of species turnover has been terribly upset by human activity.

Why should we be concerned about the extinction of species? First, the total of all species constitutes a reservoir of genetic, morphological, and biochemical diversity in our world. Therefore, the extinction of species not only represents a loss to the natural world and our ability to understand and appreciate it, but it may in a practical sense represent the loss of a chemical compound of potential value in medicine or agriculture. Second, the loss of

---

Example 14.6. The functional response has been evaluated in a number of different types of predators (see Begon and Mortimer, 1981). For example, Holling (1959) observed the functional response for three mammalian predators on sawfly cocoons. These cocoons provide a large resource population for the three summer months in Canadian pine plantations. Because each predator species opens the cocoons in a particular manner, predation rates for each species can be calculated. By using different areas with different sawfly cocoon densities, Holling was able to determine the functional response for each species (Figure 14.10b gives the data for two of these species). The short-tailed shrew, *Blarina*, increased its resource consumption very rapidly with increasing resource density, indicating a functional response curve consistent with the satiation curve in Figure 14.10a. The other species, the common shrew, *Sorex*, increased its resource consumption more slowly, and its response tended to suggest responses similar to that from search image formation. However, it is difficult to determine whether these observations are actually concave or not at the low resource density, the aspect that differentiates the two curvilinear responses given in Figure 14.10b.

Periodical cicadas (*Magicicada*) emerge only once every 17 years in the northern part of their range, and during these emergences they are extremely abundant, up to 300 cicadas/m^2. The major predators of cicadas are a number of bird species that characteristically eat the cicada body and discard the wings. In a series of populations, Karban (1982) estimated the density of emerging broods of cicadas by counting the number of cast skins, the density of bird predators, and the predation rate of birds by counting discarded wings.

Figure 14.11a gives the numerical response from these data—that is, the bird density for different cicada densities. There appears to be no numerical response. This observation is reasonable, given that most of

species is an esthetic loss to our world. What can match the intrinsic beauty of migrating whales or the majesty of the bald eagle? Last, the extinction of species is an indication of the quality of our world. The near exterminations of the brown pelican and the perigrine falcon on the west coast of the United States due to DDT were warnings to us of the deterioration of our environment.

Animals associated with humans or introduced by humans to new areas have had devastating effects on local fauna and flora. For example, nearly all of the bird fauna has been eliminated from lower elevations in the Hawaiian Islands. A number of factors, common to other island bird species, appear to have been important. These include predation on eggs by the rat and mongoose, competition by introduced bird species, and bird malaria brought in by the introduced species to which the native species were not resistant. In

---

these birds do not move about during the time of emergence, late May and June, and that a reproductive response could not occur until the next spring (or perhaps late summer). The functional response is given in Figure 14.11b, in which the predation rate per bird is given for different densities. In this case also, no relationship appears to exist between the two estimates. Because the observed cicada broods always had relative high densities, even the low-density brood had about 5 skins per m^2, and birds may be satiated at even these low densities. One interesting aspect of this study is that cicada reproductive success was greatest at high densities, the inverse of most density-dependent responses, because there was no numerical or functional response in the predators at the observed densities.

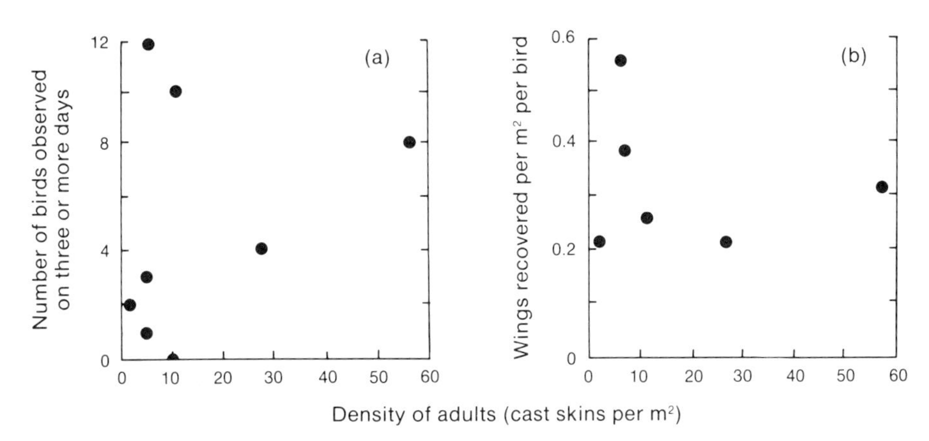

Figure 14.11. Estimates of (a) the numerical and (b) the functional responses of birds to cicada density (after Karban, 1982).

Example 14.7. Predation has been demonstrated to have an effect on the outcome of competition in sessile species, such as plants or mollusks. For example, the presence of predation by sea otters (see also Example 14.3) can virtually eliminate sea urchins from an area. Sea urchins in turn feed on a perennial kelp, *Laminaria*. When sea otters are absent, sea urchins are plentiful and the stand of kelp is dominated by annuals rather than *Laminaria* (Table 14.1, after Duggins, 1980). On the other hand, when sea otters are plentiful, they eliminate the sea urchins, and the perennial kelp dominates and virtually eliminates the annual kelp. In other words, the sea otter influences the competitive outcome of species two trophic levels below it.

One approach to measuring the diversity of species at one trophic level is to count the number of types of species. (Many other measures of diversity have been suggested, and they include additional information such as the frequencies of the different species.) In the experiment discussed in Example 14.4, a low density of the predatory starfish resulted in a reduction of the number of resource species (*Mytilus* outcompeted other species), while the species number increased when starfish were present. However, the effect of a predator on species diversity may follow different patterns.

For example, Lubchenco (1978) found that the number of algae species in tidepools was low when the density of the herbivorous snail, *Littorina littorea*, was low, and that the algal species number increased with increasing snail density to a point at which the species number declined with further increases in snail density (Figure 14.12a). At the high density, grazing is so intense that most species and individuals are eliminated. On the other hand, in nontidepool areas, the competitive relationships among the algae are different and a few species are no longer dominant. As a result, the diversity increases with lower predator density (Figure 14.12b). In other words, predators can either increase or decrease the diversity of resource species, an effect that is dependent upon the food preferences of the predator, the density of the predator, and the competitive abilities of the resource species in the given environment.

Table 14.1
The effect of sea otters on the density (number/m^2) of sea urchins and kelp in Alaska (after Duggins, 1980).

	Presence of sea otters		
	None	*< 2 yrs*	*< 10 yrs*
Sea urchins	16	0.31	0.0
Kelps			
Annuals	3	10	2
Laminaria	0.8	0.3	46

In theory, differential predation pressure on competitors could affect other types of species as well. An experimental demonstration of this effect was given for predation by newts of the genus *Notophthalmus* on three species of tadpoles (Morin, 1981). The newt is a generalist predator and occurs with tadpoles of several anuran species in areas of the eastern United States. Morin constructed a number of small artificial ponds and to them added tadpoles of each species, along with newts at four density levels. The survivals to metamorphosis of the three species were strongly affected by the density level of the predator, resulting in a reversal in the relative abundances from the zero predator level to the higher predator levels (Figure 14.13). In particular, the proportions of the *Scaphiopus* and *Bufo* tadpoles declined and the proportion of *Hyla* tadpoles increased with increasing density. The apparent reversal in competitive outcome appears to have resulted from combination of the relative poor interspecific competitive ability of *Hyla* and a predation preference by the newts for *Scaphiopus* and *Bufo* larvae.

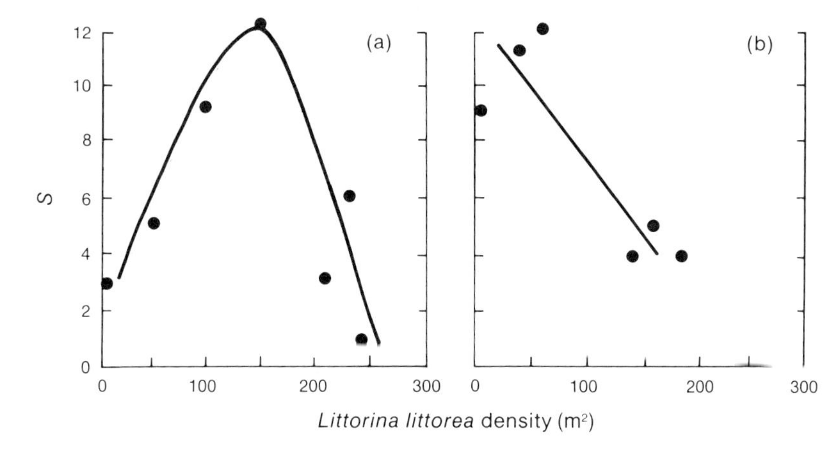

Figure 14.12. The number of algae species in (a) tidepools and (b) nontidopool areas as a function of the density of a snail (after Lubchenco, 1978).

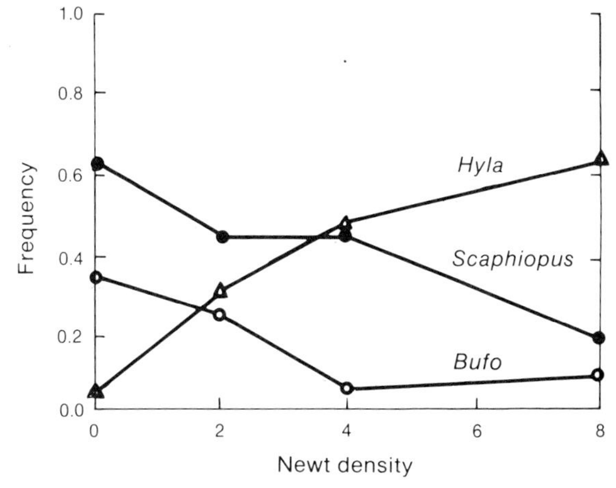

Figure 14.13. The effect of newt density on the proportions of tadpoles of three species (after Morin, 1981).

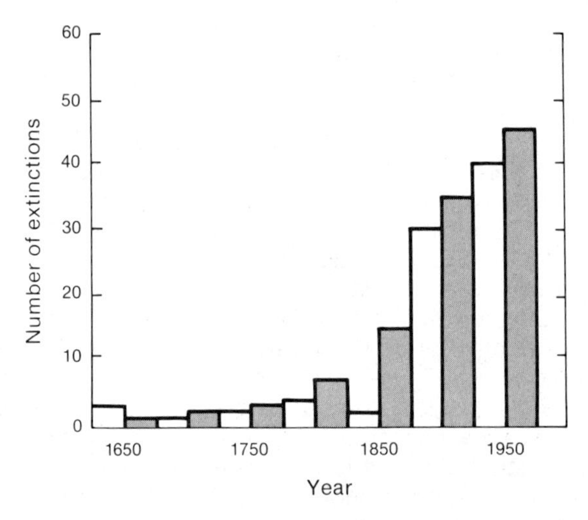

Figure 14.14. The number of bird and mammal (shaded) extinctions from 1650 to the present (after Ziswiler, 1967).

Table 14.2
The cause of extinction for 147 mammals and birds (after Ziswiler, 1967).

	Number of species	*Human predation*	*Predation by human-associated organisms*	*Habitat destruction*	*Not known*
Mammals	46	0.63	0.28	0.07	0.02
Birds	101	0.22	0.37	0.37	0.04

addition, much of the lowland vegetation was removed for agriculture or is now being replaced by introduced plant species.

Although human predation and predation by human-associated organisms appear to have been the major causes of past extinctions and will continue to be major problems, human activity displaces other species in a number of other ways. First, humans have changed the environment in many areas of the world by activities such as cutting forests and filling in marshlands, so that the natural habitat of many species is destroyed or reduced. Such large-scale habitat destruction is occurring presently in the Amazon basin, where huge expanses of rain forest are being removed for agricultural development. In this area, many species of plants and animals that have not yet been identified may go extinct. In addition, changes in the general environment by our use of synthetic chemicals, such as DDT, or fuels, such as those causing acid rain, may have extensive effects on species survival.

Many of the species most sensitive to predation and environmental change are those at the top trophic level, such as tigers or mountain lions, or the largest of their type, such as the condor or the elephant. These animals generally require large areas to forage and also have low reproductive rates. In addition, *endemic species*, those that have a very restricted range, are particularly vulnerable to extinction by predation or habitat destruction. Endemic species may either be on islands or in unusual continental habitats where evolutionary change has resulted in specially adapted species, such as the desert pupfish (see Chapter 9 and Example 14.9).

Example 14.8. A number of marine mammals have either become extinct or faced extinction because of hunting pressure. While commercial fishing pressure has influenced the abundance of some fish species, generally fish species can continue to maintain themselves at lower densities. However, it appears that the common skate, a large ray sometimes more than 2 meters in length and 50 kg in weight, faces extinction in the Irish Sea, an area where it was once abundant (Brander, 1981). Other skate species have shown a small decline in recent years but do not appear to be nearly as vulnerable to fishing pressure. For a species to maintain itself, the adult mortality rate, due to fishing pressure and other causes, must be less or equal to the recruitment rate of mature fish (see discussion in Chapter 21). The recruitment rate is dependent upon the age to maturity, the preadult mortality, and the fecundity level. Figure 14.15 gives a measure of the fecundity and the estimated age of maturity in the five major species of skates found in the Irish Sea. Of these species, the common skate has the greatest age to maturity and the second lowest fecundity. In addition, because the common skate is relatively large, it probably also suffers higher preadult mortality from fishing than the other species. It appears that the common skate will be completely fished out of the Irish Sea because of its higher susceptibility to fishing pressure.

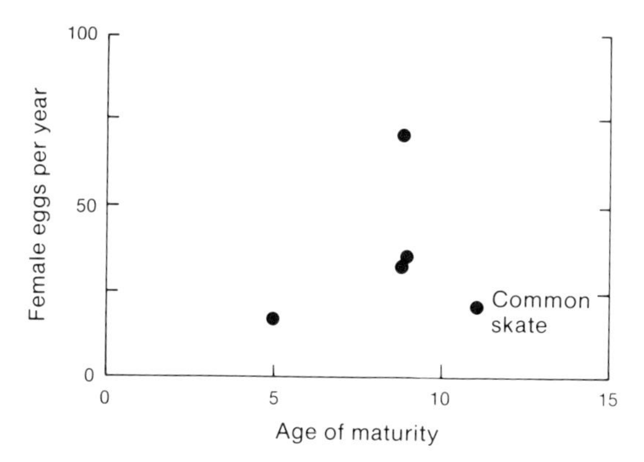

Figure 14.15. The age of maturity and a measure of fecundity for the common skate and four other species of skates in the Irish Sea (after Brander, 1981).

What can be done to keep endangered species from extinction? A number of approaches can aid species near extinction (see Frankel and Soule, 1981). First, endangered species should be protected from hunting or removal by hunters, poachers, and hobbyists. To assist in this effort, laws protecting endangered species should be strictly enforced and natural reserves set aside. For endemic species, such areas need not be large but must encompass the habitat required by the organism. For large animals, the size of the protected large area may include both a reserve and adjacent areas because of the large area necessary for foraging. Second, in areas in which a species has become extinct, reintroduction from other areas will lessen the probability of species extinction (as with the whooping crane). However, often this is difficult and requires an extensive knowledge of the biology of the organism. Last, species can be raised in zoos or other managed areas to increase population numbers and preserve the species. Of course, the zoo environment may be so different from the natural environment that reintroduction to natural areas may be quite difficult.

SUMMARY

The concept of predation is used in a general way to include host–parasite and herbivore–plant systems as well as classical predator–prey systems. The Lotka-Volterra model, which predicts change in the numbers of the predator

---

Example 14.9. Many of the large animals that face extinction were once hunted for sport or for their meat, fur, feathers, or horns. For example, the whooping crane, the tallest North American bird at 1.5 m and having a wingspan of 2.25 m, nearly became extinct in the late nineteenth century. These cranes were hunted as trophies and for their feathers and eggs, and the draining of prairie land reduced their nesting success. The whooper was thought extinct, but in 1936 a band was found wintering off the coast of Texas. Even though a refuge was provided, their numbers dwindled to 15 in 1941. With intense protection and surveillance, this band has grown to more than 80 today. The nesting site of this band was found in northern Alberta in 1954, making the flyway distance 4,000 km. Because the migration is risky and only this one band existed, a new, shorter flyway was established by introducing eggs into sandhill crane nests (Zimmerman, 1977). The sandhill crane is a smaller, gray crane that has a flyway from Idaho to New Mexico, about a third that of the whoopers. Although there has been a very high mortality in the whoopers introduced to sandhill crane nests, nearly a dozen whoopers have established themselves by living with sandhill cranes. However, none of these whoopers has bred yet, a step necessary to maintain this new population.

---

and the resource (prey) population, is discussed. A number of examples are given that illustrate the numerical and functional responses in predators or interactions among more than two species. Extinction of species, particularly as resulting from human predation, is considered.

PROBLEMS

1. How might predator–prey, host–parasite, and herbivore–plant interactions differ? How would these differences be reflected in the predator–resource equations for numerical change?
2. Using the Lotka-Volterra predator–resource equations and the values of $r = 0.2$, $a = 0.01$, $d = 0.2$, $b = 0.001$, $R = 300$, and $P = 30$, calculate the expected change in population numbers for both species. Illustrate this change in a diagram in which resource numbers are on one axis and predator numbers on the other.
3. Distinguish between a numerical and functional response in a predator population. Predators of the periodical cicada discussed in Example 14.6 did not show either type of response. Why do you think this occurred?
4. Assume that you wanted to demonstrate that the numbers in a predator and a resource species affected each other. What type of experimental perturbations would you use to show the effect of predation on resource numbers and vice versa?
5. Construct a simple mathematical model that could be used to show the effect of human predation on an endangered species (assume that human numbers are unaffected by this species). Include in this expression a term that gives a minimum critical population size and a term in which a (the attack rate) decreases as the number of the endangered species decreases.

CONCLUSIONS TO PART 2

The density (and distribution) of an organism are potentially affected by a variety of different factors (see Table C.2). The relative importance of these factors may vary with the characteristics of the organism and may vary over time or space. Expressions for the effect of these factors on the change in density, dN/dt equations, are useful in determining the situations in which different factors may be important.

Although we have focused on the effect of these factors singly, in many populations more than one of these factors may have significant effects on density. In some cases, such joint action may result in population regulation at some constant level, or in others, such as the combination of competition and predation, it may result in different effects in different situations.

Table C.2
A summary of the factors that can influence the density of a population and their general effects on density level and variation.

	General effect on density			
Factor	*Reduction*	*Increase*	*Constancy*	*Variation*
Intrinsic				
Density dependence	Yes	—	Yes	—
Dispersal	Local	Overall	Yes	—
Extrinsic				
Density independence	—	—	—	Yes
Competition	Yes	—	Yes	—
Predation-resource				
predator population	—	Yes	—	Yes
resource population	Yes	—	—	Yes

PART 3

Ecological Genetics and Evolutionary Ecology

In the first two parts, we introduced the principles of population genetics and population ecology. In this part, we will utilize those principles to examine first how environmental parameters are or may be important in determining the genetic composition of a population (Chapters 15–17) and then the genetic basis of ecologically important factors (Chapters 18–20). Many of these topics, such as speciation models, molecular variation, life-history evolution, and coevolution, are of great current interest in population biology and are areas of intensive research. Both because of the more complicated nature of these topics, which often include ideas from both genetics and ecology, and because of new developments in these areas, conclusions concerning some of these topics are often speculative. Our interest here, however, is to give a foundation of understanding for these topics that combine thought from several disciplines and that, as a result, more closely reflect the overall biology of populations. Chapter 21 discusses some areas of application of the principles of population biology to human problems. Population biology has a number of important uses, and, hopefully, greater understanding of its principles will enhance applied research.

CHAPTER 15

Ecological Models of Selection

> For the existence of a polymorphism is a sure sign of evolutionary change if it be transient, and of the action of contending selective forces if it be of the normal balanced form. That such selection is likely to be powerful, promoting sensitive adjustments to changing conditions, is a fair deduction from recent research on ecological genetics.
>
> E. B. FORD

In the development of population genetics, it is often assumed that the ecological factors significantly affecting genetic variation are constant over space and time. On the other hand, population ecology generally assumes that the population is genetically homogeneous for ecological important characteristics. In this chapter, we will concentrate on the ways in which the genetic constitution of a population may be influenced by ecological factors, such as physical or biotic factors in the environment, and the kind and number of organisms in the population. The complement of this topic, the way in which the genetic constitution of organisms in the population can affect the density and distribution of an organism, is considered in the section on adaptation in Chapter 16. The ways in which the genetic constitution of the population may affect its demographic characteristics or its interactions with other species are discussed in Chapters 19 and 20.

I. THE GENERAL EFFECT OF ENVIRONMENTAL VARIATION

In our previous discussions of fitness values we have generally assumed the relative fitness of a genotype to be some constant value. In fact, fitness values may depend not only upon a particular genotype but upon the environment in which the genotype exists. In other words, the relative fitness of a given genotype may be high in one environment and low in another. For example, the melanic forms of the peppered moth have a higher survival from predation in polluted environments, while the typical form has a higher survival in unpolluted environments (see Chapter 3).

Environmental or ecological factors that potentially affect relative fitness are such physical factors as temperature, moisture, and soil type, and such biotic factors due to other species as interspecific competitors, predators, and parasites. In addition, the composition of the population being considered may affect genotypic fitness values with respect to population numbers, age distribution, and genotypic proportions. As a first step in examining such ecological effects, we will assume that selective values are environmentally dependent and that the environment may vary over time or in space. Such an analysis is particularly appropriate in an examination of physical environment factors. In general, we will assume that the environment is coarse grained (see Chapter 9)—that is, a given individual exists throughout its lifetime primarily in an environmental patch of a given type.

As a way to understand the effect of environmental variation, assume that the fitnesses of the genotypes change as a result of the level of some environmental value (see also Example 15.1). For example, Figure 15.1 gives the relative fitness for the three genotypes at a locus with two alleles over an environmental spectrum. In this case, when the environmental level is low—say, there are low temperatures or low relative humidity—then genotype A_1A_1 has the highest relative fitness, A_2A_2 the lowest relative fitness, and A_1A_2 is intermediate (see Table 15.1). On the other hand, when the environmental

---

Example 15.1. One approach used to demonstrate an association between genotypes and an environmental factor is to examine the *in vitro* properties of electrophoretic variants in different environmental conditions (see also Chapter 17). It is assumed that the biochemical properties measured in the laboratory would be associated with fitness values in some manner. An apparent biochemical explanation for a genotype–environment association of electrophoretic variants has been demonstrated by Koehn (1969) in a small freshwater fish. In this species, there is a cline in the frequency of the two alleles at an esterase locus associated with latitude, such that the frequency of one allele—let us call it A_2—varies from 1.0 in southern Arizona to 0.17 in northern Nevada, a distance of approximately a thousand kilometers. Koehn measured the kinetic activity of esterase extracts from the three genotypes, A_1A_1, A_1A_2, and A_2A_2, at a range of temperatures and then put the enzyme activities on a relative scale (see Figure 15.2). At the highest temperature examined, 37 °C, the homozygote most common in warmer parts of the range, A_2A_2, had the highest relative activity. At the lowest temperature, 0 °C, the other homozygote, A_1A_1, the genotype that had the highest frequency in the colder parts of the range, had the highest relative activity. The heterozygote had the highest relative activity for the two intermediate temperatures. In other words, there appeared to be a strong

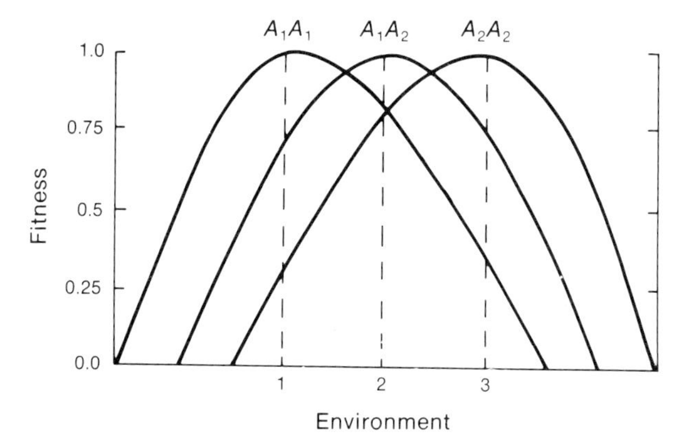

Figure 15.1. The relative fitness values for three different genotypes over an environmental spectrum.

level is high, the fitness values are reversed, with A_2A_2 having the highest fitness and A_1A_1 the lowest fitness value. Where the environmental factor is at an intermediate level, genotype A_1A_2 has the highest fitness.

If fitness values of different genotypes are related to environment variation, then in some situations a balanced polymorphism may result. In the traditional two-allele, constant-fitness model presented in Chapter 3, only when the fitness of the heterozygote is greater than that of the homozygotes is

---

association between the frequency of these alleles and the *in vitro* activity of enzymes from different genotypes in temperatures corresponding to those encountered in natural populations.

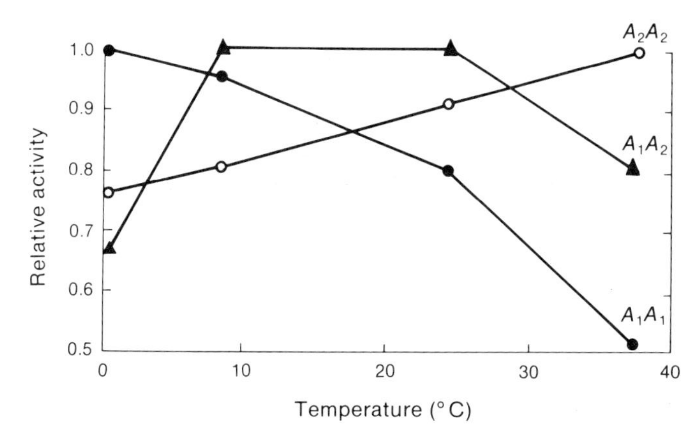

Figure 15.2. The relative activity for three esterase genotypes in a fish species (from Koehn, 1969).

Table 15.1
The fitness for different levels of an environmental effect.

Environmental level	Fitness		
	A_1A_1	A_1A_2	A_2A_2
1	1.0	0.75	0.5
2	0.75	1.0	0.75
3	0.5	0.75	1.0

Table 15.2
Variation in fitness as the result of different environments leading to a marginal heterozygous advantage.

Environment	A_1A_1		A_1A_2		A_2A_2
1	$1 + s$		1		$1 - s$
2	$1 - s$		1		1
3	1		1		$1 - s$
4	$1 - s$		1		$1 + s$
Average	$1 - \frac{s}{4}$	$<$	1	$>$	$1 - \frac{s}{4}$

there a stable equilibrium. A number of biological and environmental factors may cause fitnesses to vary, with the potential result that the relative fitness of a particular genotype may be high in one environment and low in another. If we apply the term environments generally to the factors that affect relative fitness, then over a series of environments the fitnesses may vary, as illustrated in Table 15.2. In this example, genotype A_1A_1 has the highest fitness in environment 1 and 3 and the lowest fitness in environments 2 and 4. Assuming that all environments are equally common, then we can calculate the average of the fitness values for each genotype over all environments. If the average fitness of the heterozygote is greater than that of either homozygote, then a phenomenon called *marginal heterozygous advantage* exists in which the arithmetic mean of the heterozygote is higher than that of the homozygotes. Such an average advantage may occur even though the heterozygote by itself does not have the highest fitness in a single environment.

Before we continue, it is important to realize that variation in selection over different environments does not necessarily lead to the maintenance of a polymorphism. Consider an example where differential selection affecting viability occurs within the life cycle of an organism and where there are just two morphs, the dominant morph (consisting of one homozygote and the heterozygote) and the recessive morph (Table 15.3). Assume that the relative viability is higher for the recessive morph in the larval (or seedling) stage but

Table 15.3
An example of differential viability in two life stages and the overall viability measure.

	Genotype		
	A_1A_1	A_1A_2	A_2A_2
Larval (or seedling) viability	1	1	v_L
Premating, adult viability	1	1	v_A
Overall viability	1	1	vv_Lv_A

that the recessive morph has a lower premating, adult survival. In this example, this means that $v_L > 1$ and $v_A < 1$. On the surface, it appears that balancing selection would be operating and that selection would maintain a polymorphism. However, because the variation is within a generation, the overall survival of the morphs is the product of the relative viabilities in the different life stages. Because of this property, and assuming that there are no other fitness differences for these genotypes, the overall relative fitness of the dominant form is either less than, equal to, or greater than the fitness of the recessive morph.

As an example, let us assume that $v_L = 1.2$ and that $v_A = 0.8$. The overall relative survival of A_2A_2 is $v_Lv_A = (1.2)(0.8). = 0.96$, resulting in an overall selection against the A_2A_2 genotype and consequently the A_2 allele. Therefore, when the overall relative fitness of the dominant morph is greater than the recessive, the A_1 allele will eventually become fixed, whereas if it is lower, the A_2 allele will eventually become fixed. Only in the unlikely case where the relative fitnesses are exactly equal will the two morphs remain in the population, because they are neutral with respect to each other.

If the population is structured into subpopulations, then the maintenance of a polymorphism is more likely. For example, consider the simple model in Figure 15.3 where the population has two subdivisions in different environments with limited migration m between them (see Chapter 5 for a discussion

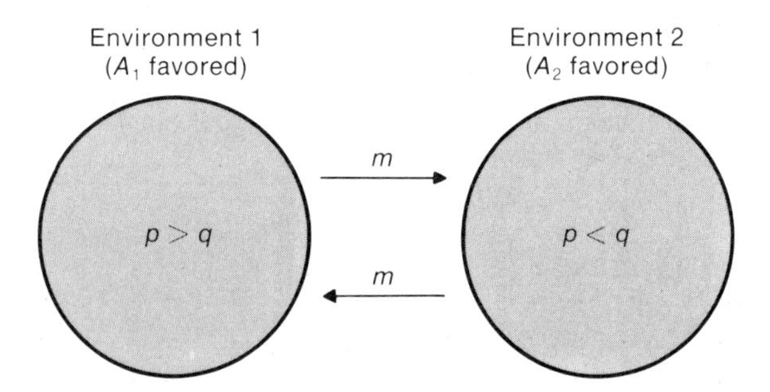

Figure 15.3. A population divided into two subpopulations, with A_1 favored in environment 1 and A_2 favored in environment 2.

of population structure). If A_1 is favored in environment 1 and A_2 favored in environment 2, then as a result of selection the frequency of the A_1 allele should increase in environment 1 and decrease in environment 2. If the amount of migration between the subpopulations is small, then the selective differences should have a cumulative effect on the allelic frequencies. However, a polymorphism will be maintained in each subpopulation, including the allele not favored, because of the constant introduction of alleles from the other subpopulation by migration. As a result, a spatially variable environment with

---

Example 15.2. Slender wild oats exist as a weed in much of the Mediterranean-climate area of California. There is generally a low migration rate in this species, with dispersal distances averaging a few meters or less, and a high proportion of self-fertilization exists, a factor that also limits migration potential. There appear to be two basic types of this species in some areas, which can be identified by morphological and electrophoretic variants and by some quantitative characteristics. Detailed mapping on a hillside showed a strong association of the two types with environmental factors within a small area (Hamrick and Allard, 1972). The "mesic" type, the type generally found in the wetter areas of California, is marked by an electrophoretic allele at the phosphatase locus that we will call A_1 (the selective differences probably occur at other loci having alleles associated with these alleles). On the other hand, the "xeric" type, the type generally found in dryer areas, is marked by allele A_2. Table 15.4 gives the frequency of the A_1 allele in several areas ranging from the moist hill bottom to the dry hilltop. There is a strong association between the allelic frequency and the environmental type, with allele A_2 fixed in the most xeric site (e) and A_1 in the highest frequency in the most mesic site (a). Presumably, the lack of fixation for A_1 at site a is partially due to the influx of A_2 alleles by migration. These patterns are consistent with an explanation based on a combination of selection and migration, although founder effect may also be important.

Table 15.4
The allelic frequencies of the phosphatase locus in five sites on a hillside arranged from the bottom (a) to the top (e) of the hill (Hamrick and Allard, 1972).

	Site				
Allele	*a*	*b*	*c*	*d*	*e*
A_1	0.65	0.15	0.21	0.41	0.0
A_2	0.35	0.85	0.79	0.59	1.0

---

limited migration among the subdivisions may be able to maintain a substantial number of polymorphisms. In a number of organisms, particularly those with limited dispersal, such as snails and many plants, small-scale spatial variations in allelic frequencies are seen (see Chapter 1 and Example 15.2). Although these patterns are consistent with the joint effects of selection and migration as just described, genetic drift that results in changes in allelic frequencies in small, semi-isolated subpopulations may also be important (see Chapter 6).

In addition to limited migration, *habitat selection*—that is, the choice of a particular habitat by different genotypes—may enhance the likelihood of a polymorphism. This can occur because a genotype can, for example, select an environment in which it has a fitness advantage (see Example 15.3). Examples of such habitat selection include melanic moths that choose dark backgrounds and lizards that pick areas that allow them camouflage protection. In these cases, the habitat selection results in increased survival from predators of the morphs in the selected habitats.

Some attempts have been made to correlate environmental variation with measures of genetic or morphological variation. The presumption is that measures of the level of genetic variation, such as heterozygosity or morphological variance, would be correlated with environmental variation. Such associations have been found in some situations, but the cause–effect relationship is difficult to substantiate.

II. FREQUENCY-DEPENDENT SELECTION

In a variety of organisms, including insects, birds, and crop plants, it appears that relative fitness values may be a function of the frequencies of different genotypes in a population, resulting in selection that is dependent upon genotypic frequencies. In essence, this frequency-dependent selection is a manifestation of the genotypic fitness being a function of an environment that includes the genetic composition of the population itself. Generally, models of frequency-dependent selection that lead to a balanced polymorphism have been of major interest. In these situations, the relative fitnesses of genotypes are some inverse function of their frequencies in the population. For example, if an allele is uncommon, the homozygote for that allele would have a higher fitness than other genotypes due to an advantage in viability, mating, or some other component of fitness. Of course, frequency-dependent selection may also lead to a faster fixation of a particular allele if, for example, an allele has a greater disadvantage when uncommon. We will not consider such models here, but situations of this sort could occur because a rare flower type is not recognized by a pollinator or an unusual color morph is picked out of a school of fish by a predator.

A general model of frequency-dependent selection assumes that the fitness of a genotype, say A_2A_2, is some function of the frequency of that genotype, q^2. Assuming that the relative fitness increases as the genotype becomes rare, the fitness of this genotype can be given as

$$w_{22} = 1 + \frac{s_2}{q^2}$$

where s_2 is the selection coefficient associated with genotype A_2A_2. If we assume that there is complete dominance—that is, that genotypes A_1A_1 and A_1A_2 are indistinguishable—then the frequency-dependent fitness values of these genotypes are

$$w_{11} = w_{12} = 1 + \frac{s_1}{1 - q^2}$$

where s_1 is the selection coefficient associated with the dominant phenotype, and $1 - q^2$ is the frequency of the dominant phenotype.

For this model, the fitness of a genotype increases quickly as it becomes rare. For example, if $s_2 = 0.1$ and $q^2 = 0.04$, then

$$w_{22} = 1 + \frac{0.1}{0.04} = 3.5$$

Example 15.3. In many cases, the genetic basis of phenotypic polymorphism is unknown. In some cases, however, phenotypic polymorphism is under genetic influence and different morphs appear to have different fitnesses in different ecological conditions. For example, the cichlid fishes of an area in Mexico have a variety of tooth, gut, and body morphologies.

In particular, the two common types of cichlids in this area consist of a small-toothed form with a long gut that feeds on detritus and algae and a large-toothed form with a short gut that feeds on snails. Although in some cases such differences in morphology can be environmentally induced, broods raised on controlled diets contained individuals of both types, suggesting genetic control of the morphology (Sage and Selander, 1975). Examination of electrophoretic variation in the two groups suggested that they were one population because there were no significant frequency differences between them.

To demonstrate directly that the two morphs were part of the same population, Kornfield et al. (1982) caught 34 breeding pairs and classified them by tooth morphology (see Table 15.5). If they were part of the same population, then the pairs should be close to random-mating proportions. For example, the expected number of pairs under random mating in which both individuals were small toothed would be the frequency of the small-toothed morph in females times the frequency of the small-toothed morph in males times the number of mating pairs, or

$$\frac{23}{34} \times \frac{18}{34} \times 34 = 11.8.$$

---

a rather high relative fitness value. Figure 15.4 gives the relative fitness values as a function of the frequency of the recessive genotype where s_1 and s_2 are both 0.1. Notice that, at $q^2 = 0.5$, the fitnesses of all the genotypes are equal, indicating the equilibrium frequency in the population.

Frequency-dependent selection appears to occur principally for differences in viability or mating. Such frequency-dependent selection is often mediated by some behavioral factor—for example, a predator preferentially attacking particular genotypes in a prey species or particular male genotypes being favored by females for mating (see Example 15.4). In other cases, the basis for frequency-dependent selection may be physiological, as in the interactions among plants or insect larvae relating either to differential use of nutrients or differential production of toxins.

Let us consider a general approach that can be employed to demonstrate that some kind of overall frequency-dependent selection is present. This technique compares the input ratio of two genotypes to their output ratio as a method of demonstrating that some kind of frequency-dependent selection is present (the basic technique was introduced in Chapter 13 when we considered

---

For all four mating types, the number of observed matings was close to that expected, and a χ^2 test with one degree of freedom gives a nonsignificant value of 0.84.

Perhaps the mechanism maintaining this polymorphism is habitat selection, in which small-toothed morphs select and have a higher fitness feeding on detritus and algae, and large-toothed morphs select and have a higher fitness feeding on snails. Such a hypothesis can be examined by measuring viability and reproductive success of the two morphs raised on both food resources and by measuring the food preference of the two morphs.

Table 15.5
The observed trophic morphology of 34 breeding pairs of Mexican cichlids and the expected numbers in parentheses assuming random mating (Kornfield et al., 1982).

	Male	
Female	*Small tooth*	*Large tooth*
Small tooth	13(11.8)	10(11.2)
Large tooth	5(5.2)	6(4.8)

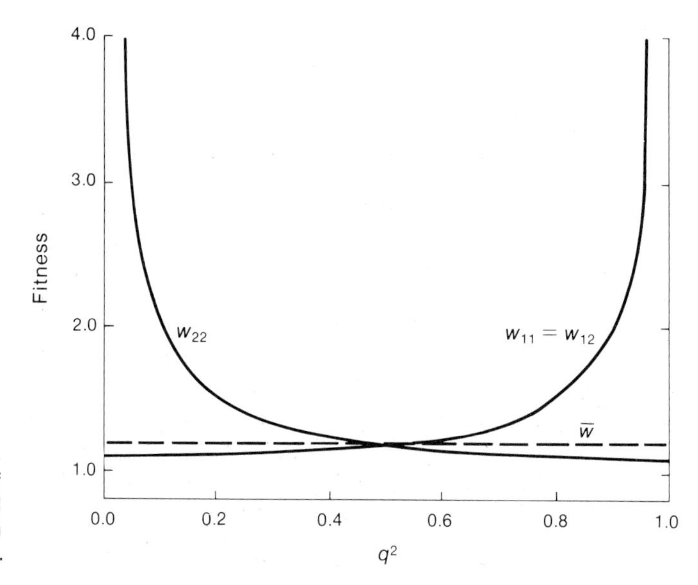

Figure 15.4. The relative fitness values as a function of the frequency of the recessive genotype determined by the frequency-dependent selection model in the text, with $s_1 = s_2 = 0.1$.

Example 15.4. It has been suggested that mating success in males may be a function of the frequencies of different males in a population. For example, in *Drosophila pseudoobscura* a number of studies have indicated that males with an uncommon genotype have an advantage in mating, often called *rare-type male mating advantage*. To examine this phenomenon, an investigator places a number of males of two types in a

Table 15.6
The input and output ratios used in Figure 15.5 for mating success in *D. pseudoobscura*.

	Ratio of males that mated (output ratio)	
Ratio of males in the chamber (input ratio)	*CH/AR* (*Ehrman*, 1967)	*AR/PP* (*Spiess*, 1968)
9	1.20	7.30
4	1.29	4.44
2.33	1.50	4.11
1.5	0.70	1.87
1	0.96	1.32
0.67	1.21	1.30
0.43	0.70	0.63
0.25	0.42	0.47
0.11	0.16	0.24

interspecific competition). If selection is not frequency dependent, then the output ratio should be similar to the input ratio for all initial frequencies.

There is some evidence that the fitnesses of plants are affected by the genotypes of the plants that surround them (Allard and Adams, 1969). Anthropomorphically, some plants are so-called good neighbors—that is, they enhance the relative fitness of other plants relative to the effect on themselves, whereas others are relatively bad neighbors and have a detrimental effect on other plants. A schematic representation of this effect is given in Table 15.7. In this case, plants of type 2 are good neighbors—that is, they increase the fitness of type 1 as compared with the situation in which type 1 is surrounded by itself. On the other hand, type 1 plants are bad neighbors, lowering the fitness of type 2 as compared with the situation in which type 2 is surrounded by itself.

Frequency-dependent selection may occur as the result of differential predation on the polymorphic forms of a species. For example, a predator may focus on a common morph (or morphs) to the relative exclusion of rare morphs. This type of focusing, which results from relying on a search image to

---

container and records their mating success. The proportions of the two types are generally varied over a range—say, from 10 to 90 percent—while the total number of males is kept constant. Data from two such experiments using different inversion types in *D. pseudoobscura* (see Chapter 1) are given in Table 15.6. Notice that, for the first experiment, the output ratio is much lower than the input ratio when the input ratio is high and vice versa. This is in contrast to the expectation of equal input and output ratios for no frequency dependence. If these data are plotted on a log–log scale, then the ratios are as given in Figure 15.5. The straight lines that indicate the overall effect of the experiments cross the no-frequency-dependence line at input ratios of about 0.667 and 7, respectively, as indicated by the arrows. Therefore, it appears that there are stable equilibria for CH and AR at 0.4 and 0.875, respectively, for these two examples (Ayala, 1972).

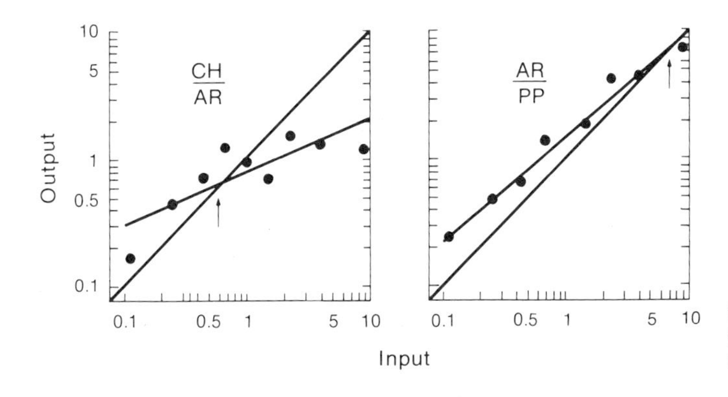

Figure 15.5. The input and output ratios of males of different inversion types (see Table 15.6).

Table 15.7
An illustration of a good neighbor effect (a) and a bad neighbor effect (b). Fitness is measured on the genotype in the center of the matrix.

(a) Type 2 is a good neighbor						
	Higher				*Lower*	
2	2	2		1	1	1
2	*1* ↑	2	versus	1	*1* ↓	1
2	2	2		1	1	1
(b) Type 1 is a bad neighbor						
	Lower				*Higher*	
1	1	1		2	2	2
1	*2* ↓	1	versus	2	*2* ↑	2
1	1	1		2	2	2

locate prey, gives the rarer morphs a higher relative survival. A number of different organisms have been shown to use search images in locating prey, including fish and birds. If the prey population is morphologically diverse, then the overall survival might be higher than in a homogeneous population because the predator is less efficient in identifying prey (see Example 15.5).

Example 15.5 Many predatory birds and mammals form search images on particular types of prey. To determine whether a population with morphological variation is subject to less predation than a homogeneous population, Croze (1970) offered predatory carrion crows artificial prey that were camouflaged mussel shells covering meat rewards. The shells were laid out 12 m from each other, in populations that were either monomorphic red, monomorphic yellow, monomorphic black, or polymorphic with equal proportions of each color. In six experiments with monomorphic shells, the average number of shells surviving (meat not taken) was only 3.5, for a survival proportion of 0.127. For the eleven polymorphic experiments the average number of shells surviving was much higher, 10.5, for a proportion of survival of 0.391. In other words, for a given morph in a monomorphic population, the proportion of survival is only about a third what it would be in a polymorphic population, suggesting that morphological variation in a prey species may reduce predation risk.

III. DENSITY-DEPENDENT SELECTION

Up to this point, we have assumed that differential selection was independent of population density. There is substantial documentation to illustrate that factors such as viability, fecundity, and mating success are affected by population density and that in general fecundity and viability decrease with increasing population size, a result documented in insects, birds, crop plants, and other organisms (see Chapter 12). However, for genetic change to occur as a result of density-dependent selection, there must be selective effects related to density that are specific to different genotypes in the population. One way to envision density-dependent selection is given in Figure 15.6, where the relative fitnesses for the three genotypes at a locus with two alleles are reduced as density increases. In this example, the effect is differential, so that the heterozygote has the highest relative fitness as measured by female offspring per female at high density. The vertical dotted lines indicate the population density (or number) at which populations consisting of only one genotype would stabilize in this environment. This density is the carrying capacity, K, of this environment, as discussed in Chapter 11. Example 15.6 gives some data that indicate a density-dependent heterozygous advantage in viability.

A general model of density-dependent selection assumes that the fitness of a genotype is some function of N such that it decreases as N increases. If this function is the logistic term, then the fitness of genotype A_1A_1 can be given as

$$w_{11} = 1 + r_{11}\left(\frac{K_{11} - N}{K_{11}}\right)$$

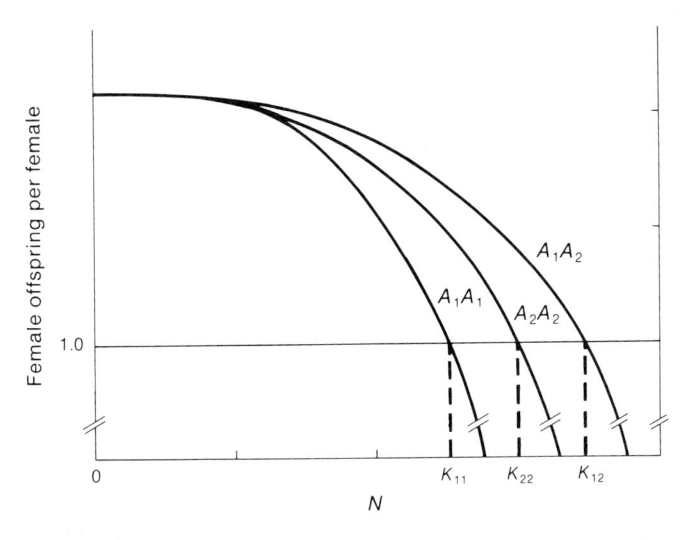

Figure 15.6. A theoretical example illustrating density-dependent selection.

where r_{11} and K_{11} are the intrinsic rate of increase and carrying capacity, respectively, associated with genotype A_1A_1.

For this model, the fitness of a genotype decreases as the density in the population increases. For example, if $r_{11} = 0.1$, $K_{11} = 600$, and $N = 200$, then

$$w_{11} = 1 + 0.1\left(\frac{600 - 200}{600}\right) = 1.067.$$

Example 15.6. A number of marine copepods exhibit polymorphisms of color patterns in natural populations. Battaglia (1958) examined the inheritance and the effects of crowding on a polymorphism in the pigment pattern of the copepod *Tisbe reticulata*, from the brackish waters of the Venice lagoon. In this population, there are three common phenotypes, which result from the genotypes V^vV^v, V^vV^m, and V^mV^m. In one experiment, Battaglia crossed $V^vV^m \times V^vV^m$ and raised the progeny under three different levels of crowding (Table 15.8). Under low crowding, the progeny are produced in close to a 1:2:1 ratio, although there is an excess of heterozygotes. Under medium and high crowding, however, there is a large excess of heterozygotes. If we weight by the expected genotypic proportions, we can obtain the viabilities of the homozygotes relative to that of the heterozygote. For example, the proportion of V^vV^v progeny produced relative to the number of heterozygotes times 2 will give the relative viability of this genotype. Under low crowding, this is (2)904/2,023 = 0.89. For high crowding, the relative viabilities of the homozygotes are only 0.66 and 0.62 that of the heterozygote. This appears to be a good example of density-dependent viability where the heterozygote advantage increases with higher density.

Table 15.8
The number of progeny and estimated viabilities (in parentheses) of a marine copepod raised at three levels of crowding (from Battaglia, 1958).

Culture condition	*N*	V^vV^v	V^vV^m	V^mV^m	q_m
Low crowding	3,839	904 (0.89)	2,023 (1.0)	912 (0.90)	0.501
Medium crowding	1,743	343 (0.68)	1,015 (1.0)	385 (0.76)	0.512
High crowding	1.751	353 (0.66)	1,069 (1.0)	329 (0.62)	0.493

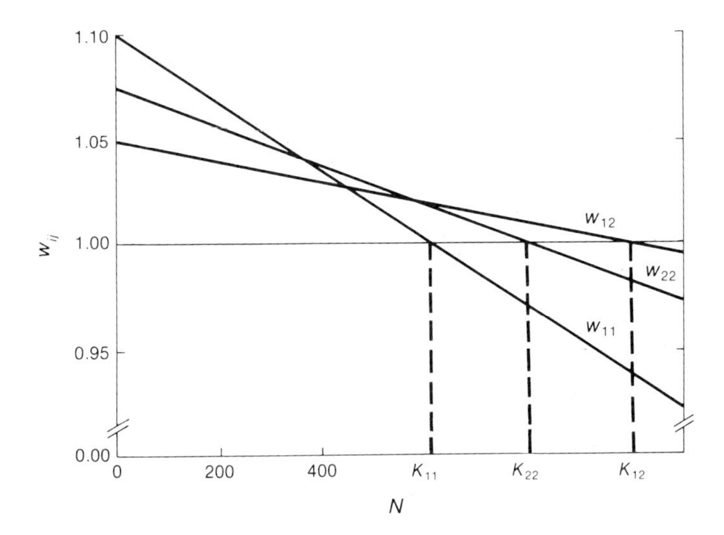

Figure 15.7. Density-dependent selection with $r_{11} = 0.1$, $r_{12} = 0.05$, $r_{22} = 0.75$, $K_{11} = 600$, $K_{12} = 1{,}000$, and $K_{22} = 800$.

If $N = 600$, then $w_{11} = 1.0$. An example of fitnesses of the three genotypes determined by the above expression is given in Figure 15.7. In this case, there is a negative correlation between r and K values so that, for example, in this environment the heterozygote has the highest carrying capacity and the lowest intrinsic rate of increase.

As with all the other ecological models of selection, density-dependent selection may occur only periodically. For example, if density-independent factors are important in a population, then in most years the density may be so low that only small selective differences exist among the genotypes. However, infrequently the density may increase to a high enough level that there is strong selection and a resultant change in allelic frequency. The consequences of the selective episode may remain for many future generations if the density returns to a low, nonselective level (see also Example 16.2).

SUMMARY

The genetic constitution of a population may be affected by a number of ecological factors. Variation in physical or biotic factors, such as temperature or food resources, may result in a stable polymorphism if there is marginal heterozygote advantage. The presence of limited migration or habitat selection increases the likelihood of such a polymorphism. In addition, variation in fitness values that are dependent on the genotypic constitution of a population—frequency-dependent selection—or on the numbers in a population—density-dependent selection—may result in genetic polymorphism.

PROBLEMS

1. How can variation in the environment—say, in temperature—result in a polymorphism? Give examples of fitness values that would result in a stable polymorphism.
2. Assume that a plant species shows genetic differentiation in adjacent areas. Design an experiment that would distinguish between the possible causes of this variation.
3. Using the model for frequency-dependent selection, with $s_1 = 0.3$ and $s_2 = 0.1$, draw a graph of the relative fitness for different allelic frequencies. Determine the equilibrium frequency.
4. Using the density-dependent selection model given in the text, $r_{11} = r_{12} = r_{22} = 0.1$, and with $K_{11} = 1000$, $K_{12} = 800$, $K_{22} = 600$, draw a graph of the fitness values for different densities. Assuming density-dependent factors were most important in this population, what would you expect the eventual frequency of A_2 to become?
5. Assume that a frog species has both brown and green morphs in a population. Design some experiments that would determine whether frequency-dependent or density-dependent selection was important in this case.

CHAPTER 16

Adaptation and Speciation

> Things cannot be other than they are.... Everything is made for the best purpose. Our noses were made to carry spectacles, so we have spectacles. Legs were clearly intended for breeches, and we wear them.
>
> FRANÇOIS VOLTAIRE

Many traits of evolutionary significance are determined by a number of genes—that is, they are polygenic traits (see Chapter 8). Phenotypic changes that result in the increased survival or fecundity of an organism in a particular environment may be due to single genes (see Chapter 15), but often such changes are affected by many genes. As we will discuss below, the process of advantageous, genetically based phenotypic change is called adaptation. In many cases, speciation appears to be an extension of adaptation—that is, two populations become so phenotypically different that they no longer interbreed. However, because the process of speciation has rarely been observed, controversy exists regarding what factors are important in speciation.

For the most part, evolutionary biology is an empirical science, which means that the observation of certain general patterns has led to hypotheses that explain these patterns in a logical and concise manner. Experimental evaluation of these hypotheses is often impossible, and in many cases it is difficult to evaluate objectively the relative importance of particular factors. In this chapter, we introduce some aspects of two processes in evolution, adaptation and speciation, and examine the potential for a genetic understanding of each.

I. ADAPTATION

As discussed in Chapter 9, the environment may vary in a number of abiotic or biotic factors. One may consider these factors challenges or problems the organism attempts to "meet" or "solve" by adaptation. In another way of thinking, *adaptation* is the cumulative genetic response in a population resulting from selective pressures posed by the environment. Adaptation may result in greater survival or fecundity rates, thereby increasing the number of individuals in the population. As a result, an adaptive genetic change may result in an increase in the density or distribution of an organism (see Example 16.3). Adaptation may take a number of forms—for example, such morphological attributes as camouflage or spines that give protection from predators, biochemical characteristics that allow survival in extreme conditions, and behavioral traits that are essential for successful foraging.

However, environmental characteristics may not remain constant over time and in fact may be continually changing. As a result, in order to remain adapted, an organism must continually respond to these changes. (The situation has been likened to that of the Red Queen in *Alice in Wonderland*, who had to continue running in order to stay in the same place). As a result, an organism may "track" the environment, changing morphologically, physiologically, or behaviorally in response to environmental changes (Figure 16.1). In the theoretical case depicted, the combination of temperature and humidity has changed over time (solid lines), with both environmental factors increasing. Here the environment is only depicted at two different times, but it is assumed that the change occurred gradually. The niche that the organism can occupy (broken lines) has also changed with the organism's adapting to the environmental change, allowing the species to continue to survive in the changing environment. Presumably, if a species cannot adapt to new conditions in a given area, then it will become extinct there.

In natural populations, selection is often divided into three different types: directional selection, in which individuals of one extreme are favored (similar to many artificial-selection experiments); stabilizing selection, in which an intermediate phenotype is favored over extreme phenotypes; and disruptive selection, in which both extreme phenotypes have an advantage compared with the intermediate types. A simple schematic representation of

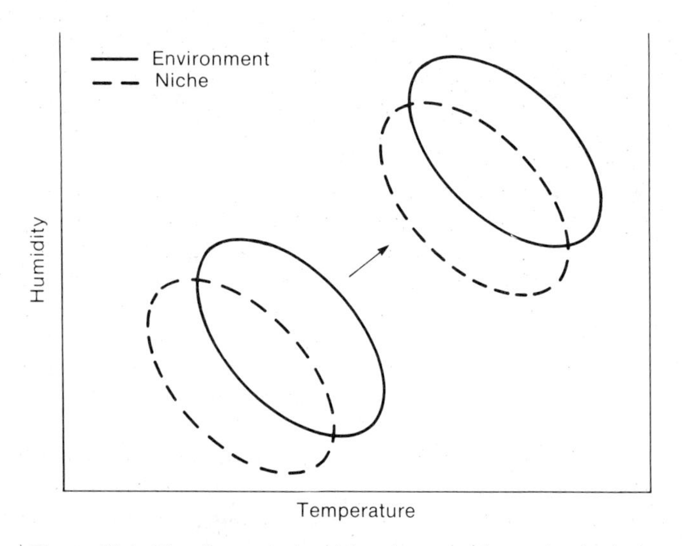

Figure 16.1. The change in the niche of an organism as it adaptively tracks a change in the environment.

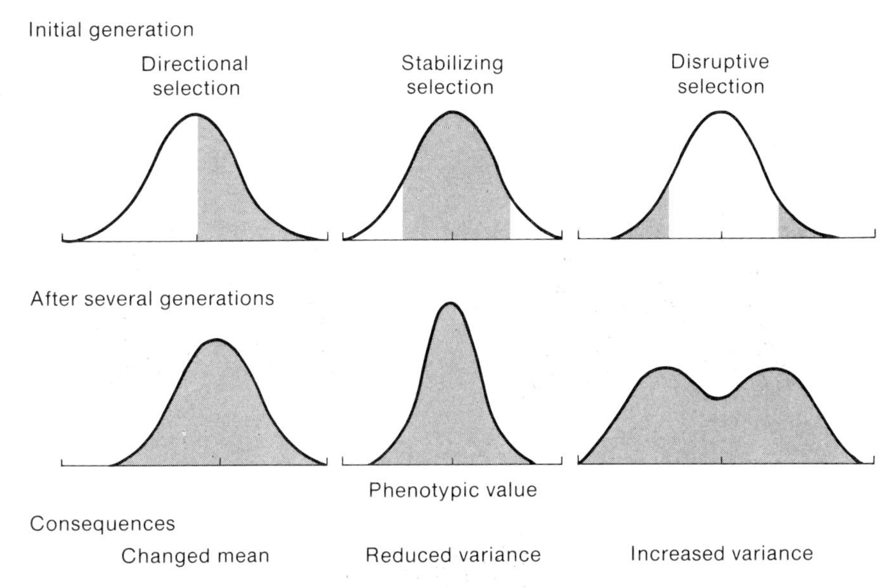

Figure 16.2. Directional, stabilizing, and disruptive selection; the shaded region is the portion of each distribution selected and the lower distributions indicate the expected change after several generations of selection.

these types of selection is given in the upper portion of Figure 16.2, in which some phenotypes have no reproduction and other phenotypes are favorably selected (indicated by shading). Of course, in natural populations, precise separation into two classes does not generally occur, but phenotypes of a particular class have a higher likelihood of survival and reproduction than others.

The expected consequences of the three modes of selection differ sharply. The changes are generally stated in terms of the mean and the variance of the phenotypic value, although the distribution of the phenotype may change in ways not described by these statistics. For these three types of selection, the mean changes for directional selection alone, remaining approximately the same for stabilizing and disruptive selection. The variance remains approximately the same, decreases, and increases for directional, stabilizing, and disruptive selection, respectively (see the lower portion of Figure 16.2).

Before we discuss some examples of adaptive change, let us briefly consider some ways in which organisms may respond more immediately to environmental challenges. More immediate responses may be categorized into three types: regulatory, acclimatory, and developmental (see Table 16.1). All these responses are themselves under genetic control, in the sense that genetic factors determine the potential for regulatory, acclimatory, or developmental responses. Regulatory responses are generally immediate physiological or behavioral changes, such as an increase in adrenaline or flight from a predator. Acclimatory responses take place over a longer period of time and may often involve some morphological change such as changes in fur thickness in cold weather. Both regulatory and acclimatory responses are temporary and are

Table 16.1
Responses of organisms to changes in their environment.

Type of response	*Time involved*	*Examples*
Regulatory	Short	Sweating, shivering
Acclimatory	Intermediate	Tanning, thickening of fur
Developmental	Longer, but within lifetime of individual	Changes in leaf morphology, changes in insect color
Evolutionary	Long, many generations	Development of melanism in moths, development of metal tolerance in plants

reversible in relatively short and somewhat longer time spans, respectively. Developmental responses to an environment are more permanent and generally irreversible (see Example 16.1). For example, leaf morphology and cuticle thickness in some plants depend upon an aquatic or terrestrial location, the development of a protective shield in some *Daphnia* results as a consequence of the presence of predators, and the occurrence of winged versus wingless forms of some insects depends upon temperature. A general term for phenotypic variation caused by environmental factors is *phenotypic plasticity*.

The phenotypic attributes of a population (or a species) generally appear to benefit individuals of the population in coping with their environment. As a result, natural selection appears to be the major factor affecting the phenotypic constitution of a population. In some instances, other factors such as mutation, migration, genetic drift, and the mating system may play major roles in determining the genetic constitution of a population, as discussed in Part 1. In a number of cases, such as pesticide resistance in insects or drug resistance in bacteria, adaptation has been thoroughly documented by a combination of experimental and empirical studies. In many other cases, similar adaptation is logical, but a direct demonstration of a selective change is difficult to document. This, of course, does not imply that an adaptive explanation is incorrect, but rather that other factors may play a role in determining genetic change or the genetic constitution of a population. For example, the genetic constitution of a population may change either because of pleiotropy or genetic hitchhiking, thereby resulting in a potentially nonadaptive evolutionary response (see Chapter 19 for a discussion; see also Lewontin, 1978; and Hedrick, 1982).

The consolidation of selection, migration, mutation, and genetic drift into a coherent theory to explain evolutionary change is by nature difficult because the relative roles of these factors, particularly over long periods, are unknown. Selection, migration, and mutation generally have directional effects on the genetic constitution of a population, but in all these factors, and in the mating system as well, random variation may occur. Variation in these factors may be either insignificant or of major importance, depending upon the magnitude of

Example 16.1. For most vertebrates, sex determination is controlled genetically. However, recently it has been discovered that developmental responses determine sex in many reptiles. In particular, the temperature during embryogenesis influences the sex ratio in a specific manner, depending upon the organism. For example, higher temperatures produce females in green turtles and males in alligators, and low temperatures have the opposite effect (Mooreale et al., 1981; Ferguson and Joanen, 1982). Obviously, when reptiles, and particularly endangered species such as the green turtle, are raised in captivity, it is quite important to maintain an intermediate temperature to ensure that both sexes are produced in substantial proportions. The reliance on temperature to determine the sex ratio is puzzling in many respects from an adaptive viewpoint, because it is assumed that a nearly equal number of both sexes is optimum for continued breeding success. In alligators, however, eggs incubated at higher temperatures are larger hatchlings and become female. Female hatchlings occupy marsh areas, while male hatchlings occupy more open-water habitats, suggesting that this early size differential may have some adaptive significance.

In the Atlantic silverside, a small marine fish, the proportion of male progeny is higher with warmer temperatures (Conover and Kynard, 1981). However, this proportion is highly dependent upon the female parent, suggesting the involvement of genetic factors in determining the sex of progeny (Table 16.2). Some female parents produce similar proportions of male progeny at both high and low temperatures (A, B, and D in Table 16.2), while other females produce nearly all males in warm temperatures and more similar proportions of males and females in cold temperatures (C, E, and F). Conover and Kynard suggest that this largely developmental response to temperature differences is related in this species to the adaptive significance of sexual dimorphism in size.

Table 16.2
The proportion of males observed in progeny raised from six female Atlantic silversides under two temperature regimes (after Conover and Kynard, 1981).

	Water Temperature	
Female	*Warm*	*Cold*
A	0.52	0.63
B	0.0	0.04
C	0.02	0.34
D	0.50	0.52
E	0.0	0.54
F	0.01	0.32

the variation. For example, selection may be a strong factor affecting genetic variation in one generation, in which there is an extreme environmental fluctuation, and of much less significance in other generations (see Example 16.2). Furthermore, unique events, such as periodic intense selection pressures, the swamping of a population by an excess of immigrants, or the near extinction of a population resulting in extreme genetic drift, may drastically affect the genetic constitution of a population, overriding the importance of either directional or random processes.

Populations that are separated from each other for an extended period of time would be expected to accumulate genetic differences resulting from selection, genetic drift, and mutation. Such genetic changes may result in directional selection, that is, local adaptation of different populations when selection is the major factor. This adaptive process produces particular *ecotypes*—that is, population phenotypes that are the result of adaptation to

---

Example 16.2. Periodic changes in weather may result in extreme ecological conditions and, consequently, in intense selection. One such example was documented in a population of finches from an islet in the Galápagos Islands. More than 1,500 of these birds were individually banded and measured for a number of morphological characters (Boag and Grant, 1981). In 1977, an extreme drought occurred—only 2.4 cm of rain fell—and the population declined by approximately 85 percent, with the smaller females suffering even higher mortality. During the drought, the small seeds ordinarily eaten by this species declined sharply in abundance and only larger and hardier seeds were available. The morphological characteristics of the survivors differed from the predrought individuals in their greater body and beak sizes (Table 16.3). These directional phenotypic changes are consistent with the observed directional change in seed resources—that is, a larger beak size as an adaptive change to eat larger seeds. Estimates of the heritability of these traits are quite high, and the phenotypic changes probably therefore reflect a substantial genetic change in the population, one that may have long-term effects.

Table 16.3
The mean body weight and bill depth before and after a drought in a population of ground finches (from Boag and Grant, 1981).

	Before	*After*
Body weight (g)	15.79	16.85
Bill depth (mm)	9.42	9.96

---

local ecological factors. There are many examples of such adaptations. Particularly impressive are the adaptations in plants to unusual environmental factors such as peculiar soil types or the presence of heavy metals in the soil (see Example 16.3). As a result of such adaptive changes, the distribution of an organism may expand into areas that the organism may have been unable to occupy previously.

Species that range over large areas may show geographic variation in phenotypic characters. The geographic trends of some traits appear to be consistent with changes predictable from physiological considerations. For example, many warm-blooded vertebrates tend to be larger in colder areas (a tendency called Bergman's rule) and to have longer limbs relative to body size in warmer areas (called Allen's rule). Bergman's rule is consistent with the physiological explanation that the body core of a large animal conserves heat more efficiently than that of a small animal (the ratio of the volume to surface area is greater than that in a smaller animal), and Allen's rule is consistent with the observation that thermoregulation at high temperatures is more efficient in animals with relatively long limbs.

Geographic trends in body size or other phenotypic characteristics may be due in part to developmental responses. In order to determine critically how much of a geographic trend is genetic, one must raise individuals from populations that differ in a phenotypic value in a common environment(s) (see Chapter 8). For some species, however, transplant or common-garden experiments may be difficult to carry out, making the extent of the genetic basis of the geographic variation difficult to ascertain (see the latter part of Example 16.3).

One of the best established and best understood examples of adaptation is mimicry. Mimicry can be divided into three basic categories: Batesian, Müllerian, and cryptic (Wickler, 1970). In *Batesian mimicry*, one species, the model, is repellent to predators because of its taste, sting, or bite, and another species, the mimic, is similar to the model in morphology and sometimes behavior and is consequently also avoided by predators (see Example 16.4). In *Müllerian mimicry*, all the species involved are noxious and resemble each other (all are models), often forming what is termed a mimicry complex—that is, a group of morphologically similar species. As a result, predators may avoid all of the species of such a group after encountering only one species. In most cases of Batesian and Müllerian mimicry, both the models and the mimic are generally aposematically colored—that is, brightly colored in a distinctive pattern that enhances recognition by predators. In *cryptic mimicry*, the mimic resembles a background object such as a stick, pebble, leaf, or tree branch. In general, mimicry is an example of directional selection, although in some cases disruptive selection appears to have occurred, in which one morph in the species is part of a mimicry complex and is brightly colored, while another morph is cryptically colored.

Although adaptation due to directional selection is the most widely cited type of selection in natural populations, stabilizing selection is probably of equal importance in many cases. Stabilizing selection is static, in that the phenotypic mean does not change, but it may result in genetic change that

Example 16.3. The mining of heavy metals, such as lead, nickel, zinc, and copper, has contaminated the surface of the earth with soil containing large amounts of these metals. Generally, this soil is toxic to plants, but various species, particularly grasses, are gradually able to colonize these habitats. Populations of the bentgrass, *Agrostis tenuis*, have developed tolerance to lead, nickel, zinc, and copper and have been examined in detail in England (Bradshaw et al., 1965).

Often, the boundary of metal-tolerant plant populations is quite sharp, with plants just off the mine areas lacking tolerance to the heavy metal. An example is given in Figure 16.3 for *Agrostis tenuis* on and near a lead mine in Wales. These samples were only 7 m apart, with the mine boundary between them. The mean lead tolerance (as measured by an index of root growth in lead-contaminated soil) is approximately 75 in the population from the mine and only 35 just outside the mine boundary. Frequently, such tolerant populations also differ from nearby populations in growth form, flowering time, proportion of self-fertilization, and other traits, which may themselves be adaptive in that they are important in the maintenance of the metal-tolerance characteristics.

Species that become established in a new geographic area similar to their ancestral home may expand their ranges very quickly. For example,

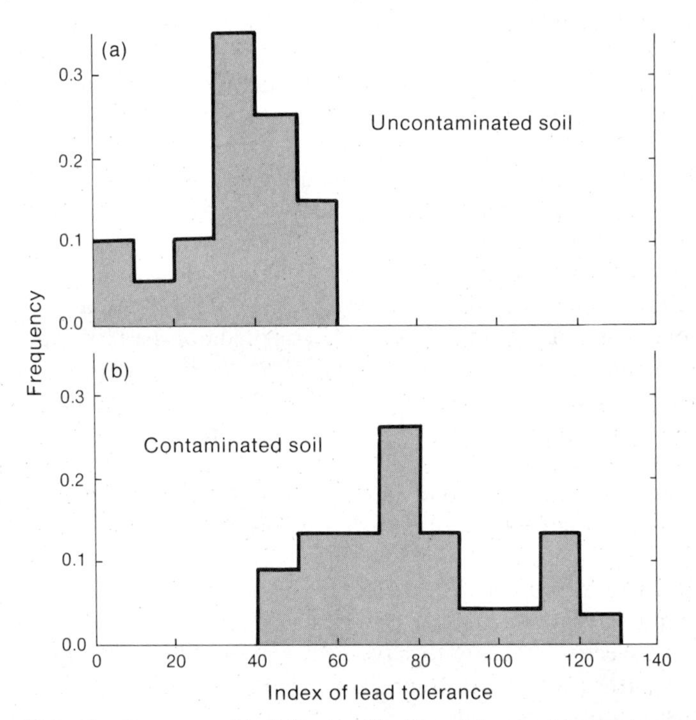

Figure 16.3. The frequency of individuals with different levels of lead tolerance in an *Agrostis tenuis* population sampled from contaminated and uncontaminated soil (after Bradshaw et al., 1965).

starlings (see Chapter 9), house sparrows, and pigeons spread throughout North America within a century after their introduction from Europe. In addition to its rapid spread after its introduction in the period 1852–1860, the house sparrow already exhibits geographic variation in a number of phenotypic traits in North America. Figure 16.4 gives the distribution of a measure of general size in house sparrow males (after Johnston and Selander, 1971). There are significant geographic trends in size and these trends are generally associated with winter temperatures—that is, the largest sparrows are in areas with coldest winters and the smallest are in the areas with the warmest winters. These trends are consistent with Bergman's rule, and in fact the geographic variation is nearly as large as that found in Europe. Because young birds are full size by the end of their first summer, most of these trends in phenotypic variation appear to be due to genetic differences rather than developmental responses.

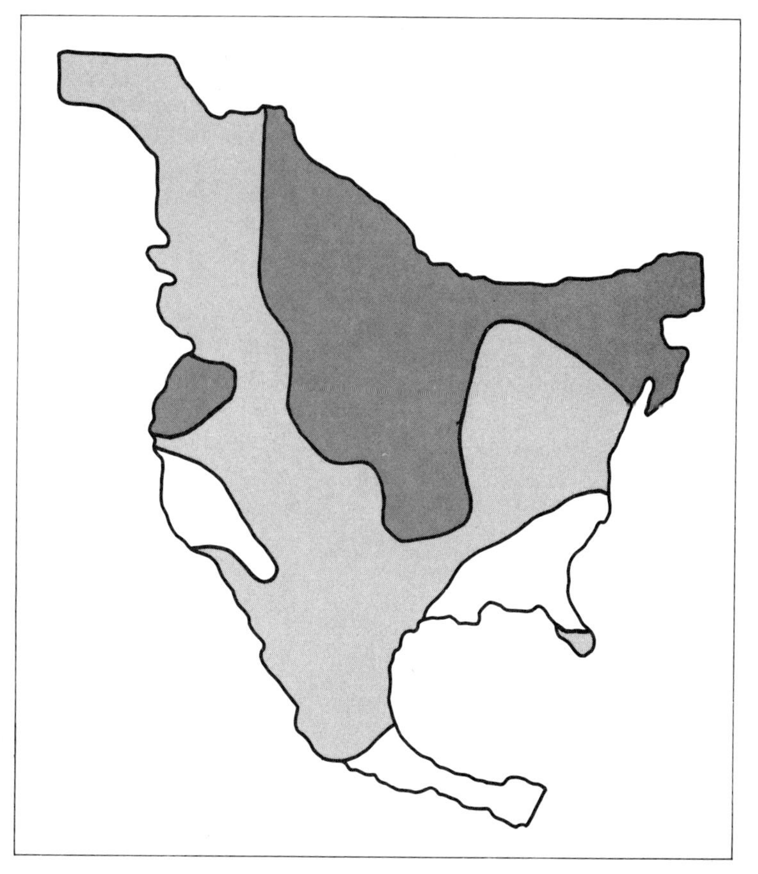

Figure 16.4. Geographic variation in a measure of body size for male house sparrows (after Johnston and Selander, 1971). The darkly shaded areas contain the largest birds, and areas with no shading contain the smallest birds.

leads in turn to a particular array of genotypes yielding the least phenotypic variability around the phenotypic optimum. Often, the combination of genetic factors and biotic and abiotic environmental considerations results in selection favoring an intermediate optimum (see Example 16.5). In a related discussion in Chapter 19, we cover some of the genetic constraints that are important in determining the demographic aspects of a population.

Example 16.4. One of the most thoroughly examined examples of mimicry is that involving the monarch butterfly of North America (Brower, 1969). In this case, the monarch is a model and is generally quite noxious to most avian predators, exemplified by the blue jays that have been used extensively as a potential predator. Naive blue jays that eat an unpalatable monarch quickly become ill and regurgitate the butterfly. This reaction is caused by the presence of compounds in the butterfly, called cardiac glycosides, that both disrupt the heartbeat and influence the area of the brain that controls vomiting. Cardiac glycosides are acquired by monarch larvae feeding on certain species of milkweed that contain the toxins in their leaves. Figure 16.5 illustrates the unpalatable monarch

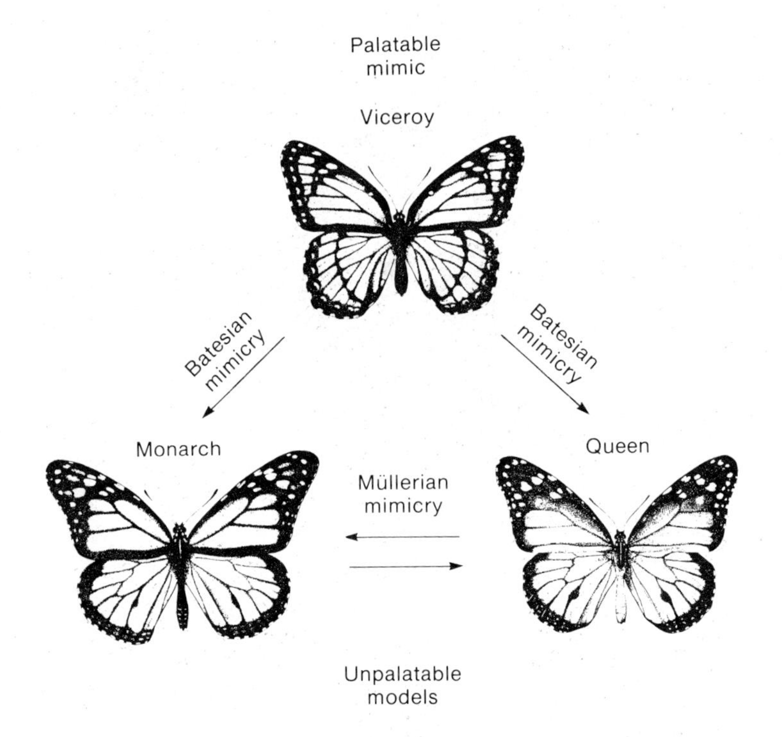

Figure 16.5. Three species of mimetic butterflies (from Brower, 1969).

When the environment is heterogeneous, as, for example, when there are soils of several types, then the different parts of the population living on the distinct patches may become differentially adapted to them. Over time, particularly when migration between patches is low or habitat selection is high, the parts of the population may diverge phenotypically (for a discussion in relation to single-gene models, see Chapter 15). Such disruptive selection for

(lower left) and two morphologically similar butterflies, the palatable viceroy and the unpalatable queen butterfly. The viceroy is a Batesian mimic of the monarch in much of its range and a Batesian mimic of the queen in Florida. Because the monarch annually migrates south, it coexists with the more tropical queen, making these two unpalatable species Müllerian mimics.

If a palatable mimic is very abundant relative to its model, then its survival should be lower than when it is rare relative to its model. Such frequency-dependent selection should occur because predators must learn (and be reinforced in) what are noxious prey, a process that should occur most quickly when most individuals are noxious models. To examine this prediction, Brower et al. (1967) released prometheus moths that had either been painted black (making them cryptic, because black is the general color of their resting background) or painted aposematically to mimic the noxious species *Parides*, which is red, black, and green. When the two types of painted moths were released at a high density (all at one locality), the cryptic moths had higher recapture and relative survival rates than the aposematic moths (see Table 16.4). On the other hand, when the moths were released at low density (at different times in many localities), the aposematically colored moths had a higher relative survival. These results are consistent with the hypothesis, given above, that the survival of a mimic is a function of its abundance relative to its model.

Table 16.4
The recapture and relative survival rates for prometheus moths that were painted either cryptically or aposematically and were released at two densities (after Brower et al., 1967).

	Cryptic		*Aposematic*	
Density	*Recaptured/ released*	*Relative survival*	*Recaptured/ released*	*Relative survival*
High	129/414	1.0	95/414	0.74
Low	37/194	0.70	53/199	1.0

different forms is thought by many to be a potential first step for the development of new species (see Example 16.9).

In addition, selection for different traits in males and females leading to sexual dimorphism can be considered disruptive selection. In this case, the two forms are part of a random-mating population and have basically the same genetic makeup, but they have different characteristics due to the hormonal impact on development and behavior. Sexual dimorphism is generally thought to be the result of *sexual selection*—that is, selection that occurs as the result of intraspecific competition for mates. The main forces involved in sexual selection are male-to-male competition for mating opportunities and female choice of males. The consequence of these selective factors are male

---

Example 16.5. Stabilizing selection favors intermediate phenotypes over extreme phenotypes. For example, for most bird species, the number of eggs in a clutch appears to be strongly selected for a given number. Clutch size is presumably determined polygenetically, although there are specific countable categories—for example, clutches with two, three, or four eggs. Clutch size above and below this optimum number produces fewer fledglings—for the smaller clutches because the potential number of fledglings is small, and for the larger clutches because fewer young survive, presumably because the parents cannot adequately feed and care for so many young.

The swift normally has clutches of two or three eggs, and in most years the survival of young to fledging is high. In four different years in Oxford, England, eggs were transferred to some three-egg clutches to make four-egg clutches, and the survival of these young was monitored (from Lack, 1966). Table 16.5 gives these results, illustrating that the individual survival in the four-egg clutches is low, 0.44, and that the average number of fledglings produced from these clutches, 1.76, is smaller than that for three-egg clutches. From these data, it appears that the optimum clutch size is three and that selection acts against clutches of other sizes.

Table 16.5
The individual survival and number of fledglings per clutch over a four-year period for swifts (after Lack, 1966).

Clutch size	*Individual survival*	*No. fledglings/ clutch*
2	0.96	1.92
3	0.85	2.54
4	0.44	1.76

---

phenotypes such as the bizarre plumages of the male birds of paradise and the incredible structures built by male bowerbirds (see Example 18.5).

II. SPECIATION

Different populations of a species may become adapted to particular environments, but, given the opportunity, they may interbreed and produce a hybrid population. However, it is possible for populations to diverge to the point of being unable any longer to interbreed, thus actually becoming separate species. In many cases, this type of species formation is related to, or is an extension of, the adaptation process discussed above.

Let us begin our discussion of the process of speciation with a definition of the term species. A simple and generally useful definition of a *species* is a group of actually or potentially interbreeding populations. In contrast to a population as defined earlier, a species does not necessarily comprise individuals existing together in time and space. For example, the starling populations now existing in Europe, those extant in the nineteenth century, and those found today in North America all belong to the same species.

Related species usually share many phenotypic characteristics but differ in some aspects that distinguish them as species. Traditionally, species have been identified on morphological grounds, with a "type specimen" illustrating the phenotypic characteristics typical of a particular species. Presumably, such morphological characteristics are in large part genetically based and are indicative of important biological differences between species. For example, behavioral differences in mating or habitat preferences are often correlated with morphological differences between species. An exception is found in many sibling species—different species that are almost identical morphologically but that can exhibit quite different behavioral patterns or habitat preferences.

What types of biological changes are necessary for the establishment of a new species? Do the changes that result in a new species occur quickly and result from a few genes? Although simple answers are tempting, such questions are difficult to answer explicitly. In fact, it is doubtful that simple or single answers to these questions exist at all, and more likely that a new species becomes established via many avenues. Furthermore, the factors important in a particular situation depend upon the characteristics of the organism, the environment, the genetic variants available, and chance.

Several different approaches to the process of speciation have been used. For example, closely related species have been examined both in terms of historical (climatic or geological) information and of genetic differences between the species. Using these data, we can reconstruct the process of speciation in a logical, albeit hypothetical, manner (see Examples 16.8, 16.9, and 16.10). A more direct approach is to examine populations undergoing speciation (see the first part of Example 16.9), although finding a good situation to examine is sometimes difficult. Last, certain aspects of speciation, such as the development of reproductive isolation, (discussed below) have been mimicked in artificial populations (see Example 16.6).

Using the population genetics context we have considered, one can assume that new species appear and others disappear over a period of time. In certain periods, these processes may be accelerated—for example, the rate of speciation may be increased if there are many relatively uninhabited areas, and the rate of extinction may be increased by a rapid deterioration or change in the environment. It should be noted that changes seen in geological time take place over thousands (or perhaps millions) of generations. Obviously, in view of our understanding of population and quantitative genetics, large phenotypic changes in a species, *macroevolution*, could easily take place over such a time span. Examination of the fossil record suggests that for many species periods of no morphological change may be followed by rapid (in

---

Example 16.6. Several experiments have shown that prezygotic reproductive isolation can be quickly generated. For example, experiments in *Drosophila* have been able to generate ethological isolation between phenotypes in a population in a relatively short time. The basic design used in these experiments is to have two lines that are homozygous for different mutant alleles so that progeny from matings between individuals of different phenotypes may be identified and selected against, resulting in selection for positive assortative mating (Table 16.7). In the notation used in Table 16.7, all such progeny of matings between different types are aa^+bb^+ and phenotypically wild type. In one such experiment, Knight et al. (1956) were able to establish nearly complete ethological isolation between two mutant strains of *D. melanogaster* in only 15 generations.

A similar experiment was carried out in two varieties of maize, yellow sweet and white flint, that had different mutant markers (Paterniani, 1969). In five generations, the two varieties were almost completely isolated from each other. The results are given in Figure 16.6 as the proportion of

Table 16.7
A scheme that allows selection against wild-type progeny and increases positive assortative mating.

	Females	
Males	aab^+b^+	a^+a^+bb
aab^+b^+	aab^+b^+ (mutant)	aa^+bb^+ (wild type)
a^+a^+bb	aa^+bb^+ (wild type)	a^+a^+bb (mutant)

---

geological time) morphological change. This pattern, called a *punctuated equilibrium* and consisting of alternative episodes of morphological stasis and change, is also consistent with our understanding of quantitative and population genetics (see Charlesworth et al., 1982).

Before discussing speciation models, let us consider the means by which species identity is maintained. The separation of species is maintained by biological barriers termed *reproductive isolating mechanisms.* These isolating mechanisms are most easily categorized as *prezygotic isolating mechanisms,* those important prior to the fertilization of gametes from two different species, and *postzygotic isolating mechanisms,* those that act after fertilization. Table 16.6 distinguishes three major types of prezygotic and postzygotic isolating mechanisms, respectively (see Example 16.6).

The first prezygotic isolating mechanism—potential mates do not

---

heterozygotes on ears from the two varieties, a value that decreased from an average of more than 0.4 (0.5 would be the Hardy-Weinberg expectation) to less than 0.05. The increased isolation was found to be the result of a change in the time of maturation of the two varieties, temporal reproductive isolation.

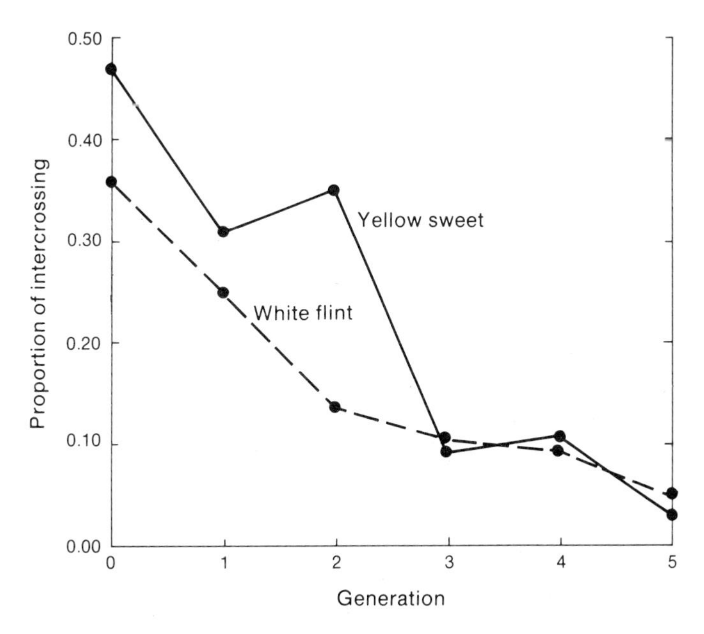

Figure 16.6. The change in proportion of heterozygotes over time in two varieties of maize (after Paterniani, 1969).

Table 16.6
Categorization of reproductive isolating mechanisms.

Prezygotic isolating mechanisms	*Postzygotic isolating mechanisms*
1. Potential mates do not meet (temporal or ecological isolation)	1. Hybrid zygotic mortality or low hybrid viability
2. Mates meet but do not mate (ethological isolation)	2. Hybrid sterility
3. Mating but no fertilization occurs (mechanical isolation or gametic mortality)	3. Hybrid breakdown

meet—may occur when organisms are reproductively mature or active at different times. For example, temporal isolation may occur if two species of plants flower at different times of the year or if two species of fish are active at different times of the day. If the two species exist in different habitats—that is, exhibit ecological isolation—then they may even fail to encounter each other, making interspecific mating impossible. The second prezygotic isolating mechanism—mates meet but do not mate—is more important in animals (or animal-mediated pollination in plants) and becomes more important where elaborate courtship patterns occur. Last, fertilization may fail to take place, either because of mechanical constraints in the genitalia or because of an antigenic response to the sperm or pollen. In this case, fertilization may be prevented or the effectiveness of fertilization greatly reduced.

In some cases, particularly in plant populations, prezygotic isolating mechanisms are poorly developed and, as a consequence, postzygotic factors are often more important in maintaining species identity. The first postzygotic isolating mechanism, hybrid mortality or low viability, presumably occurs because the two parental genomes are somehow genetically incompatible. For example, genetic control of development may differ, causing abnormal morphology in the hybrids. Second, the hybrids may be sterile even though they are as viable as either of the parental types. Hybrid sterility is often caused by such chromosomal differences between the species as translocations or pericentric inversions. Individuals heterozygous for chromosomal differences of these types often produce a high proportion of gametes that have extra or missing chromosomes or chromosomal segments. The last postzygotic isolating mechanism, hybrid breakdown, occurs when the offspring of hybrids have reduced viability or fertility. Such reduced fitness presumably results from segregation and recombination among the genomes of the two species, which produce abnormal development, genetic regulation, or meiosis.

Of course, not all these reproductive isolating mechanisms are usually operating simultaneously between a pair of species, although in most cases more than one type of isolating mechanism is important. Prezygotic isolating mechanisms are obviously more efficient, since gametes are not generally wasted on progeny that will not propagate the species. In fact, hybrid progeny may use resources that otherwise would be utilized by the parental species. As a result, it is generally assumed that species separated by a postzygotic isolating mechanism but in contact will eventually develop some prezygotic isolation.

The traditional classification of modes of speciation is based upon the location of the speciation event in the species distribution. Speciation may occur when the incipient species are geographically separate (*allopatric*), co-existent (*sympatric*), or adjacent (*parapatric*) populations. In some ways this classification is artificial, because, for example, geographically sympatric populations may be separated ecologically, and the classification of sympatric versus parapatric is a function of dispersal ability. Noting these problems, we still find the geographic framework to be useful in examining speciation models. Allopatric speciation is a widely accepted model of speciation and in general is an extension of differential adaptation in different areas. On the other hand, no consensus exists about the relative importance of sympatric and parapatric speciation. In part, this controversy is due to the lack of critical data and in part it is probably due to differences in the relative importance of the modes among different organisms.

In all of these geographic modes, the process of speciation is generally assumed to be gradual, although it may occur in a relatively short time, and based upon divergent selection pressures. It is possible that nonselective forces can reorganize the genome in such a manner that a new species is created instantaneously. For example, individual polyploids between two species are sometimes formed. Such individuals, because of their hybrid qualities, may have a fitness advantage over the parental species and may be reproductively

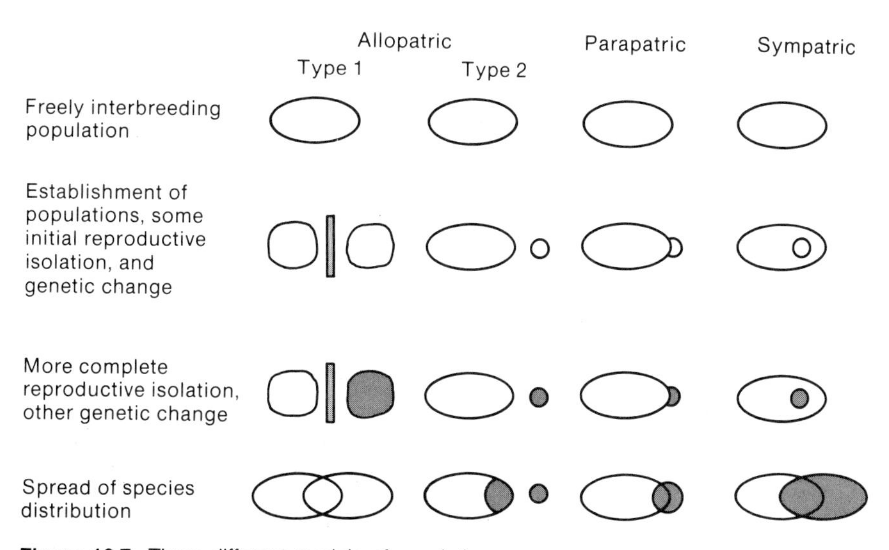

Figure 16.7. Three different models of speciation.

isolated because backcross progeny have an odd number of chromosomal sets. Furthermore, genetic drift may generate new combinations of alleles and thereby result in novel genotypes.

A simple illustration of these three geographic models of speciation is given in Figure 16.7. Initially, for all models it is assumed that the ancestral species exists in a given geographical area as a freely interbreeding population. There are two basic models of allopatric speciation. In type 1, some *geographic barrier*, a physical obstacle such as a glacier or a river, divides the population into two separate parts. In type 2, the populations are separate geographically

---

Example 16.7. The plant genus *Clarkia* appears to have evolved rapidly into a large number of species. The differences among the species generally involve morphological changes, particularly in floral characteristics and large amounts of chromosomal change. For example, two primarily insect-pollinated species, *C. nitens* and *C. speciosa*, which exist in the oak woodlands in central California, differ by floral color (see Table 16.8). *C. nitens*, the northern species, has yellow flowers, and *C. speciosa*, the southern species, has purple flowers (Bloom, 1984). However, the zone of chromosomal transition between the two species (they differ by a minimum of six to seven reciprocal translocations) begins 120 km south of the narrow species border and is itself 50 km wide.

Because the climate in this area has become increasingly xeric in the last 5,000 years, and because *C. speciosa* is a relatively xeric species with other close relatives to the south, it seems reasonably certain that this is a zone of secondary contact, in which *C. speciosa* has invaded and is moving north. The *C. speciosa* morphological characteristics have trasversed the chromosomal barrier and are now 120 km north in the *C. nitens* (or north) chromosomal type. Furthermore, in this region the flower color purple is dominant, while it is recessive further south. This suggests that the introgression of *C. speciosa* characteristics may be aided by a new dominant flower-color allele and that the pollinators have a fidelity to a particular flower color.

Table 16.8
The flower color and chromosomal characteristics of two *Clarkia* species as they vary in their north–south distributions.

Species	*Flower color*	*Chromosomal type*
C. nitens	Yellow	North
C. speciosa	Dominant purple	North
C. speciosa	Recessive purple	Transition
C. speciosa	Recessive purple	South

because a small founder population colonizes a previously uninhabited disjunct area. In the parapatric and sympatric models, the incipient species continues to share either a border or its entire range, respectively, with the ancestral species. During the initial part of the speciation process, it is generally assumed that some genetic change occurs and that reproductive isolation begins to develop between the two groups. Over time, genetic differentiation accumulates, leading by chance to more reproductive isolation. In addition, when the geographic barrier in the allopatric models disappears, and throughout the process for parapatric and sympatric speciation, further reproductive isolation itself should be selectively advantageous owing to the disadvantages in producing hybrid progeny.

When two similar species share a common border on their geographic ranges, then in many cases there is a *hybrid zone*, an area in which there are individuals having characteristics of both parental species. As we discussed earlier, species hybrids may be viable and fertile to some extent. In such a case, it is difficult to distinguish between a hybrid zone that is due to secondary contact—that is, contact of two formerly disjunct differentiated species—and clinal variation, in which the populations on either side of the putative hybrid zone show differential adaptation. When there is historical information (past weather, water, or glacial patterns) or information about other related species, then secondary contact and clinal causation of hybrid zones can be inferred (see Example 16.7).

Hybrid zones due to secondary contact may have important evolutionary significance. For example, there are suggestions that individuals that are species hybrids may have higher chromosomal and individual gene mutation rates. Furthermore, F_2 and backcross progeny may have new combinations of genes not present in either parental species. Although most of these new mutations and gene combinations are probably detrimental, it is possible that some new advantageous type may be generated in hybrid zones. In addition, hybrid zones can potentially allow gene flow of alleles from one species into another, a phenomenon called *introgression*.

a. Allopatric Speciation

As indicated in Figure 16.7, there are two basic models of allopatric speciation. In both, the genetic change involved in the formation of species occurs in basically two steps. First, while the populations are separated by a geographic barrier or by distance, genetic change may take place in one or both of the populations. This genetic change may be the result of differential adaptation to different environments, genetic drift, or incorporation of new mutants in one or the other of the populations.

It has been suggested that a genetic "revolution" may take place during the process of allopatric speciation. In the type 2 model of allopatric speciation, because the population that establishes itself in the new area may be small, genetic drift may play a particularly important role in genetic change.

It is possible to indicate the extent of genetic change by comparing electrophoretic variants from different species. In both some fishes and some *Drosophila* species (Avise, 1976), there are indications that genetic divergence increases for electrophoretic variants in a rather gradual and regular fashion as the relationship between species becomes more distant. However, because the role of electrophoretic variants in speciation is unclear, these loci may not be good indicators of the genetic variants important in the process of speciation (see Example 16.8).

b. Sympatric Speciation

Sympatric speciation assumes that the incipient species are coexisting in the same geographic area. However, most of the genetically based models of sympatric speciation assume that there is some ecological diversity within this area. One genetic model for sympatric speciation is an extension of the model that allows polymorphism due to diversity in the habitat (see Chapter 15 and Example 16.9). Once the population has become subdivided due to selection

---

Example 16.8. Some of the most spectacular examples of species complexes are on islands—for example, the finches on the Galápagos Islands, which were important in Darwin's formulation of natural selection (see Grant, 1981). Recently, species on the Hawaiian Islands have been thoroughly studied, particularly many of the more than 500 species of *Drosophila*. The Hawaiian Islands are the most isolated islands in the world and have been formed sequentially by volcanic action. The oldest islands are in the northwest end of the archipelago, while on the island of Hawaii, the youngest island, there is still volcanic activity.

Two closely related species of picture-winged *Drosophila* from Hawaii that have been examined in detail are *D. silvestris* and *D. heteroneura*. These two species are related to another species on the next youngest island, Molokai, and they are thought to be descended from founder migrants from Molokai, the type of speciation event described in the type 2 allopatric speciation model. *D. silvestris* and *D. heteroneura* are chromosomally homosequential (that is, they have the same chromosomal banding patterns), their larvae utilize the decaying bark of the same tree species, they are genetically very similar for alleles from 25 protein loci, and their courtship patterns exhibit no qualitative differences. However, the morphology of the head is strikingly different for the two species (Figure 16.8), with *D. heteroneura* having an extraordinary mallet-shaped head (Kaneshiro and Val, 1977). It appears that

---

and perhaps to limited migration, other genetic changes such as habitat selection, differential reproductive maturity, and positive assortative mating may occur that could serve to isolate the subpopulations biologically.

c. Parapatric Speciation

Parapatric speciation occurs when the incipient species share a common border. In many ways, parapatric speciation models are a combination of the allopatric and sympatric models, and distinguishing between them appears to be difficult. The theoretical initial stage of parapatric speciation often involves invasion of a new habitat and subsequent adaptation to it. As a result, a change in allelic frequencies may be established over the border between the two populations. Consequently, it is thought that parapatric speciation is most important in species with relatively low mobility—for example, flightless grasshoppers or grasses (see Example 16.10). Subsequent to the establishment of the cline, further genetic change may take place that eventually results in reproductive isolation.

---

this difference is important in the determination by a female of the species of a courting male and is important in the ethological isolation of the two species. The two species will produce fertile F_1, F_2, and backcross progeny in the laboratory, allowing the genetic determination of the head shape to be evaluated. From crosses between the two species, it has been estimated that approximately ten genes are responsible for this morphological difference.

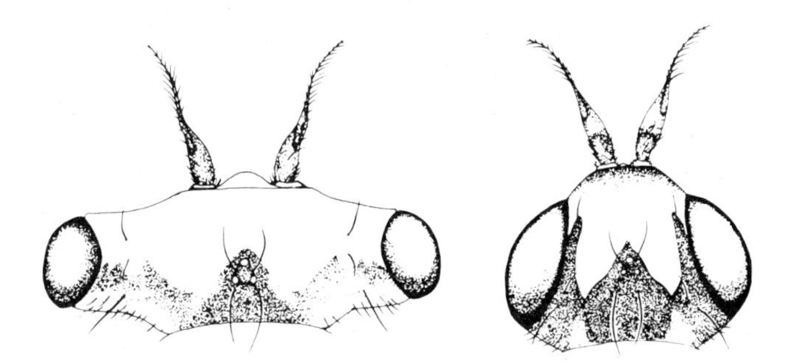

Figure 16.8. The head shape in two closely related species of *Drosophila* (Kaneshiro and Val, 1977).

PROBLEMS

1. The grass plants in Figure 16.3 taken from contaminated and uncontaminated soil had somewhat overlapping distributions of lead tolerance. What might explain this overlap? Design an experiment to examine your hypothesis.
2. The experiment that generated the data in Table 16.4 suggests that some maximum density exists for a mimetic species relative to the density of its model. What qualities of the model, mimic, and predator might be important in determining this optimum?

Example 16.9. Sympatric speciation may occur in monophagous species—that is, species that feed on only one resource species. Because many monophagous species use their hosts as a site both for feeding and mating, the choice of mates is strongly associated with the choice of hosts. Differential host selection can lead to ecological isolation between groups, and postzygotic isolation need not be involved.

Bush (1975) has studied the host races of the hawthorn fly, *Rhagoletis pomonella*, a fly that shifted hosts from the hawthorn tree to apple trees in 1864 and to cherry trees in 1960. These host races are morphologically indistinguishable, have nearly identical frequencies at a number of enzyme loci, and can be readily hybridized, producing fertile F_1, F_2, and backcross progeny in the laboratory. There is some evidence from crosses between the host races of *Rhagoletis* that only one or a few genes are responsible for host-selection and host-survival differences (however, see Example 18.3).

Let us discuss a simple scenario that could account for the development of the cherry host race. First, the ancestral race is homozygous for the host-recognition alleles for hawthorn, genotype A_HA_H, and homozygous for host-survival alleles for hawthorn, genotype B_HB_H (see Table 16.9a). The change of hosts begins with the establishment of a new host-recognition allele for cherries, A_C (the transition stage in Table 16.9a). Because the species is not adapted to survive on cherries, alleles that increase survival, such as B_C, would have a selective advantage on the new host. Therefore, the cherry host race may differ only by alleles at two loci from the ancestral host race.

Another possible case of sympatric speciation is for two sibling species, *Chrysopa carnea* and *C. downesi*, of lacewings (Tauber and Tauber, 1978). These two species hybridize and produce fully fertile and viable F_1 and F_2 offspring. However, they differ in morphology, with *C. carnea* being light green (reddish-brown in the winter) and *C. downesi* dark green. In addition, *C. downesi* is univoltine (one generation a year), in early spring, and is restricted to conifers, while *C. carnea* is multivoltine (several generations in the summer) and occurs mainly in grassy areas and meadows.

---

3. In the process of speciation, is there a general temporal order to chromosomal, electrophoretic, and morphological change? Is there a general temporal order to prezygotic and postzygotic isolating mechanisms? Support your answers with examples and a logical argument.
4. What is a hybrid zone? How would you determine experimentally whether a hybrid zone is due to secondary contact?
5. Suggest organisms in which the three basic types of speciation models might be important. What are the characteristics of these organisms that make them likely candidates for a particular mode of speciation?

From matings between the two species, it appears that allelic difference at one locus determines the color patterns, and that two loci are involved in the change in reproduction. Because *C. downesi* has a restricted habitat range and geographic distribution compared with *C. carnea*, it appears to be descended from *C. carnea*, or from an ancestral form of *C. carnea*. Therefore, a possible sequence of steps leading to *C. downesi* from the ancestral *gg* (light green) and $D_1D_1D_2D_2$ (multivoltine) form are given in Table 16.9b. First, a transition form with a dark-green phenotype (*GG*) began living in a conifer habitat; the color was protective from predation. Second, the reproduction changed to univoltine and early spring breeding ($d_1d_1d_2d_2$) to coincide more closely with resource availability in the conifers. As a result of the changes at these three loci, the species are separated by habitat and by temporal differences in reproduction, a divergence that may well have occurred sympatrically.

Table 16.9
The possible sequence of events leading to sympatric speciation through (a) host races in *Rhagoletis* and (b) reproductive change in *Chrysopa*.

(a) *Host race*	*Genotype*	*Host recognition*	*Host survival*
Hawthorn	$A_HA_HB_HB_H$	Hawthorn	Hawthorn
Transition	$A_CA_CB_HB_H$	Cherry	Low on cherry
Cherry	$A_CA_CB_CB_C$	Cherry	Cherry

(b) *Species*	*Genotype*	*Color*	*Reproduction*
C. carnea	$ggD_1D_1D_2D_2$	Light green	Multivoltine
Transition	$GGD_1D_1D_2D_2$	Dark green	Multivoltine
C. downesi	$GGd_1d_1d_2d_2$	Dark green	Univoltine

Example 16.10. Large numbers of cases of parapatric incipient species exist, but, as we mentioned in our discussion of hybrid zones, it is difficult to determine whether these cases are due to clinal variation or secondary contact. Often, adjacent species differ in some chromosomal rearrangements, such as translocations or inversions. Chromosomal differences may result in reduced hybrid fertility, a type of postzygotic reproductive isolation. It is not clear whether chromosomal differences are merely associated with speciation or whether they are a major cause of speciation.

Related rodent species often differ by a number of chromosomal rearrangements. For example, four chromosomal forms of the mole rat, *Spalax ehrenbergi*, are parapatrically distributed in Israel (see Figure 16.9; after Nevo et al., 1975). These incipient species (they may actually be sibling species) have different chromosomal numbers and also differ by several inversions. An examination of electrophoretic variation indicated that the four chromosomal species were very similar at these loci (Nevo and Shaw, 1972). However, from laboratory experiments there does appear to be more aggression between contiguous types (58 and 60), suggesting the development of a prezygotic isolating mechanism (ethological isolation).

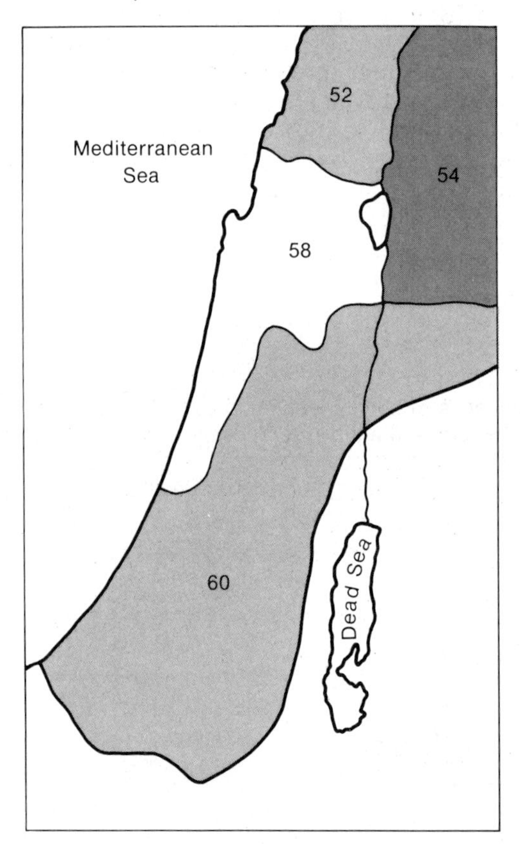

Figure 16.9. The geographic distribution of the four chromosomal forms of the mole rat, *Spalax ehrenbergi*, in Israel (after Nevo et al., 1975).

CHAPTER 17

Molecular Variation and Evolution

> In the last decade, however, evolutionary biology has been revitalized and enriched through contact with molecular biology.... Considering the magnitude of this effect, we may not be overfanciful to think that future historians will see molecular biology more as the salvation for than, as it first seemed, the nemesis of evolutionary biology.
>
> ROBERT SELANDER

An important recent development in evolutionary genetics is the accumulation of new information on the extent and nature of genetically determined molecular variation within and between species. As we discussed in the first chapter, in the late 1960s electrophoresis of proteins gave a measure of the extent of molecular variation. Recently, the amino acid sequence of proteins and the nucleotide sequence of DNA have given more detailed information about the nature of molecular variation. In this chapter, we will first consider the theories that have been proposed to account for the levels of molecular variation observed within a species and then examine some data relating to these hypotheses. Second, we will discuss molecular evolution between species and construction of phylogenetic trees.

I. THEORIES REGARDING THE MAINTENANCE OF MOLECULAR VARIATION

The finding that large amounts of genetic variation exist for loci that encode for proteins has generated a controversy that has endured for more than a decade concerning the factors responsible for this variation. The conflicting opinions can be simply stated as two dichotomous views: *selectionism*, which holds that the large majority of variants are maintained by some form of balancing selection, such as heterozygote advantage or other related selection models (see Chapter 15), and the *neutrality theory*, which holds that the large majority of molecular variants are neutral with respect to each other—that is, they are selectively equivalent—and that changes in frequency are due to a combination of genetic drift and mutation (see Chapter 6). Both views are more complex than this simple explanation suggests, and below we will discuss both hypotheses in more detail. Of course, it is quite likely that some combination of these hypotheses may actually explain the large amounts of molecular variation.

The simplest selectionist explanation for the presence of large numbers of molecular variants is that the variants are maintained by an advantage of the heterozygote at each of the polymorphic loci (see Chapter 15 for other balancing selection models). However, a heterozygous advantage model for many loci may lead to an intolerably low fitness for most populations, suggesting that such a simple model is not a feasible hypothesis (Lewontin and Hubby, 1966). To examine this argument, let us assume that the fitness values for the three genotypes A_1A_1, A_1A_2, and A_2A_2 are $1 - s$, 1, and $1 - s$, respectively. The mean fitness is then

$$\begin{aligned}\bar{w} &= p^2(1 - s) + 2pq + q^2(1 - s) \\ &= 1 - s(p^2 + q^2).\end{aligned}$$

At equilibrium, $p = q = 0.5$ and $\bar{w} = 1 - \frac{1}{2}s$. If there are m such loci and we assume that the mean fitness over all such loci, W, is the product of the mean fitnesses at each locus, then

$$W = \bar{w}^m.$$

Because $\bar{w}$ is smaller than unity, and if the number of such loci is large, the product of the fitnesses of the individual loci quickly approaches zero. For example, if there are 1,000 such loci and $s = 0.02$, then

$$\begin{aligned}W &= (1 - 0.01)^{1,000} \\ &= 0.00004\end{aligned}$$

a very low fitness indeed. Assuming that selection occurs in preadult viability, then this fitness value implies that only 1 out of 25,000 zygotes would survive to adulthood.

The theory of neutrality assumes that the majority of polymorphic molecular variants are affected primarily by the combination of mutation generating new variation and genetic drift leading to the eventual fixation of these variants. For these alleles, selection plays a minor role in determining frequency change because different genotypes have almost identical relative fitness—that is, they are neutral with respect to each other. This of course means not that the different allelic forms have no function but that they are functionally equivalent. As a result, neutrality theory suggests that there are allelic substitutions that have no selective advantage over those that they replace; hence the term *non-Darwinian evolution* is sometimes used to describe this process. On the other hand, neutrality theory does not claim that allelic substitutions that are responsible for evolutionarily adaptive traits are neutral.

As we discussed in Chapter 6, when mutation and genetic drift are the only factors affecting the genetic variation, the equilibrium heterozygosity in the population is

$$H_e = \frac{4N_e u}{4N_e u + 1}.$$

However, because the production of new mutants is generally quite slow and the population sizes in most species are relatively large, making the effect of genetic drift small, it takes an extremely long time to reach this equilibrium heterozygosity. In addition, low population sizes or bottlenecks some time in the past could reduce the heterozygosity to below the equilibrium expected from a large contemporary population size, and migration from a larger source population could increase the heterozygosity to above the expected equilibrium value.

II. EVIDENCE RELATING TO THESE THEORIES

A number of approaches have been suggested or tried to resolve the controversy about the cause of molecular variation. For the most part, these studies have either involved experimental examination of the properties of different allozymes or the statistical association of genetic variation with environmental variation or population genetic parameters. Because both the selectionist and neutrality hypotheses have become somewhat diffuse, it has been difficult in many cases to falsify one or the other hypothesis, and in fact many data are consistent with a form of either the selectionist or neutrality views.

One approach to comparing these views is to consider the amount or pattern of electrophoretic variation observed for a number of loci and to compare it to that expected from the neutrality model (see Example 17.1). This approach illustrates one of the advantages of the neutrality model—that is, it provides equilibrium predictions about the amount or pattern of genetic variation. For example, using the expression above, we can predict the amount of heterozygosity at equilibrium for the neutrality model. On the other hand, predictions based on a selectionism model vary depending upon the type of balancing selection assumed.

Other studies have endeavored to establish an association between particular electromorphs and selective factors. Overall, it appears that only limited success has been achieved in demonstrating selective difference between naturally occurring electromorphs, even though some of the investigations have been extensive. Several experimental approaches have been used to demonstrate that selection is occurring for particular alleles. For example, genetic perturbation studies, in which the allelic frequency is perturbed above and below an equilibrium frequency and returns from both above and below, can demonstrate that balancing selection is present (see Example 3.3). On the other hand, if allelic frequencies change at different rates in different environments, then this indicates the importance of these environmental factors (see Example 17.2).

Another approach is the examination of the gene products (allozymes) for particular biochemical properties that result in balancing selection. Generally, extracts of the allozymes are measured for their *in vitro* activity under different environmental conditions, such as temperature and salinity (see Example 15.1). It appears that many allozymes differ in some biochemical properties, but these differences are often small and their selective importance in

---

Example 17.1. Ayala et al. (1972) measured the heterozygosity for a number of electrophoretic loci for four closely related species of the *D. willistoni* group. They found that the heterozygosity ranged from 0.152 to 0.195, with a mean of 0.173 for the four species. These *Drosophila* are very abundant tropical insects, so they suggest that the effective population number (actually the number in the species) is 10^9 or greater. Assuming that the mutation rate is about 10^{-6}, the approximate value estimated in laboratory experiments, then H_e is

$$H_e = \frac{4(10^9)(10^{-6})}{4(10^9)(10^{-6}) + 1} = 0.999$$

a value close to the maximum heterozygosity possible. Obviously, this value, predicted from neutrality, is much larger than the values observed, and Ayala et al. consider it a falsification of the neutrality theory. However, it is not clear whether estimates of N_e and u are accurate and whether the populations are at equilibrium.

A second example utilizing this approach is in populations of the univoltine spittlebug, *Philaenus spumarius*, which live on islands between Finland and Sweden (Saura et al., 1973). These islands are young, having arisen by volcanic action in the past several hundred years, and the age and approximate number of generations of spittlebugs on these islands is therefore known. Furthermore, the female spittlebug lays eggs in an easily identifiable mass, so a good estimation of the breeding population size can be obtained by counting egg masses. The island age, population size, and heterozygosity for an average of about 20 electrophoretic loci per population are given in Table 17.1. Using the above expression and the observed values of N_e and H, we can estimate the mutation rate in these populations. For example, on island 1

$$0.023 = \frac{4(1100)u}{4(1100)u + 1}$$

natural populations is not clear. However, differences between species are often consistent with a selective hypothesis (see last part of Example 17.2).

III. DNA VARIATION

One of the most exciting developments in genetics in recent years is the knowledge of gene structure obtained from DNA sequencing. For some viruses, the nucleotide sequence of the whole genome is known (all 5,400 nucleotides of the ϕX174 virus), and, for some bacteria, the nucleotide

---

so that

$$u = 5.4 \times 10^{-6}.$$

In general, the mutation rates necessary to explain these data are high (last column of Table 17.1). Furthermore, the young age of the islands might suggest that the populations have not reached equilibrium heterozygosity. Because H would presumably be higher at equilibrium than these observed values, the necessary u value would be even higher than those given in Table 17.1. These results are inconsistent with neutrality in a way opposite to those of Ayala et al.—that is, if we assume a mutation rate of 10^{-6}, then the predicted heterozygosity is smaller than that observed. However, in this case the heterozygosity may be affected by recurrent migration from the mainland, making it greater than expected from the estimate of the population size. These two examples illustrate the problems associated with resolving the relative importance of selectionism and neutrality using indirect means.

Table 17.1
The age, population size, heterozygosities, and estimated mutation rate for six island populations of a spittlebug (after Saura et al., 1973).

Island	*Maximum age of population (years)*	*Population size*	*H*	*Estimate of u*
1	200	1,100	0.023	5.4×10^{-6}
2	100	90	0.071	1.8×10^{-4}
3	300	500	0.063	3.4×10^{-5}
4	200	1,000	0.099	2.7×10^{-5}
5	400	3,000	0.096	8.8×10^{-6}
6	200	200	0.112	1.6×10^{-4}

Example 17.2. Probably the most thoroughly studied naturally occurring allozymic polymorphism is the variation at the *alcohol dehydrogenase* (*Adh*) locus in *Drosophila melanogaster*. The two major alleles, *F* and *S*, are present in substantial frequencies in populations from Japan, Japan, the United States, England, and Australia, and exhibit similar north-south clines in all continents. Because alcohols are an important component of the rotting fruit in which *Drosophila* live, breakdown of alcohols by alcohol dehydrogenase is probably necessary for fly survival. A number of researchers have found that the frequency of the two alleles is affected by the presence of alcohols. Furthermore, the rate of change appears to be affected by the type of alcohol present in laboratory experiments, as illustrated in Figure 17.1 for the change in allele *F* for different alcohols (van Delden et al., 1975). In this case for nearly all the alcohols tested, the *F* allele quickly increases in frequency.

The amino acid sequences for both allozymes are known, and the only difference is one amino acid of the 255 total. In position 192, the

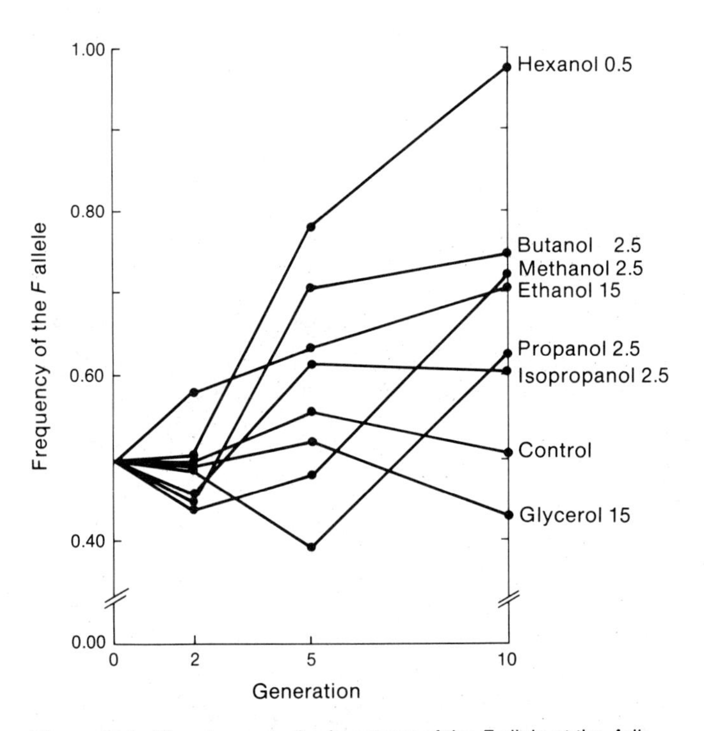

Figure 17.1. The change in the frequency of the *F* allele at the *Adh* locus in the presence of different alcohols (van Delden et al., 1975).

F allele has the amino acid threonine and the *S* allele has lysine (Thatcher, 1980). This single amino acid difference appears to have a number of effects on the molecule; for example, the allozymes show significant differences in enzyme activity, substrate specificity, heat stability, and other biochemical properties. It is difficult to know, however, which of these factors may be related to selection in natural populations and how they might give an overall balance of selective forces that could maintain the stable polymorphism.

Another approach to evaluating the importance of selective factors on allozyme variation is the comparison of closely related species that have similar habitats but differ in a particular environmental aspect that is likely to affect enzyme function. For example, three species of barracudas of the genus *Sphyraena* living in the eastern Pacific are quite similar except that the water temperature in their ranges differs (Graves and Somero, 1982). The three species are monomorphic for different alleles at a *Ldh* (*lactate dehydrogenase*) gene, the protein product of which is important in muscle contraction. The kinetic activities of extracts of the allozymes for each species were determined over a range of temperatures (Figure 17.2). There was an inverse relationship between the relative activity of the allozyme and the temperature of the habitat; for example, *S. argentea* has the highest relative activity but the lowest habitat temperature. However, the relative activity is nearly identical for the three species at the temperature midrange for each species (indicated by closed circles in Figure 17.2). These results suggest that there is selection for an optimal enzyme activity that may result in different allozymes in different temperatures.

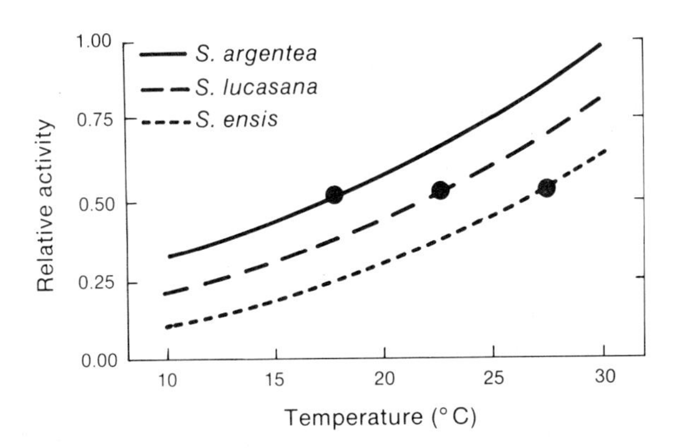

Figure 17.2. The relative activity of different LDH allozymes from three species of barracudas at different temperatures (from Graves and Somero, 1982). The closed circles indicate the midrange temperature for each species.

Table 17.2
Two restriction endonucleases and the nucleotide sequences that they recognize. The arrows indicate where the strands of DNA are cut.

Enzyme	*Source*	*Recognition sequence*
Eco R1	*E. coli*	↓ GAATTC CTTAAG ↑
Hin dIII	*Hemophilus* influenza	↓ AAGCTT TTCGAA ↑

sequence of sizable regions of the genome is known. In eukaryotes, some genes and adjacent areas have been sequenced, as has the human mitochondrial DNA, which consists of more than 16,000 nucleotides.

One of the surprising findings is that in many eukaryotic genes some internal regions of the gene, called *intervening sequences* (or *introns*), are transcribed into RNA but are not translated into polypeptides. During the processing of RNA, the protein-coding regions are spliced together to form the complete messenger RNA sequence to be used for translation into the polypeptide sequence. Preliminary studies seem to indicate that there is a greater likelihood of nucleotide differences in the regions that are spliced out than in the coding regions of a gene, making it appear that selective constraints are greater in the coding regions. In other words, these preliminary results suggest that nucleotide variation may be consistent with neutrality expectations in some parts of the genome and not in others. It is not clear how important balancing selection is in maintaining nucleotide variation.

Another approach used to measure variation in DNA utilizes enzymes present in bacteria that recognize foreign DNA and degrade it. These enzymes, called restriction endonucleases, recognize specific sequences in the DNA and cleave the DNA molecule at those sequences. For example, enzyme Eco R1, found in *E. coli*, recognizes a portion of DNA with the sequence GAATTC and cleaves it unevenly (between G and A on both strands), so that each end has an unpaired "sticky" region AATT (Table 17.2). Restriction enzymes have been used to characterize variation in mitochondrial DNA, mtDNA (see Example 17.3), or cloned segments of nuclear DNA. This technique appears particularly appropriate for analysis of relationships of closely related individuals because the rate of change in restriction sites appears rapid. Furthermore, it is assumed that the four- or six-base sequences recognized by the enzymes are probably unrelated to selective factors.

Recently, the finding in bacteria, yeast, and *Drosophila* of small transposable genetic elements, called "jumping genes" by some, has raised speculation as to their role in evolution. Particular transposable elements,

Example 17.3. Because mtDNA is maternally inherited (mitochondria are always obtained from the female gamete) and there appears to be no or very little exchange of DNA between mitochondria, it is thought that individuals sharing the same mtDNA patterns have a common ancestry from a single female some time in the past. It is possible as a result of this maternal inheritance that variation in mtDNA and variation in nuclear genes may show different patterns over space or time.

Two species of mice form a hybrid zone across Denmark, an area 50 kilometers wide with a marked anatomical difference between the species. In addition, a number of electrophoretic differences distinguish these species, which are *Mus domesticus*, occupying the southern portion of the zone and present in most of western Europe, and *M. musculus*, the species in the northern portion of the zone and present in northern Denmark, the rest of Scandinavia, and eastern Europe. Using eight electrophoretic loci that were earlier shown to be different in these two species, Ferris et al. (1983) found that populations of *M. musculus* from northern Denmark were similar to those in other areas but that both groups of *M. musculus* differed sharply from populations of *M. domesticus* (see Table 17.3).

When the same populations were examined for mtDNA variation, a very different pattern emerged. The populations of *M. musculus* from northern Denmark were found to be more similar to *M. domesticus* than to the other *M. musculus* populations (see Table 17.3). On the other hand, other populations of *M. musculus* were quite different from *M. domesticus*. It appears from these results that mtDNA from *M. domesticus* crossed the hybrid zone and, probably because of some selective advantage, invaded and spread northward. In all likelihood, the variants observed using restriction enzymes were carried along—hitchhiked—as this advantageous mtDNA spread. Ferris et al. conclude that, because of such apparent interspecific transfer of mtDNA, mtDNA may not be a good indicator of the evolution of genes in the nucleus.

Table 17.3
The average difference in frequencies for eight protein loci and the average difference in mtDNA nucleotide sequence for populations of *M. Musculus* (*M. m.*) from northern Denmark, *M. musculus* from other areas, and *M. domesticus* (*M. d.*) (after Ferris et al., 1983). The numbers in parentheses are standardized differences for the two types of variation.

	Northern Denmark M. m. and other M. m.	*Northern Denmark M. m. and M. d.*	*Other M. m. and M. d.*
Electrophoretic difference	0.258 (0.284)	0.909 (1.0)	0.892 (0.981)
mtDNA difference	0.059 (1.0)	0.023 (0.390)	0.052 (0.881)

either small insertion sequences of 800 to 5,000 bp (base pairs) or larger elements of 10,000 bp, have been found in multiple copies in a given genome. Furthermore, the same elements may be in different locations in other individuals. Because these elements can both insert new copies elsewhere and leave a copy in the old position, they have been termed "selfish DNA" (Orgel and Crick, 1980). These properties make transposable elements potentially important in a number of aspects of evolutionary genetics, such as influencing the amount of DNA and causing mutations (see Example 17.4).

Example 17.4. A class of transposable elements of about 3,000 bp in size, found in *Drosophila melanogaster*, is called *P* elements. In strains having *P* elements, up to 30 to 50 elements may be dispersed throughout the genome, while other strains, called *M* strains, may have no *P* elements. When *P* type males are crossed to *M* type females, offspring manifest a syndrome of effects called *hybrid dysgenesis*, which includes sterility, higher mutation rates, and chromosomal breakage.

Kidwell (1983) examined *D. melanogaster* strains collected over fifty years and determined whether they were *P* or *M* strains. The older strains were nearly all *M* strains, while almost 90 percent of the strains collected in the 1970s were *P* strains (see Figure 17.3). Nearly all of the strains collected lately in the United States are *P* (or *P* related), while both European and Asian strains often are *M*. It appears that the *P* element originated in recent decades and has rapidly spread to *D. melanogaster* populations throughout the world. Because of the association of the *P* element with the multiple phenotypic effects of hybrid dysgenesis, it has been suggested that this or other similar elements may be important in formation of reproductive isolation.

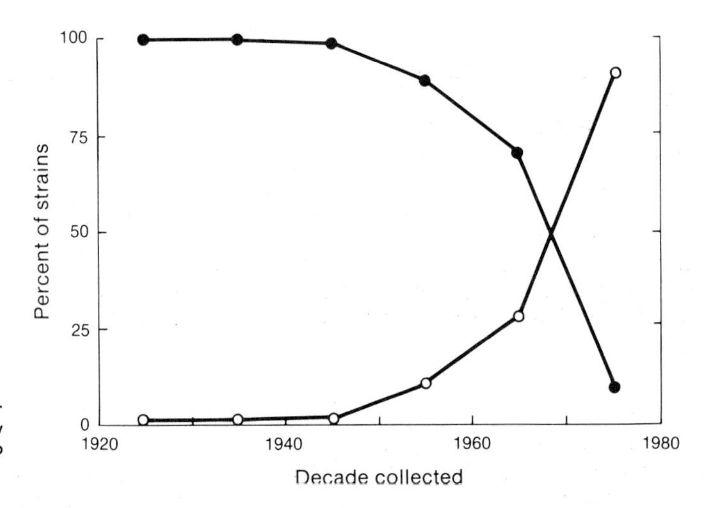

Figure 17.3. The percent of *D. Melanogaster* strains collected over fifty years that are *M* (closed circles) or *P* (open circles) (after Kidwell, 1983).

IV. THE MOLECULAR CLOCK

Data from amino acid sequences for a number of proteins from different organisms appear to indicate that substitutions accumulate at a regular rate for a given protein. Of course, this apparent regular replacement is an average of a number of events over a long period of time. Assuming that there is such a regular replacement, then the differences in amino acid between two species may serve as a *molecular clock*, indicating the time since two species diverged from a common ancestor. In addition, using comparisons of homologous proteins (or nucleotide sequences) from different species, an estimate of the rate of substitution of molecular variants can be made, given the approximate time since divergence. Below we will briefly develop the logic used for these estimates when comparing amino acid sequences.

Assume that the number of differences, d_{aa}, in two homologous amino acid sequences with n_{aa} amino acids of a particular protein for two species is known. When the amino acids at a particular position are different, this indicates that at least one amino acid replacement has occurred in the course of evolution (there could have been more than one, but it would not be possible to detect the exact number). If we let K_{aa} be the mean number of amino acid substitutions per site and assume independence of replacement of amino acids at different sites, the probability of no substitutions is

$$\Pr(0 \text{ substitutions}) = e^{-K_{aa}}$$

from the Poisson distribution (see Chapter 10). Therefore, the probability of at least one substitution is

$$\Pr(1 \text{ or more substitutions}) = 1 - e^{-K_{aa}}$$

and then

$$d_{aa} = n_{aa}(1 - e^{-K_{aa}}).$$

Assuming that the proportion of sites that are different is $p_{aa} = d_{aa}/n_{aa}$, then

$$p_{aa} = 1 - e^{-K_{aa}}$$
$$e^{-K_{aa}} = 1 - p_{aa}$$
$$K_{aa} = -\ln(1 - p_{aa}).$$

To make this measure a function of time, the rate of substitution per amino acid site per year can be found as

$$k_{aa} = \frac{K_{aa}}{2T}$$

where T is the estimated number of years (from paleontological or other

Table 17.4
A hypothetical array of amino acids from two species where the abbreviations indicate different amino acids.

Species	*Position*					
	1	*2*	*3*	*4*	*5*	...
a	lys	glu	ala	asp	arg	...
b	lys	glu	lys	asp	iso	...

evidence) since the divergence of the two species from their common ancestor. The 2 in the denominator of this last expression occurs because molecular change occurs in both lineages back to the common ancestor (see discussion in next section about phylogenetic trees).

As an example, let us consider the hypothetical array of homologous amino acids from two species in Table 17.4 (see also Example 17.5). Of the five amino acids compared, they are identical at positions 1, 2, and 4 and different at positions 3 and 5. Therefore, the proportion of sites that are different is

$$p_{aa} = \tfrac{2}{5} = 0.4.$$

The mean number of amino acid substitutions per site is then

$$K_{aa} = -\ln(1 - 0.4) = 0.51.$$

K_{aa} is larger than p_{aa} because it is possible that there may have been two or more changes at the sites that differed. If we assume that the two species diverged 200 million years in the past, then the estimated rate of substitution per year is

$$k_{aa} = \frac{0.51}{400{,}000{,}000}$$

$$= 1.26 \times 10^{-9}.$$

The hypothesis of a molecular clock is consistent with the neutrality theory. In other words, if the replacement of molecular variants over time is a function of genetic drift and the mutation rate, it will result in a relatively regular turnover (replacement) of molecular variants over time. To illustrate this, assume that genetic drift and mutation are only determinants of changes in frequencies of molecular variants. Let the mutation rate to a new molecular variant be u, so that in a diploid population of size N there are $2Nu$ new mutants per generation. From our earlier discussion of genetic drift in Chapter 6, we know that the probability of fixation of a new neutral mutant is equal to its initial frequency. In this case, because each new mutant is assumed to occur only once, the initial frequency, and therefore the probability of

Example 17.5. The α chain of the hemoglobin molecule has been sequenced in a number of different species. For example, the α chains of humans and carp differ at 68 sites and are the same at 72 sites; therefore $p_{aa} = 68/140 = 0.486$. Using the equation above, $K_{aa} = 0.666$, and, because the time since divergence of human and carp is estimated to be about 375 million years, $k_{aa} = 8.9 \times 10^{-10}$.

The rate of amino acid substitution appears to differ for different protein molecules. For example, Dickerson (1971) has compared the amount of divergence for several molecules that have been sequenced in a number of organisms (Figure 17.4). From these comparisons, it appears that the rate of amino acid substitution is very fast for fibrinopeptides, intermediate for hemoglobin, and very slow for cytochrome c. The estimated rates of amino acid substitutions for fibrinopeptides, hemoglobin, and cytochrome c are 9.0×10^{-9}, 1.4×10^{-9}, and 0.3×10^{-9}, respectively. These differences may result from the relative importance of the three-dimensional structure of these molecules and the proportion of amino acids involved in active enzymatic sites. For example, histone IV, which has a k_{aa} value of 0.006×10^{-9}, fits in tightly with the DNA molecule, and therefore nearly the whole amino acid sequence is critical for the three-dimensional structure. At the other extreme, the fibrinopeptide molecule appears to be hardly more than a spacer molecule with no enzymatic sites.

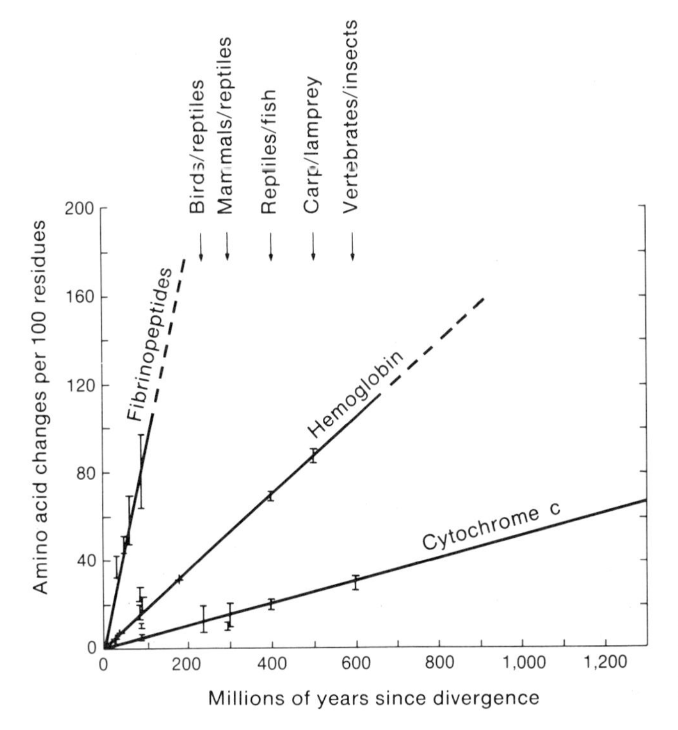

Figure 17.4. The number of amino acid changes for three different molecules in organisms that diverged at different times (after Dickerson, 1971).

fixation is $1/(2N)$. Assuming an equilibrium situation between mutation and genetic drift, the rate of allelic substitution (k) is the product of the number of mutants generated per generation and their probability of fixation or

$$k = 2Nu\left(\frac{1}{2N}\right)$$
$$= u.$$

In other words, the average rate of allelic substitution in this simple model is equal to the mutation rate at the locus.

If we assume that the mutation rate per gene is about 10^{-6} and that there are about 100 amino acids in a typical protein, then the mutation rate per amino acid should be about 10^{-8}. With neutrality over the whole molecule, this should be an approximate estimate of the rate of amino acid substitution. Except for fibrinopeptides, this rate is much higher than the ones given above. One explanation is that, as suggested above, the selective constraints may vary for different molecules and among sites within a molecule. Where constraints are greatest, the rate of amino acid substitution is reduced. In other words, the rate for fibrinopeptides is close to that expected from neutrality for the whole molecule, while the rate for histones is much lower, presumably because of the functional constraints on a large part of the histone molecule.

V. PHYLOGENETIC TREES

The differences among related species may be measured for morphological, biochemical, immunological, or other traits, with more closely related species presumably being more similar. Thus, one of the important uses of genetic or phenotypic measures of differences is the organization of related populations or species in a biologically meaningful way—that is, the arrangement of such groups to illustrate ancestral relationships. Relationships are usually given in the form of a *phylogenetic tree*, in which the root or ancestral type is at the top and the branches or evolutionarily derived types are at the bottom. The similarity and differences in molecular structure among different organisms may be used to understand evolutionary relationships. Presumably, organisms that have similar molecules are closely related, and those that have quite different molecules are only distantly related.

To illustrate a simple approach used to construct a phylogenetic tree, let us assume that there are a number of groups indicated by a,b,c . . .that could be populations, species, or other biological groups. Assume also that the genetic differences between these groups have been measured in some quantitative way using genetic data, such as allelic frequencies, amino acid differences, or restriction map information, so that a matrix of values as shown in Table 17.5a is available. Here we will use D_{ab}, a measure of the genetic

Table 17.5
The genetic distance measures (a) among four groups and (b) among three groups after groups *a* and *b* have been clustered. A numerical example of genetic distance values illustrating the above, with (c) the four groups and (d) the three groups after *a* and *b* have been clustered.

	(a)				(b)	
	a	*b*	*c*		*(ab)*	*c*
b	D_{ab}	—	—	*c*	$D_{(ab)c}$	—
c	D_{ac}	D_{bc}	—	*d*	$D_{(ab)d}$	D_{cd}
d	D_{ad}	D_{bd}	D_{cd}			

	(c)				(d)	
	a	*b*	*c*		*(ab)*	*c*
b	0.12	—	—	*c*	0.24	—
c	0.21	0.27	—	*d*	0.36	0.18
d	0.33	0.39	0.18			

differences between groups *a* and *b* known as the *genetic distance*. Genetic distance values are zero when the two groups are identical and increase as the groups become more different.

We would like to organize all the groups in a way that reflects biological relationships. (The approach used here is actually a phenogram that is only congruent with a phylogenetic tree if the rates of evolution are the same among all species.) The first step of a simple method of determining this organization is putting together, or clustering, the two groups that have the smallest genetic distance—that is, combining them into a single new group. We then calculate new estimates of genetic distance between this new composite group and the remaining groups by taking the arithmetic average of their distances. For example, if populations *a* and *b* have the smallest genetic distance, then they can be combined into a new group (*ab*). The genetic distance between this unit and *c* is

$$D_{(ab)c} = \tfrac{1}{2}(D_{ac} + D_{bc})$$

and the distance between this unit and *d* is

$$D_{(ab)d} = \tfrac{1}{2}(D_{ad} + D_{bd}).$$

Now we have a new matrix of distance values (given in Table 17.5b) using the three remaining groups, (*ab*), *c* and *d*. Assume that D_{cd} is the smallest genetic distance remaining so that a second cluster of *c* and *d* is formed. The last step in this example is to calculate the average distance between the two clusters, which is

$$D_{(ab)(cd)} = \tfrac{1}{2}(D_{(ab)c} + D_{(ab)d}).$$

It is useful to give a numerical example of this procedure so that the branch lengths have a particular value and the phylogenetic tree can be drawn to proper scale. Some hypothetical data reflecting the above discussion are given in Table 17.5c for the genetic distance among four groups. First, D_{ab} is the smallest genetic distance value, and *a* and *b* are therefore clustered and the reduced matrix in Table 17.5d is calculated. In this case, the distance between *ab* and *c* is

$$D_{(ab)c} = \tfrac{1}{2}(0.21 + 0.27) = 0.24$$

and the distance between *ab* and *d* is

$$D_{(ab)d} = \tfrac{1}{2}(0.33 + 0.39) = 0.36.$$

Of the distance values in this reduced matrix, D_{cd} is the smallest. The distance between the two clusters *ab* and *cd* can then be calculated:

$$D_{(ab)(cd)} = \tfrac{1}{2}(0.24 + 0.36) = 0.30.$$

The phylogenetic tree for these data is given in Figure 17.5, in which the vertical scale indicates branch length. Notice that this scale is half the genetic distance, because evolutionary change is assumed to occur equally in all lineages. For example, the distance between *a* and *b* is 0.12, but the vertical branch length from the time at which they split is 0.06 in each branch. Because

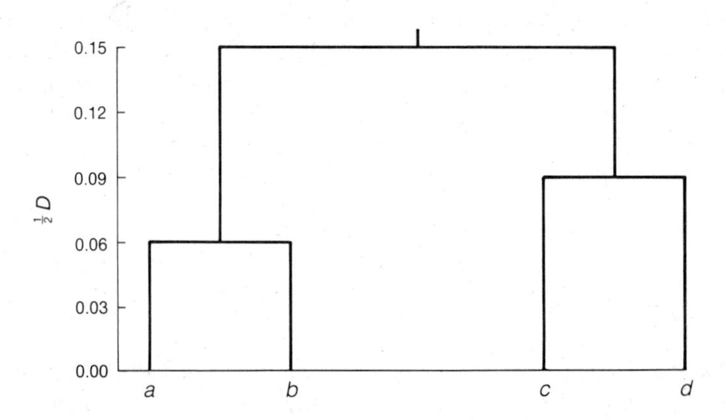

Figure 17.5. The phylogenetic tree obtained from the data in Table 17.5.

the genetic distance between the two clusters is 0.3, the branch length back to their common ancestor is 0.15. An application of this approach to allelic frequencies is given in Example 17.6.

Comparisons of phylogenies based on molecular data with the traditional phylogenies based on morphological differences are often quite interesting. For example, based on electrophoretic or immunological distances, humans, chimpanzees, and gorillas are relatively quite similar, suggesting that they only diverged approximately five million years ago. However, morphologically the differences are much greater than would be expected on this basis, leading to a great deal of speculation to explain this rapid morphological divergence. One hypothesis is that the relatively fast morphological change is the result of adaptive change at *regulatory genes*—that is, genes whose major function is to control the amount of product from other genes.

Another example of the inconsistencies between morphological and molecular data is that of the Tasmanian wolf, an extinct marsupial carnivore. Because in morphology it is much more similar to an extinct South American marsupial rather than Australian marsupials, it has been suggested that it is more closely related to the South American form. However, immunological comparison shows that the genetic distances between the Tasmanian wolf and some Australian marsupials are small, supporting the hypothesis that the Tasmanian wolf is most closely related to the Australian marsupials (Lowenstein et al., 1981).

Although changes in molecular traits may not exactly follow a molecular clock—that is, occur at a regular rate over time—it is even less likely that morphological changes occur at a regular rate. For example, human morphology from paleontological evidence appears to have changed at a very rapid rate, and the morphology of the Tasmanian wolf appears to have converged rapidly on a morphology present in a separate lineage. Selection is presumably important in these rapid morphological changes, but the details of the genetic change underlying them are not known.

Example 17.6. The frequencies of alleles at blood group and electrophoretic loci have been calculated for many human populations. Nei and Roychoudhury (1980) have used these data to calculate both genetic distance values and phylogenetic trees for a number of human populations. A sample of their distance values is given in Table 17.6 for five different populations. The phylogenetic tree obtained using these data and the method described above is given in Figure 17.6. The phylogenetic tree clusters first the Japanese and the Australian Aborigine populations together and then the English and the Brazilian Indian together. These four appear to form a group separate from the Bantu. However, this tree is based on relatively small genetic distances among the groups, because these groups share many alleles in common, so that the clustering may not reflect close biological kinship.

Proteins are still intact in some specimens of some extinct animals. Comparison of the immunological responses of these proteins to those of related extant animals gives a measure of the relationships among these organisms. For example, a 40,000-year-old baby mammoth was discovered in the USSR in 1977. Comparison of an extract of the protein albumin from the mammoth, two species of elephants, and a manatee, using immunological techniques, gave the genetic distances in Table 17.7 and the phylogenetic tree in Figure 17.7. The differences between

Table 17.6
The genetic distance among five human populations based on 22 polymorphic blood group and enzyme loci (from Nei and Roychoudhury, 1980).

	a	*b*	*c*	*d*
a (English)	—	—	—	—
b (Japanese)	0.034	—	—	—
c (Brazilian Indian)	0.036	0.037	—	—
d (Australian Aborigine)	0.059	0.031	0.057	—
e (Bantu)	0.044	0.106	0.138	0.131

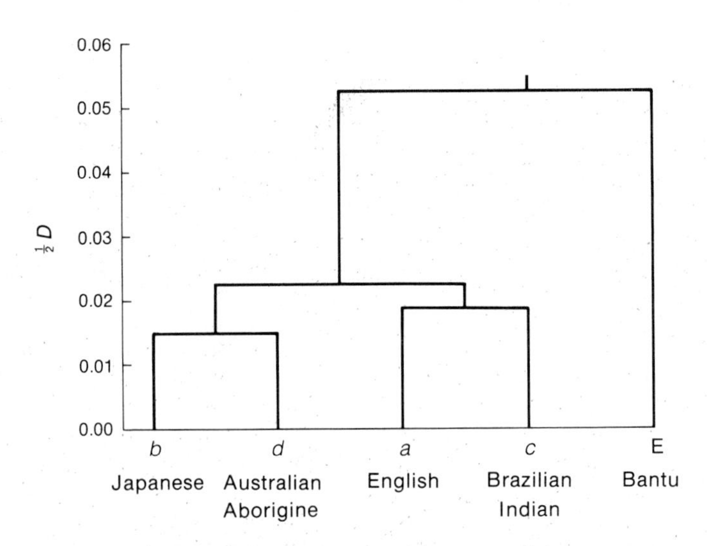

Figure 17.6. The phylogenetic tree of human racial groups constructed from the genetic distance data in Table 17.6

the mammoth, African elephant, and the Indian elephant are small, corresponding to current ideas that the three species are closely related. The similarity of the genetic distances suggests that they diverged from each other at about the same time.

Table 17.7
The immunological genetic distances between the extinct mammoth, the two extant species of elephants, and the manatee (after Lowenstein et al., 1981).

	a	*b*	*c*	*d*
a (Mammoth)	—	—	—	—
b (Indian elephant)	0.027	—	—	—
c (African elephant)	0.036	0.032	—	—
d (Manatee)	0.523	0.444	0.377	—

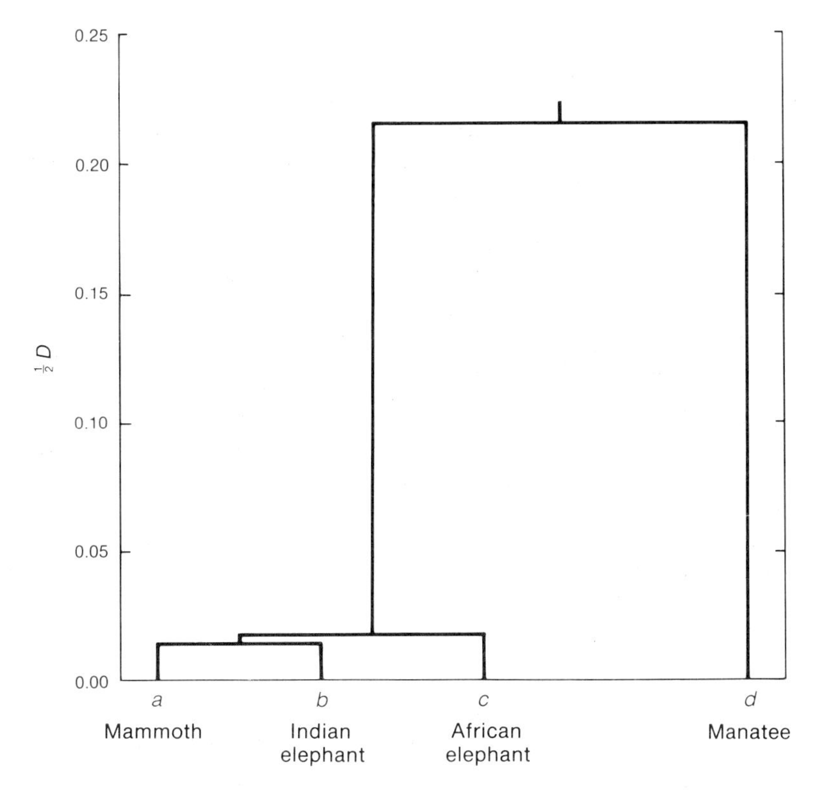

Figure 17.7. The phylogenetic tree constructed from the immunological genetic distance data in Table 17.7.

SUMMARY

There is a large amount of molecular variation in natural populations. Two different views have been espoused to explain this variation, selectionism and neutrality, although it is likely that a combination of these hypotheses may be most accurate. Molecular differences between groups can be used to infer divergence times between species and their phylogenetic relationships. In some cases, these constructions are inconsistent with rates of morphological change or phylogenetic trees based on morphology and suggest that relationships based on morphology should be cautiously evaluated.

PROBLEMS

1. Compare selectionism and the neutrality theory and suggest an experiment that might differentiate between them as explanations for variation at a given locus.
2. What is the molecular clock? What is the rate of amino acid substitution? Would you expect the rate of nucleotide substitution to be faster or slower than the rate of amino acid substitution? Why?
3. If two species, thought to have diverged a million years ago, differ at 40 out of 90 amino acid positions for one protein and at 20 out of 80 amino acid positions for another, what are K_{aa} and k_{aa} for the two proteins? Why might these estimates differ for the two proteins?
4. Using the following genetic distance values between groups, construct a phylogenetic tree. What assumptions do you think are implicit in the construction of such a tree? If other data gave a different tree, how would you resolve the differences?

	a	*b*	*c*
b	0.2	—	—
c	0.15	0.3	—
d	0.55	0.7	0.65

5. If we knew the complete nucleotide sequence and the complete amino acid sequence for the proteins in a group of organisms, what could we discover about selectionism and neutrality and the relationship of the organisms in the group?

CHAPTER 18

The Evolution of Social Behavior

Evolution is the most powerful and the most comprehensive idea that has ever arisen on Earth.

JULIAN HUXLEY

Particular behavioral patterns distinguish species from one another much as different morphological phenotypes do. Behavioral patterns can generally be understood in adaptive terms as important for acquiring food, finding and keeping mates, raising young, or defending against predators. Often these patterns manifest themselves on an individual level— for example, in the prey-searching behavior of a hawk or in a male frog's call to a mate. However, here we will be primarily concerned with behavior that involves groups of individuals. Group behavior is often consistent with the adaptive advantage in a group of foraging, predator avoidance, and reproduction. In particular, we will consider groups that exhibit some form of organization due to the behavior of individual members. As an introduction, let us briefly discuss social dominance and territoriality as examples of behavior.

In a number of organisms that forage, defend, or raise young as a group— wolves, chickens, and apes, for example—a *dominance hierarchy* often exists. This social organization consists of an ordered array of dominant and subordinate individuals (see Example 18.1). Generally, this ranking is determined in a series of confrontations, with the lower ranking individuals losing to the more dominant ones. After the dominance hierarchy is established, it is generally maintained only by threat and subsequent submission. Individuals that are high in the dominance hierarchy because of their success in confrontations are generally thought to have higher relative fitness values, since the rank of individuals may determine access to food or mates and the role of an individual in the group. In addition, such a social organization appears to reduce the energy and time spent in intraspecific conflict and consequently to increase the efficiency of the population's resource utilization.

In many vertebrates, suitable habitat is divided among individuals in some manner (see Example 13.1) or among groups of individuals (see Figure 10.3). If these areas are actively defended either for use in foraging, mating, or breeding, they are called *territories*. For example, such diverse organisms as *Drosophila* (see Example 16.8), the heath hen (Example 11.5), and antelopes form *leks*, areas that males defend during mating times. On the other hand, if these areas are not actively defended and are used primarily for foraging, they are called *home ranges*. The home range of an organism is generally larger than its territory, but because the aggressive interactions used in defining territorial limits are often difficult to observe, home ranges are

---

Example 18.1. The social organization of many primate populations exhibits definite social hierarchies. Position in the hierarchy is determined by a number of factors, including sex, age, and size. As an example, let us briefly discuss the social organization of a captive population of Japanese macaques (Eaton, 1976). This troop has been observed for a number of years since it was transferred, essentially intact, from Japan. The general pattern of dominance categories and their associated social roles are given in Table 18.1. At any time, only one alpha male, five or six subleader males, and approximately twenty peripheral males exist in the troop.

From Table 18.1, it is obvious that both sex and age are important in determining these general social categories. However, within these categories, and in the determination of the alpha male, other factors may be of great significance. For example, the dominance of a male macaque is strongly affected by the rank of his mother. The alpha male in this troop, Arrowhead, is an older male, but he is relatively small and has no canine teeth and only one eye. It appears that Arrowhead's rank results primarily from the dominance of his mother, because when he was a juvenile his mother defended him in fights with other juveniles (and also defended him from the mothers of other juveniles). After such intervention occurred several times, the male offspring of lower ranking females ran away from Arrowhead, resulting eventually in his dominance. In this troop, Arrowhead's dominance results in preferential access to food, but a lower ranking male called Big V was observed to have mated much more frequently.

Table 18.1
The different dominance categories and their social roles in a troop of macaques.

Dominance category	*Social role*
Alpha male	Directs movement, defense, and policing of troop
Subleader males	Police troop; defend troop against predators
Adult females	Raise and protect offspring; defend allied females
Juveniles	Groom adults
Peripheral males	Warn troop against predators; defend troop

---

more practical to examine (see the latter part of Example 18.2). Behavioral interactions among individuals that result in the division of habitat into territories or home ranges give the successful individuals greater access to

Example 18.2. Territories may vary in size, encompassing many square kilometers, as in the foraging territories of wolf packs, or remaining small, as in the mating territories for male grouse. An individual's occupancy of a territory may be affected by factors such as age, size, or first entry into the area, which would permit the establishment of priority. In territorial birds, a particular individual or flock may occupy the same territory for a number of subsequent seasons. Figure 18.1 gives an example of the territories at two different times of flocks of an African, insect-eating bird, the green woodhoopoe (*Phoeniculus purpureus*; Ligon and Ligon, 1982). In 1975, there were eleven territories, indicated by letters, with different numbers of individuals in the flocks. In 1981, seven of these territories were still present in the same general areas, while four of the territories had been taken over by new flocks.

The size of a home range may be affected by a number of factors, such as the mobility of the organism, the density of the population, and the level of resources. For example, experiments with chipmunks have shown that the level of resources is of major importance in determining home-range size (Mares et al., 1982). Here it is assumed that home-range size, estimated by live trapping, is correlated with territory size. If an excess of food resource is added to an area in the form of seeds, then the home-range size is reduced from an average of 978 to 633 m^2. On the other hand, reduction of density by removal of individuals does not increase home-range size, suggesting that density may play a secondary role in determining home-range size in this species.

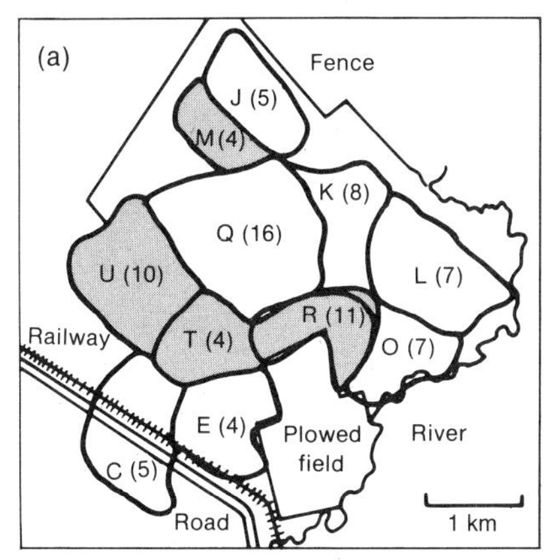

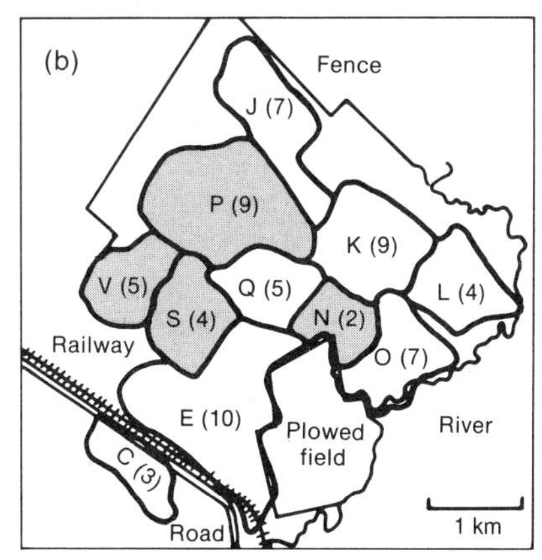

Figure 18.1. The territories of the green woodhoopoe (a) in 1975 and (b) in 1981 on a farm in Kenya (after Ligon and Ligon, 1982). The shaded areas indicate the flocks that lost territories in (a) and the new flocks in (b); the numbers are the individuals in the flocks.

resources or mates. In addition, the establishment of territories or home ranges is thought to result in less intraspecific conflict and a more efficient use of resources by the population.

I. BEHAVIOR AS A PHENOTYPIC TRAIT

Behavior involved in determining success in dominance hierarchies, territoriality, or other social behavior may be influenced by both genetic and environmental factors. For example, the rank in a dominance hierarchy or possession of a territory may be affected by the size of an individual, with size being the result of a combination of both genetic and environmental factors. Furthermore, specific behavioral patterns, such as the aggressive behavior occurring in the establishment or maintenance of rank or a territory, may be

---

Example 18.3. Let us define learning in a broad sense as behavioral change resulting from experience, in contrast to instinctive behavior that occurs without previous experience. As we discussed in Example 16.9, the fruit fly, *Rhagoletis pomonella* occurs in several different host races. Prokopy et al. (1982) examined the oviposition preference of some of these flies taken from an apple orchard. In their initial experiments, they found that flies taken from the field all chose apples as a host, while flies raised in the laboratory showed some preference for hawthorn, the ancestral host. These results suggested that learning by exposure as larvae to apples influences host preference.

To investigate this hypothesis further, the investigators trained female flies to accept one host or another by repeated exposure and then presented them with the same or the alternate host (see Table 18.2). More than 90 percent of the flies trained on a host attempted to oviposit on it when it was presented again. On the other hand, flies trained on one host rejected the second host a large proportion of the time. To further support their hypothesis that host preference is learned, they showed that the preferred host could be switched through retraining. Flies initially trained on hawthorn and retrained on apple, when presented with hawthorn, accepted it only 6 percent of the time (see Table 18.2). These observations suggest that host selection in this species may be modified by learning—that is, particular chemical or physical cues experienced previously may be used in host selection.

In general, cultural traits are considered only in terms of humans. However, in other species particular behaviors may be passed on from generation to generation through learning and imitation in a kind of precultural behavior. For example, a troop of Japanese macaques, starting with a single individual, learned to wash sweet potatoes in water

influenced by the genotype of the individuals involved. In other words, many of the same problems encountered when discussing quantitative traits and their adaptive significance are particularly important when examining behavior (see Chapters 8 and 16).

First, the determination of many behavioral traits (as well as some morphological traits) may be extensively affected by ecological factors (or by genotype–environment interaction; see Chapter 8). As an example, let us consider the dichotomy between pack hunting and solitary hunting in mammals. In some species, such as coyotes, the size of the available prey can determine the type of hunting behavior. If we assume that prey may vary from a small size, such as rodents, to a large size, such as deer, then the proportion of pack hunting could consequently vary from near zero, when only small prey are available, to near unity, when only large prey are available. In this case, the genotypes of the individuals remain constant, and the change in the behavioral phenotype, from solitary to social hunting, is thus ecologically induced.

(Kawai, 1965). Another interesting example of the transmission of a behavioral trait is the milk-pilfering behavior of several songbird species in England. The opening of the tinfoil tops and removal of some cream from milk bottles delivered to doorsteps was noticed first in one community in the 1940s. Initially, bottle opening was localized and due only to one bird species, the great tit. However, within a few years the behavior had spread over a large part of England and was imitated by other bird species. Because of the extremely rapid spread of this behavior in great tits and its extension to other species, this behavioral change could not be genetic. These behavioral changes in macaques and songbirds appear to have been extensions of normal behavior, since macaques in other troops brush off dirt from tubers and great tits often peck vigorously for insects.

Table 18.2
Oviposition choice in fruit flies when trained on one host and then presented the same or a different host (after Prokopy et al., 1982).

Flies trained on	*Offered*	*Proportion attempting oviposition*
Hawthorn	Hawthorn	0.91
Hawthorn	Apple	0.52
Apple	Apple	0.92
Apple	Hawthorn	0.20
Hawthorn, Apple	Hawthorn	0.06

In other words, the proximate or immediate stimulus that elicits behavior is an environmental factor, prey size.

Viewed in a different way, the flexibility of individuals or groups of individuals that enables them to exhibit variable behavior is genetically determined. In other words, the ability to switch behavior, depending upon the environmental cues, is a function of the genotypes present (see Example 13.6 for an example of change in prey utilization). Some genotypes may not have such plasticity and may result in a given behavior no matter what the environment. In other words, a particular genotype may predispose learning of a certain behavior. On the other hand, a population that consisted only of a genotype that exhibited flexible behavior would have genetically determined phenotypic variability. However, using the concepts of quantitative genetics, there would be no genetic variation for the trait and, therefore, a heritability of zero in this case.

In a similar vein, many organisms have the potential to learn a variety of behaviors. As a result, behavioral variability in a learned trait may have little relationship to the extent of genetic variation in a population. Only the presence of a genotype that can learn different behaviors given different stimuli is necessary for such behavioral variability. Because learned behavior is generally more important in higher organisms, the importance of genetic factors may be smaller in higher organisms. For example, because learned behavior is of great importance in humans, the proportion of genetic variation in a behavioral trait should be low relative to that in other vertebrates or invertebrates (however, see Example 18.3).

Geneticists have long studied behavioral traits, generally by using quantitative genetics approaches or by examining the pleiotropic effects of mutants. When studying mutants, whether they be single-gene or chromosomal, researchers assume that these variants will help dissect the determination of normal behavior. It is noteworthy that mutants are often used because very few large, behavioral differences are known that are inherited in a simple, single-gene fashion in natural populations. Although it is obviously difficult to document the existence of such variants, it can be assumed that naturally occurring variants at loci with mutants will also affect the same behavior, although in a more subtle manner. While it is probable that most types of behavior are affected by genetic factors, genetic analysis using quantitative genetic techniques (see Example 18.4) generally indicates that the genetic determination of variation in many behavioral traits within a population may often be relatively low.

While the behavioral differences among individuals in a population may often be primarily environmentally induced, the difference in behavior between individuals of different species is probably primarily genetically determined. (A similar general statement could be made for morphological or physiological traits, but see the discussions in Chapters 16 and 19.) In other words, the courtship patterns that separate different species of *Drosophila* or different species of birds appear to be caused by fixed genetic differences between species. Presumably, individuals of the same species from geographically isolated populations would be intermediate between these two and differ

Example 18.4. Orientation in insects may be determined by a number of factors, such as light, odors, and gravity. The extent of genetic variation affecting orientation due to the different factors in *Drosophila pseudoobscura* has been determined by directional selection experiments for phototaxis, chemotaxis, and geotaxis. One approach is to use a maze in which a number of sequential choices are available, either toward or away from a stimulus (Figure 18.2). If an individual makes all choices toward a stimulus, say light, it is positively phototactic. In this case, the individual would end up in vial 16 in Figure 18.2, while an individual that made all choices away from light would end up in vial 1.

In an experiment using such an approach to select for positive phototaxis over 15 generations, the mean score for females was changed from 8.7 to 13.4 (Dobzhansky and Spassky, 1967). The cumulative selection differential, the sum for all generations of the difference between the selected individuals and all individuals scored, was 47.2. Therefore, an estimate of heritability is the ratio of the total response over the cumulative selection differential (see Chapter 8), or 4.7/47.2 = 0.10, a rather low value of heritability. Heritabilities estimated in the same manner in populations selected for geotaxis were even lower, about 0.03.

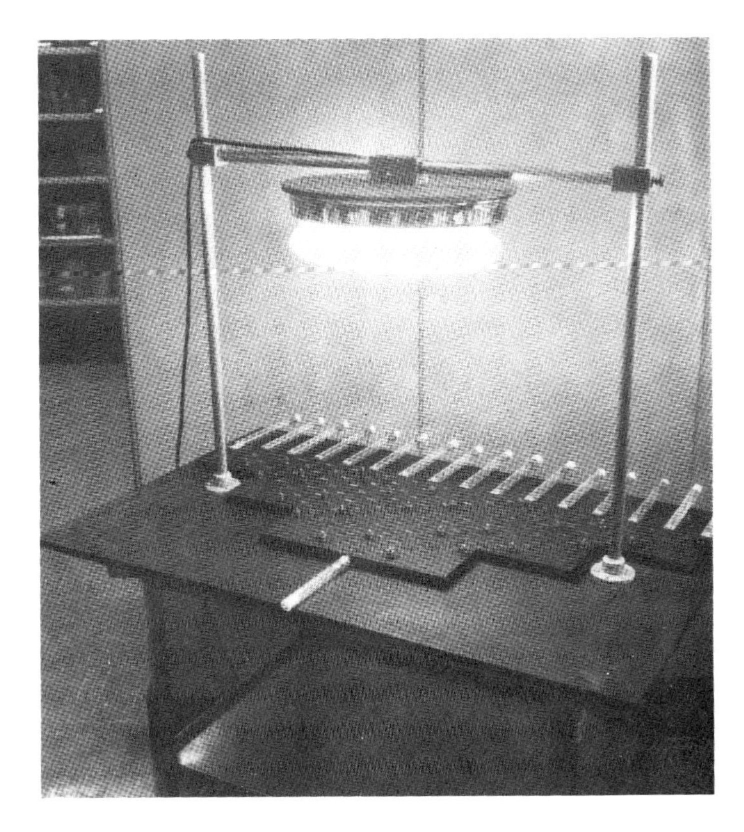

Figure 18.2 A maze for determining phototaxis in *Drosophila* (from Dobzhansky and Spassky, 1967).

Example 18.5. In many closely related species, even those with similar morphology, male courtship behavior often differs sharply. Such behavioral variation may be important in species recognition by females and in male mating success (see the brief discussion of sexual selection in Chapter 16). Within a species, male behavior is quite stereotypic, suggesting that differences between species have a large genetic component. A fascinating example of differences in male courtship behavior among related species is the bowerbirds of New Guinea and Australia. The males of these species build bowers (mating arenas), which are often quite large and may have fresh flowers, berries, or other objects grouped by color, and with sides painted in blue, black, or green. Figure 18.3 illustrates the bowers of two related species, *Amblyornis inornatus* and *A. macgregoriae*, from New Guinea (Diamond, 1982). The bower of *A. inornatus* may be nearly 3 meters long and 1.5 meters high and have a mossy lawn decorated by more than a hundred objects grouped by color. On the other hand, the simple bower of *A. macgregoriae* is less than a half meter high and has few or no decorations. In general, species of bowerbirds in which the male is not brightly colored construct elaborate bowers, and those species in which the male has distinctive plumage build simple bowers. For example, *A. inornatus* does not have a crest on its head, while *A. macgregoriae* has an orange crest three inches in diameter.

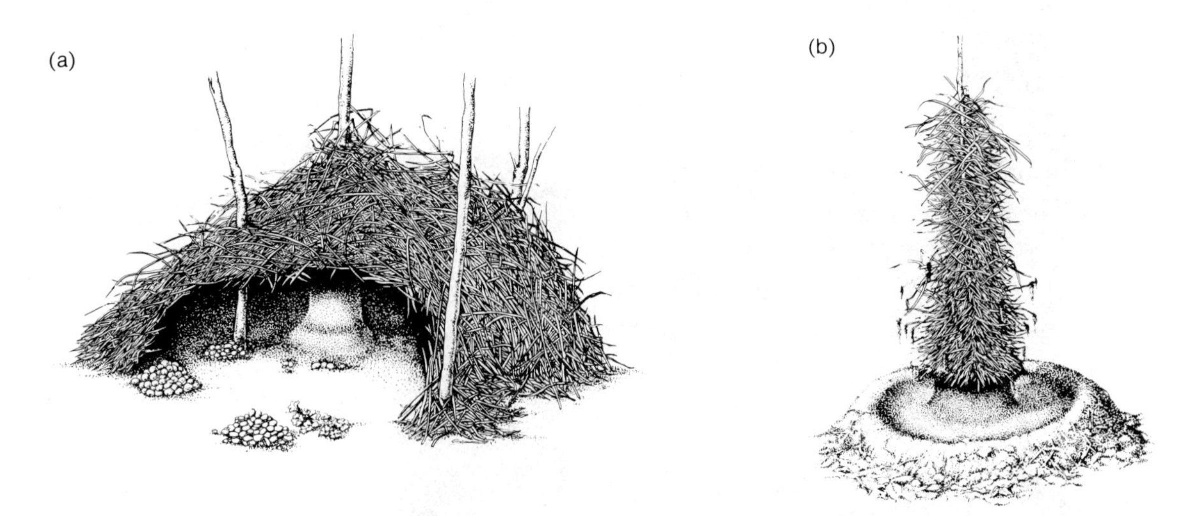

Figure 18.3. The bowers of two related species of bowerbirds from New Guinea, (a) *Amblyornis inornatus* and (b) *A. macgregoriae* (from Diamond, 1982).

because of both environmental and genetic factors. In other words, genetic factors determining the difference in behavior of individuals should be increasingly important for individuals taken from the same population, different populations, or different species.

In the past decade, both the scientific community and the general public have shown substantial interest in evolution and behavior, particularly social behavior. Much of this interest has been generated from the synthesis of evolution and social behavior into *sociobiology*. The central pursuit of sociobiology is to examine the biological factors affecting social behavior and its evolution. Obviously, this is an aspect of the study of evolution, and only the overextension of sociobiological interpretations to human behavior has made sociobiology controversial. For example, when sociobiological interpretations are used to explain such human behavior as homosexuality and aggressiveness, they suggest that these behaviors are inherent (genetic) and that an individual may have little control over his or her behavior (for a reasoned appraisal of the factors affecting human behavior see Montagu, 1980). As discussed previously, while it is probable that most types of behavior are affected by genetic factors, the demonstration that a trait has a genetic component does not imply that behavioral differences among individuals within a population are genetic.

Often, in sociobiology, it is assumed that the behavior observed in a population or species is the result of evolutionary history based on selective factors. In other words, the observed behavior is an example of a selectively advantageous phenotype and is present because it is adaptive. Although this may be true, one must consider that other nonadaptive factors may be responsible for the presence of a behavior (or other phenotypic traits). From a genetic viewpoint, the chance effects of genetic drift may have resulted in the fixation of a disadvantageous behavioral trait in the population, or a negative behavioral pattern may be present because it is a pleiotropic manifestation of an allele with a positive effect on another trait (see Chapter 19). Furthermore, a particular behavior may have become fixed in a previous environment and, because of morphological or physiological constraints, has not adapted to the present environment. In other words, when we examine variation in behavior as for any other phenotype, we should first evaluate the variants for genetic determination and, second, demonstrate the adaptive significance of the variants.

II. KIN AND GROUP SELECTION

Until now we have been considering models of individual selection—that is, natural selection based on the phenotypes of each individual, assuming that these phenotypes in turn determine such aspects of fitness as the fecundity, survival, and mating success of that individual. It is possible, however, that selection may act primarily through either relatives of individuals, that is, kin selection or on groups of individuals. We will briefly discuss kin and group selection because they have been suggested as important in the evolution of social behavior.

When the frequency of an allele is influenced by the effects of the allele on the fitness of relatives, then this type of selection is termed *kin selection*. When related individuals interact with one another in a nonrandom way, these actions may affect the fitness of kin differently from that of unrelated individuals. For example, individuals often interact differently with their sibs than with other individuals and, as a result, increase the fitness of their sibs. To describe this effect, a useful concept is that of *inclusive fitness*, which defines the fitness of an individual as its own intrinsic fitness plus the effects of relatives on the individual, weighted by the expected proportion of alleles they share in common. For kin selection to operate, the population must be structured into kin groups, such as families, but these kin groups need not be

---

Example 18.6. Although kin recognition is not essential to kin selection (see discussion above), kin recognition should facilitate kin selection. There have been few reports of kin recognition that have not been confounded with learned recognition. Learned recognition may occur early in development if sibs are not separated until after birth or hatching. However, a recent report on tadpoles of the Cascades frog, *Rana cascadae*, suggests that this organism may be able to distinguish between its sibs and nonsibs (Blaustein and O'Hara, 1981).

In this case, frog eggs were collected from nature, and in one group individual frog eggs were placed in isolated containers and allowed to mature. In a second group, tadpoles were allowed to develop with their sibs. Individuals from each of these groups were placed in the center of a tank that had sibs at one end and nonsibs at the other to determine their preference for sibs or nonsibs. Two measures of preferences, the proportion of individuals spending most of their time on the sib side of the tank and the proportion of time on the sib side of the tank, both demonstrated sib recognition and preference (see Table 18.3). This species does form aggregations as larvae, so kin selection in this species appears feasible.

In colonies of the primitively social bee, *Lasioglossum zephyrum*, one individual usually acts as guard bee and protects the nest from intruders. However, these guards usually recognize nestmates and allow them to enter whether they be relatives or unrelated individuals that are artificially introduced. When individuals from a laboratory colony were presented to the guard in a colony that had only related bees, the probability of acceptance directly corresponded to the coefficient of relationship (see Figure 18.4). These observations are consistent with the hypothesis that odor composition is genetically controlled and that guard bees are able to learn the odor of nestmates, thus allowing discrimination among odors of bees of different relationship.

---

separate physical entities and may exist as part of a large population. In other words, kin selection does not require any discontinuities in population breeding structure as does group selection (see discussion below).

The importance of kin selection in evolutionary thought is that it allows the formulation of the conditions for the development of altruistic behavior. *Altruism* is apparent concern and sacrifice for the welfare of others. For example, a worker honey bee will sacrifice her life by stinging an invading animal when defending the hive. However, a great deal of such altruistic behavior occurs between relatives, and as a result, one can consider such acts selfish because they may increase the probability of the propagation of alleles shared by the donor (the altruist) and the related recipient of the altruistic act.

The development of altruistic behavior via kin selection does not actually require that relatives be able to recognize each other (although recognition can

Table 18.3
Sib recognition in frog tadpoles measured by two different criteria (from Blaustein and O'Hara, 1981).

Treatment	*Proportion of individuals spending most time with sibs*	*Proportion of time spent with sibs*
Raised in isolation	0.90	0.58
Raised with sibs	0.80	0.57

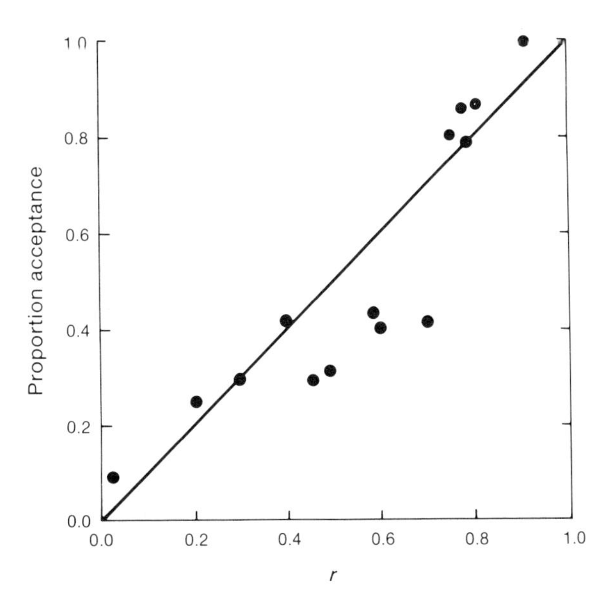

Figure 18.4 The proportion of acceptance by the guard bee of a colony in a primitively social bee of bees with different coefficients of relationship (after Greenberg, 1979).

facilitate kin selection; see Example 18.6) but without kin recognition it assumes that the probability of encounter (and the subsequent altruistic act) is directly related to the closeness of kinship of the individuals involved. Altruistic behavior assumes that an individual or individuals (the altruists or donors) sacrifice some fitness in order to increase the fitness of another individual or individuals (the recipients). Obviously, individuals practicing altruistic behavior may decrease their own fitness, thereby resulting in selection against altruistic behavior. Only if there is some counteracting selection on the alleles causing the altruistic behavior, will they increase in the population.

Let us assume for the sake of argument that there is such an altruistic allele and that it has a predictable phenotypic effect. The conditions for the genetic origin of altruistic behavior have been given in terms of the cost c in fitness to the donor, the benefit b in fitness to the recipient, and the coefficient of relationship r between the donor and the recipient. The coefficient of relationship is generally twice the inbreeding coefficient (see Chapter 7) and indicates the proportion of alleles shared in common between two individuals. For example, the r value between individuals A_1A_2 and A_1A_3 is 0.5 because they have half their alleles in common, assuming that A_1 came from the same ancestral allele. For an allele that results in altruistic behavior to increase,

---

Example 18.7. Eusocial insects exhibit cooperative care of young, overlapping of generations within a colony, and division of labor and reproduction into morphologically distinguishable castes. For example, the castes in honeybees are composed of workers, nonreproductive females that gather food and defend against predators; drones, males that inseminate the queen; and the queen, the reproductive female in the colony.

The naked mole rat, *Heterocephalus glaber*, is the only vertebrate that appears to be eusocial. These mole rats live in extensive systems of foraging tunnels in the hot, arid regions of east Africa. A colony was collected in Kenya and was subsequently observed under laboratory conditions (Jarvis, 1981). The captured colony consisted of 40 individuals that were separated into several types of workers: those that performed nest building and foraging tasks; nonworkers, including breeding males; and individuals that assisted in the care of the young. The queen was not captured, but one of the workers soon became the reproductive female. Naked mole rats have division of labor and reproduction, and exhibit size differences among the individuals that perform different tasks; they also have cooperative care of the young, only one breeding female, and overlapping generations. However, like the termites, the naked mole rat is diploid, demonstrating that haplo-diploidy is not essential for the development of eusociality.

$$\frac{b}{c} > \frac{1}{r}.$$

In other words, because r is always smaller than unity, for such an allele to increase in the population the ratio of the benefit to the cost must be greater than unity, and for smaller values of r this ratio must be quite large. As an example, r between two sibs is 0.5. Therefore, for an altruistic allele to increase in frequency, the benefit would have to be more than twice the cost. Dramatically emphasizing this concept, J. B. S. Haldane is reported to have said that he would "consider laying down his life for two brothers or eight cousins."

Sociality has arisen in arthropods a number of different times among bees, ants, and wasps, organisms in which males are haploid and females diploid, and only once in diploid arthropods (in termites) (see Example 18.7). For such haplo-diploid organisms, the r value between two sisters is 0.75, a very high value (assuming, perhaps unrealistically for many hymenoptera, a single fertilization of their mother). The basis for this is illustrated in Table 18.4, in which it is assumed that each parental allele is unique. Because the haploid male parent always contributes the same allele, every female offspring will have allele A_3. In addition, half of the time female offspring will have the same allele from the female parent and, consequently, identical genotypes—for example, members of a sister pair may both have genotype A_1A_3. The overall coefficient of relationship for sisters is then $\frac{1}{2}(1.0 + 0.5) = 0.75$, while that for the mother–daughter pair is 0.5, because the two always share only one allele. Therefore, the condition for an altruistic

Table 18.4
(a) The parental and offspring genotypes, assuming each parental allele is unique, and (b) the coefficient of relationship for sister–sister and mother–daughter pairs in haplo-diploids.

(a)		*Female*		*Male*
	Parents	A_1A_2	×	A_3
	Female offspring	$\frac{1}{2}A_1A_3$	and	$\frac{1}{2}A_2A_3$

(b)	*Sister–sister*	*Sister*		
	Sister	$\frac{1}{2}A_1A_3$	$\frac{1}{2}A_2A_3$	
	$\frac{1}{2}A_1A_3$	1.0	0.5	$r = 0.75$
	$\frac{1}{2}A_2A_3$	0.5	1.0	
	Mother–daughter	*Daughter*		
	Mother	$\frac{1}{2}A_1A_3$	$\frac{1}{2}A_2A_3$	
	A_1A_2	0.5	0.5	$r = 0.5$

allele to increase is that the benefit be 4/3 the cost, a larger allowable cost than for diploids. Furthermore, a female has more genes in common with her sisters than with her daughters, suggesting that for the maximum promulgation of her alleles, she can do better by caring for sisters that are emerging in a colony than by starting a new colony. This genetic hypothesis appears to be a reasonable hypothesis for the multiple origin of social behavior in hymenoptera, although there are other possible explanations.

The term *group selection* has been used in a general sense to describe any selection in which there is differential survival or reproduction (splitting) of groups of individuals (see Figure 18.5). In this example, group (a) has expanded to two groups over time and group (b) has become extinct. In many cases, the differential success or extinction of groups may be related to individual selective values in the groups. Here we will mention the potentially novel aspect of group selection—that is, the situation in which particular alleles or traits are advantageous to a group as a whole but are detrimental (or neutral) to individuals that have them.

An oft-cited example of a trait that is detrimental to individuals but advantageous to the group is alarm-call behavior in birds and other animals (although this can be explained by kin selection if the members of the group are related). In this case, the individual that warns of the presence of a predator is potentially exposed to a greater mortality risk. Other members of the group, because they are warned, have a greater probability of survival. Group selection has also been suggested as important in the regulation of population growth (see Chapter 12). If the growth of a population is unregulated, then there is a high probability that a population will outgrow its resources, crash, and become extinct. Some behavioral traits, such as

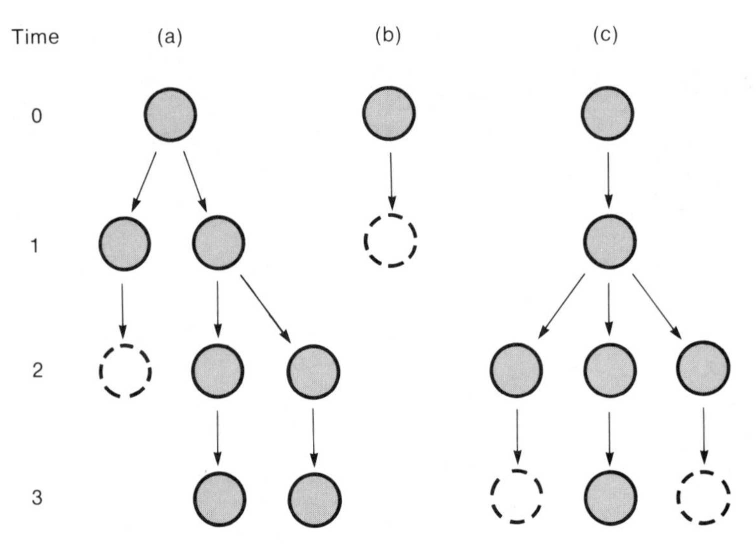

Figure 18.5. Differential survival and splitting of groups over time indicating group selection.

establishing territories or dominance heirarchies, may serve to prevent wide oscillations in population numbers, and groups possessing these attributes could, therefore, have higher survival. However, such traits may also have individual advantages—for example, having a territory should enhance individual survival and offspring survival, and group selection may not be necessary to explain such behavioral characteristics.

SUMMARY

Individual behavior and social behavior, such as those in dominance hierarchies or territoriality, are phenotypic traits. However, behavioral traits are particularly susceptible to environmental influences, and in fact, behavioral variation may often result from individual flexibility or learning. Sociobiology combines the study of social behavior and evolution, often assuming that variation in behavior traits is genetically based and adaptive. When examining behavioral variation within a population, one should evaluate these assumptions carefully. Kin selection and group selection are sometimes suggested as the basis of the evolution of altruism. The increase of an altruistic allele due to kin selection is a function of the cost of the altruistic behavior to the donor, benefit to the recipient, and the relationship between the two individuals.

PROBLEMS

1. Discuss dominance hierarchies and territoriality and how one might determine the adaptive difference between individuals and populations with and without such behavior.
2. Can behavior be considered a phenotypic trait? What problems might be encountered using a quantitative genetics approach?
3. What are the basic assumptions of sociobiology? Give an example where they may not be true. How useful do you think adaptive explanations of behavior are as applied to human behaviors such as aggression?
4. Calculate the coefficient of relationship for two daughters when both are diploid. Assume that the mother is A_1A_2 and the father is A_3A_4. If the ratio b/c is 1.25, would an allele for altruism increase in frequency, given this coefficient of relationship?
5. How do individual, kin, and group selection differ? Design an experiment that would differentiate among these types of selection.

CHAPTER 19

Demographic Genetics and Life-History Evolution

> We must realize that the possibilities for selection in a population are not unlimited, that the genetic structure of a population is the result of a compromise among various selective exigencies so that it is not possible, in any short time, for natural selection to cause an increase in fitness with respect to every stage of the life cycles and every metabolic activity.
>
> RICHARD LEWONTIN

When we discussed selection in Part 1 of this book, we assumed that each genotype could be given a constant relative fitness value determined by its relative viability. Obviously, the relative fitness of a genotype may also be determined by other factors, such as fecundity and mating success, that affect the genotype's ability to pass on its alleles. As discussed in Chapter 12, relative viability (or fecundity) may be dependent upon age. When the genotype as well as the age (or life stage) of the individuals is considered, so that there is both genotype-dependent and age-dependent selection, then this is termed *demographic genetics.*

For a given environment, selection may favor change in a particular aspect of fitness. For example, when there is a short growing season selection may favor an early initiation of reproduction—that is, a shifting of fecundity to earlier ages. Such a selective change may also result in a shift in other aspects of fitness, such as reduced longevity. Over long periods, such selective changes may result in one species having a relatively early and high fecundity and low viability, and in another, related species that lives in a different environment perhaps having a lower fecundity and higher viability. These and other related evolutionary responses can result in different species having different reproductive and survival patterns, often termed *life-history evolution.*

I. DEMOGRAPHIC GENETICS

Selection may occur at all or any of the different ages or life stages, such as gametes, juveniles, or adults. In addition to this general age specificity, selection may take different forms during various aspects of the life cycle of an organism. To reflect this complexity, we can graphically arrange the life cycle to reflect those stages in which selection can occur, and divide the total amount of selection in a particular population into four major selection (or fitness) components—zygotic, sexual, gametic, and fecundity. We can then use this general scheme for virtually any organism, such as trees, mammals, or insects, by focusing the emphasis on the aspects of the life cycle that seem most important in that organism. The total effect of selection is then the sum of these different selection components.

Figure 19.1 is a diagram outlining this division of selection into four components. Until now we have focused on selective differences resulting from different zygotic survival patterns. However, it is clear that each of these types of selection may be quite important in a given situation, suggesting that adaptive genetic change or stasis may be the net result of several of these selection components.

a. Zygotic Selection

Let us begin our examination of selection by assuming that there is a given composition of zygotes at the top of Figure 19.1. The initial proportions (or numbers) of the different genotypes in different age classes are modified by differential zygotic survival at different ages and in various life stages and differential rates of development. Zygotic selection includes age-specific

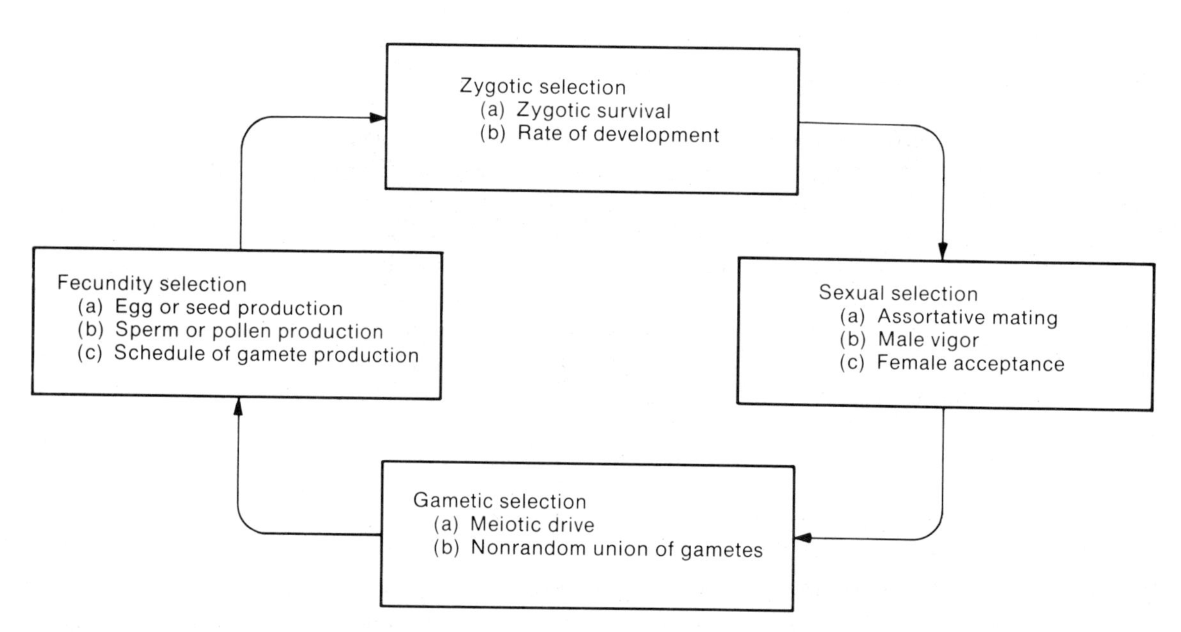

Figure 19.1. A diagram illustrating the division of selection into four components. Selection generally proceeds in a generation in a clockwise manner.

survival of different preadult stages—for example, in many insects there is differential hatchability (egg survival), larval survival, pupal survival, and premating adult survival. In plants, there are analogous stages, such as differential germination of seeds, seedling survival, and prereproductive survival. This is the type of selection we have discussed in previous chapters.

Zygotic selection can be estimated for a given life stage (or a given age interval). For example, to estimate the amount of selection that occurs due to preadult survival differences, we can use the ratio of the genotypic frequencies in an early stage of census, such as seedlings, to the genotypic frequencies of adults. We are assuming here that we are following a cohort through time to obtain these frequencies, making this a modification of the cohort approach for estimation of a survival schedule (see Chapter 12).

Let us assume that P_{ij} and P'_{ij} are the genotypic frequencies of genotype A_iA_j before and after zygotic selection and that the relative viabilities of genotypes A_1A_1, A_1A_2, and A_2A_2 are v_{11}, 1, and v_{22}, respectively. Therefore, the genotypic frequencies after selection are

$$P'_{11} = \frac{P_{11}v_{11}}{\bar{v}}$$

$$P'_{12} = \frac{P_{12}}{\bar{v}}$$

$$P'_{22} = \frac{P_{22}v_{22}}{\bar{v}}$$

where

$$\bar{v} = P_{11}v_{11} + P_{12} + P_{22}v_{22}.$$

If we divide the above expressions for the two homozygotes by that for the heterozygote, the ratios of the genotypic frequencies of the homozygotes over the heterozygote are

$$\frac{P'_{11}}{P'_{12}} = \frac{P_{11}v_{11}}{P_{12}}$$

$$\frac{P'_{22}}{P'_{12}} = \frac{P_{22}v_{22}}{P_{12}}.$$

These two equations can then be solved for estimates of the relative viability of the two homozygotes as

$$v_{11} = \frac{P'_{11}P_{12}}{P'_{12}P_{11}}$$

$$v_{22} = \frac{P'_{22}P_{12}}{P'_{12}P_{22}}.$$

This approach has been widely used when the initial genotypic composition is known because the parental mating is a cross between a homozygote and a heterozygote or between two heterozygotes (see Example 15.6). Example 19.1 is a study in which the genotypic frequencies were accurately estimated before and after zygotic selection by use of large sample sizes.

The above procedure could be extended to estimate zygotic selection at several different ages, as long as the zygotic frequencies can be estimated

---

Example 19.1. A composite cross of 30 varieties of cultivated barley has been propagated at Davis, California, since 1940. For nearly every generation, a random sample of seed was stored so that seeds from many generations are available. Extensive analysis of electrophoretic variants, primarily at several esterase loci, have been carried out in this population. In an effort to measure the amount of selection occurring at regions marked by these loci, Clegg et al. (1978) grew plants from several generations simultaneously and estimated the relative zygote-to-adult viabilities of different genotypes. Some of these data are presented in Table 19.1 for the *esterase A* locus, one of three tightly linked loci they analyzed, in plants from generations 8 and 28. The data are standardized against the A_1A_1 homozygote, the most common genotype in both generations. Using the above expression, the relative viabilities can be estimated for the different genotypes. In generation 8, none of the relative viabilities is substantially different from that of A_1A_1, with A_2A_2 slightly greater than unity and A_3A_3 slightly smaller than unity. In generation 28, the viabilities of A_2A_2 and A_3A_3 are significantly smaller than that of the A_1A_1 genotype. However, because of the high amount of selfing in this population, Clegg et al. caution that these viability differences may not be due to allelic differences at the electrophoretic loci but to other associated loci (see discussion below and Table 19.6). In this case, the genotype with the overall highest viability, A_1A_1, had the highest frequency, suggesting that viability may be an important fitness component in this population.

Table 19.1
The zygotic and adult frequencies and estimates of viability for the *esterase A* locus in a barley population (Clegg et al., 1978).

	Generation 8			*Generation 28*		
	Zygote	*Adult*	*Viability*	*Zygote*	*Adult*	*Viability*
A_1A_1	0.496	0.470	1.00	0.582	0.650	1.00
A_2A_2	0.350	0.388	1.17	0.354	0.306	0.77
A_3A_3	0.143	0.134	0.98	0.047	0.030	0.56

---

accurately. For example, the proportion of the different genotypes in a cohort can be observed several times as the cohort ages and, as a result, changes in selective values can be measured with age. If a population cannot be easily followed over time, then for organisms with known ages, such as humans and trees with annual growth rings, a modification of the static approach to estimation of a survival schedule is possible (see Chapter 12). In this case the mortality of individuals of different ages and genotypes could be monitored for a given time interval.

In an age-structured population, the genotypic frequencies for different age classes may also be estimated at one point in time. In this situation, inferences about selection assume that the older cohorts began with the genotypic frequencies observed for the younger cohorts. Because initial genotypic frequencies may vary due to genetic drift, inbreeding, or migration, such an assumption may not always be valid (see Example 19.2). This approach also assumes that selection pressures for a given age have not changed over time.

b. Sexual Selection

Assuming that zygotic selection is essentially complete before reproductive maturity and mating, the frequencies of the zygotes that contribute to the next generation depend upon the factors that determine the mating genotypes. In sexually reproducing organisms, the premating genotypic proportions of males and females may not be the same as the genotypic proportions among the mating individuals because of sexual selection (see also Chapter 16). For example, some male and female genotypic combinations may be favored by assortative mating either between like or different types, males may have different vigor in mating, or females may have different receptivities to males (see Example 19.3). For example, we discussed an example of rare-type male mating advantage in Example 15.4. Although some particular patterns of positive assortative mating do not result in changes in allelic frequencies, most types of assortative mating cause changes in allelic frequencies. As a result, here we will include such patterns as types of sexual selection.

c. Gametic Selection

The types and proportions of gametes produced by these mating genotypes depend upon gametic selection. For example, in some cases, the two types of gametes produced from heterozygotes may not be in equal proportions because of a phenomenon called *meiotic drive* (see Example 19.4). Of particular importance in a number of plants are factors that inhibit fertilization by particular types of pollen. These incompatibility systems are generally a function of either the pollen genotype or the genotype of the pollen parent. In addition, there may be competition by either sperm or pollen for fertilization.

d. Fecundity Selection

The number of zygotes produced by the mating individuals is determined by fecundity selection factors. Included in this category are all factors that

---

Example 19.2. Human populations have been examined for differences in allelic frequencies or heterozygosity among different age classes, and in general there are only small, if any, differences with age. These results may be due in part to the fact that the opportunity for viability selection is small in most contemporary human populations—that is, the proportion of the population surviving to reproduction is high. On the other hand, many plants produce large numbers of seeds and seedlings, only a few of which generally survive to adulthood.

The simplest estimate of zygotic selection from one-time sampling is to compare progeny and adult frequencies. For example, Moran and Brown (1979) estimated allelic frequencies and heterozygosity for three enzyme loci in adult alpine ash (*Eucalyptus delegatensis*) and seeds found on these adults (progeny). There were no differences in the allelic frequencies of parents and progeny, but the heterozygosity was much higher for the parents (see Table 19.2). Because there was approximately 25 percent self-fertilization in the population, the progeny heterozygosity was less than the 0.45 expected from random mating. The 0.55 heterozygosity in the parents suggests that there is viability selection that counteracts the decrease in heterozygosity due to the mating system.

A more detailed examination was done by Schaal and Levin (1976) on the perennial herb *Liatris cylindracea*. They divided a population into six "age" classes based on the number of rings in the corm (this classification has been questioned by Werner, 1978) and estimated allelic frequencies and heterozygosity at 14 loci. In this population, both allelic frequencies and heterozygosity were associated with these classes.

Table 19.2
The frequency of the most common allele and the observed heterozygosity for three loci in alpine ash in progeny and parents (from Moran and Brown, 1979).

	Allelic frequency		*Heterozygosity*	
	Progeny	*Parent*	*Progeny*	*Parent*
Mdh-1	0.57	0.58	0.45	0.57
Mdh-2	0.75	0.72	0.33	0.50
Aph-3	0.67	0.65	0.30	0.57
Mean	0.66	0.65	0.36	0.55

---

contribute to the zygote pool for the next generation. For example, if the contribution varies with particular male–female mating types, then there is the potential for selection via either egg, seed, sperm, or pollen production. Fecundity selection is often the result of differences in the gametic potential of

---

One way to measure heterozygosity is with the *fixation index*

$$F = 1 - \frac{H}{2pq}$$

where H is the observed heterozygosity and p and q are the frequencies of alleles A_1 and A_2, respectively. When there are Hardy-Weinberg proportions of the genotypes, then $F = 0$. If the only factor affecting genotypic frequencies is inbreeding, then this expression is between zero and unity and gives an estimate of the inbreeding coefficient (see Chapter 7). However, selection, migration, or genetic drift may also substantially affect genotypic frequencies, and the fixation index by definition includes these factors.

The observed heterozygosity was lowest, and the fixation index was highest for the lower classes in this population (see Table 19.3), suggesting at first glance that heterozygotes have an advantage. However, this result is somewhat confusing because *L. cylindracea* is thought to be self-incompatible, so that zygotic frequencies should be near Hardy-Weinberg proportions, making F near 0. These data show that there are only 24 percent of the Hardy-Weinberg expectations of heterozygotes in class 1. Although it is obvious that there are large differences between these classes, more extensive studies using either a static or cohort approach would reveal the dynamics of genetic change in this population.

Table 19.3
The expected and observed heterozygosity and the fixation index for *Pgi* and the average fixation index for 14 loci in different classes determined by the number of rings in *Liatris cylindracea* (from Schaal and Levin, 1976).

	"Age" class					
	1	2	3	4	5	6
$2pq$	0.50	0.50	0.50	0.48	0.50	0.50
H	0.12	0.27	0.28	0.28	0.42	0.34
F	0.77	0.46	0.45	0.42	0.17	0.31
$\bar{F}$(14 loci)	0.77	0.59	0.56	0.45	0.35	0.26

females because sperm and pollen production are generally thought not to be limiting (see Example 19.5). However, differences in pollen production may vary with genotype and affect the number of eggs fertilized. In addition, the schedule of egg or seed production over time may be an important factor in determining the contribution to the next generation.

Of all these selection components, selection affecting zygotic survival has been emphasized most often, probably in part because it is one of the easiest to quantify and to use in theoretical modeling. However, it appears from a number of studies that sexual, gametic, and fecundity selection may be as

---

Example 19.3. Although for many years the snow goose and blue goose were categorized as different species, today they are considered polymorphic forms of the same species. Analysis of segregation of these color types in progeny from different mating types suggests that the polymorphism is determined primarily by a single pair of alleles, with genotypes *BB* or *Bb* being blue and genotype *bb* white (Cooke and Cooch, 1968). The white phase predominates in flocks found in the western United States and the blue phase predominates in the east.

In mixed flocks, there appears to be positive assortative mating which results from males selecting mates that are similar in plumage to one of their parents. In other words, there are more white × white or blue × blue matings than expected if there were random mating (see Table 19.4). Here, in 771 mating pairs, 0.67 of the individuals were blue and 0.33 were white, resulting in the expected numbers of mating pairs given in the bottom row of Table 19.4. The cause of this sexual selection appears to be the imprinting of young goslings to the color of their parents. Experimental tests with young placed with foster parents of different colors have shown that imprinted geese will choose the color of their foster parent (Cooke et al., 1972). If the imprinting were complete, then one would expect that gene flow between the two phases would cease and that the potential for further differentiation and possible speciation would be present.

Table 19.4
The observed numbers of mating pairs among the blue and white phase of the snow goose and the expected numbers based on random mating (from Cooke and Cooch, 1968).

	Blue × blue	*Blue × white*	*White × white*
Observed	470	98	203
Expected	349.3	339.2	82.5

---

important as zygotic survival in many situations. Although we have not emphasized the age-specific nature of sexual, gametic, or fecundity selection, there are indications that all of these selection components may be age dependent. Obviously, a full accounting of selection and its effect on these different components, as well as age-dependent considerations, requires an extensive and detailed study. Potentially, overall selection that is a combination of these selection components may explain the maintenance of genetic polymorphisms in natural populations or the development of particular life-history characteristics.

Example 19.4. In a number of animals, heterozygous individuals do not produce equal proportions of their two different alleles because of meiotic drive, a phenomenon thought to result from the interaction of the different chromosomes when they are synapsed in meiosis. For the *t* alleles in the house mouse, and in most other cases of meiotic drive, the distortion from normal segregation takes place only in males. Most *t* alleles are lethal (or sterile) when homozygous, but heterozygous males generally produce a large majority of sperm with the *t* allele. In other words, the viability disadvantage of *t* alleles as homozygotes balances their gametic advantage in male heterozygotes. Lewontin and Dunn (1960) reported the estimated proportion of *t* carrying sperm for 16 *t* alleles from wild populations (Figure 19.2). Heterozygotes for the wild *t* alleles produced an average of 0.952 *t* alleles, and all the alleles had a segregation ratio greater than 0.85. However, many new mutant alleles have a segregation ratio near 0.5, suggesting that selection has favored the establishment of *t* alleles that result in meiotic drive.

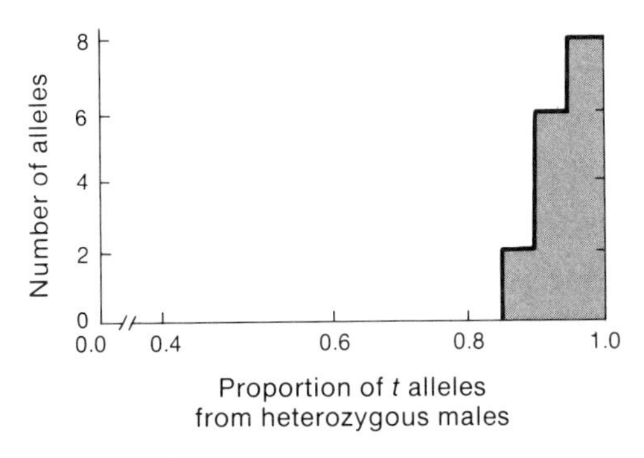

Figure 19.2. The segregation ratio found for 16 *t* alleles occurring in natural populations of the house mouse (Lewontin and Dunn, 1960).

II. LIFE-HISTORY EVOLUTION

A variety of types of survivorship and fecundity has been observed in different natural populations. Generally, these patterns, resulting from zygotic and fecundity selection, are called life-history traits (or "strategies"). In Chapter 12, we discussed the different types of survivorship curves, which range from high juvenile mortality (Type III curve) to a low juvenile mortality and higher mortality in the older age classes (Type I curve). Fecundity schedules may also be separated into different basic types—repeated reproduction (*iteroparity*) and single reproduction (*semelparity*). Many organisms are iteroparous, such

---

Example 19.5. The eelpout, *Zoarces viviparus*, is a live-bearing fish distributed along the northwestern coast of Europe that is very abundant in some areas. Because the young of a brood can be obtained from a female in late fall and early winter, so that mother–offspring combinations may be examined, this organism has been used to study in detail selection in different life stages. In particular, a polymorphism at an esterase locus has been examined to determine what selective factors are important at, or associated with, this locus. For example, the relative fecundities of the three esterase genotypes in two different years are given in Table 19.5 (from Christiansen et al., 1973). The basic data are the total number of young carried by a pregnant female of a particular genotype. In 1969, the genotype A_2A_2 had the highest fecundity of the three genotypes, whereas in 1970, A_1A_1 had the highest fecundity. However, the overall fecundity estimate for the two years suggests that there are no significant fecundity differences among these genotypes. The complete analysis of selection at this locus also suggests that there is no gametic selection and no sexual selection but some slight zygotic selection.

Table 19.5
The relative fecundities of three esterase genotypes in the eelpout (Christiansen et al., 1973).

	Genotype	*Number*	*Mean number of young per female*	*Mean relative to A_1A_2*
1969	A_1A_1	111	106.2	0.98
	A_1A_2	357	108.6	1.00
	A_2A_2	314	118.2	1.09
1970	A_1A_1	54	102.0	1.14
	A_1A_2	158	89.1	1.00
	A_2A_2	165	84.8	0.95

---

as most birds, primates, and trees, and reproduce a number of times. However, some species, such as salmon and bamboo, are semelparous and reproduce only once and then die. Among populations of a species there may be great variation in fecundity. For example, the clutch size in bird populations from the tropics is often much lower than in populations of the same species from temperate zones.

Based on the extensive variation in fitness components in different related species or populations of the same species, a theory of life-history evolution has been developed to explain these combinations of traits. (The other selection components, sexual and gametic selection, are generally not considered in terms of life-history evolution.) In a general way, one would expect an organism to evolve towards both a high and early fecundity and also towards a high survival rate. However, two factors appear to influence this trend. First, the environmental pattern may be such that either fecundity selection or zygotic selection is more important, thereby resulting in the major effect of selection on one of these fitness components (see discussion below).

Second, for most organisms there appears to be a limited amount of energy available, and if a large amount is expended on reproduction, then relatively less is available for other activities. In a simple way, we can assume that there is a fixed total amount of energy available for the organism and that this amount must be used for reproduction, growth, and maintenance. If most of this energy is used for reproduction, then less is available for growth and maintenance, thereby resulting in a lowered viability. The results of such energy allocation can be understood intuitively for organisms with parental care. If an organism, say a bird, expends a large amount of energy on reproduction, producing a clutch of eggs and tending and protecting its nestlings, its own survival may be negatively affected. In other words, there appears to be a negative association (a tradeoff) between fecundity and survival because of the allocation of resources (energy) either to reproduction or to growth and maintenance. To demonstrate this phenomenon experimentally, it is best to examine variation of individuals within a species in a given environment (see Example 19.6). Comparison of different species or of individuals in different environments may confound the observations either with species or environmental differences.

Here we have discussed fecundity and survival and general components of fitness. Before we continue, as a general background for an examination of the value of life-history characteristics, let us consider the potential relationships of various phenotypic traits with overall fitness, what determines the rate of change in fitness and traits related to fitness, and what is the genetic basis for associations between traits.

We can arbitrarily define three basic types of phenotypic traits. For the first type of trait (as given in Figure 19.4a), the fitness is essentially the same for all phenotypic values, with a slightly higher fitness for intermediate values. Peripheral traits, such as bristle number in *Drosophila* or scale count in snakes, generally fit this category, although extreme values of both these traits may have significantly lowered fitness. For the second type of trait, as illustrated in Figure 19.4b for body size, an intermediate phenotype has a definite optimum fitness. More extreme phenotypes in both directions have lowered fitnesses. Last, the relationship of a fitness component and fitness, as given in Figure 19.4c, shows a marked increase in fitness as the fitness component

Example 19.6. To determine whether there was a tradeoff between survival and fecundity, Snell and King (1977) measured the age-specific fecundity and survival in a large number of individuals in a predatory rotifer. To standardize the environment, they kept the rotifers in large Petri dishes containing a small amount of water, fed them 250 paramecia per day, and kept them at a constant temperature. The individuals initiating a cohort were born within a three-hour period and were examined each 12 hours to record any offspring or deaths. A summary of the age-specific fecundity, the m_x values for a group of ten individuals, versus the survival probabilities in these groups to age class $x + 2$ is given in Figure 19.3. From these data, it appears that high fecundity is detrimental to future survival and that low-fecundity individuals have relatively higher survival, data consistent with the hypothesis that allocation of resources to reproduction reduces the resources available for growth and maintenance.

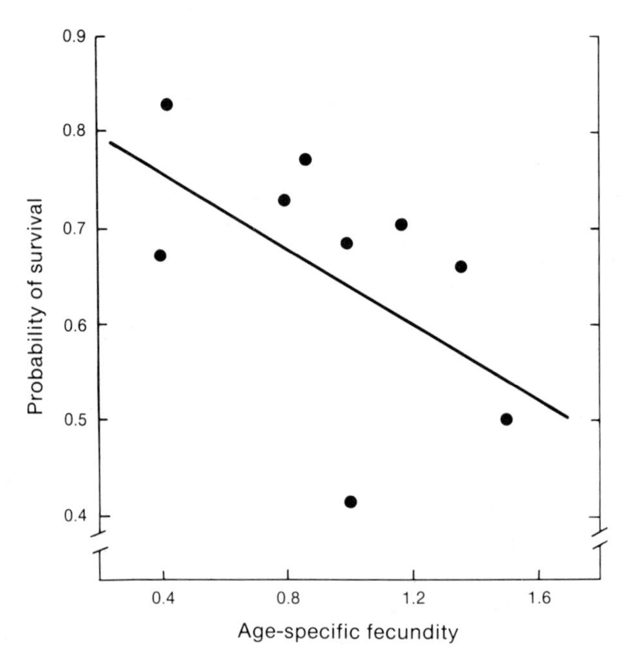

Figure 19.3. The relationship between fecundity and subsequent survival in a rotifer (Snell and King, 1977).

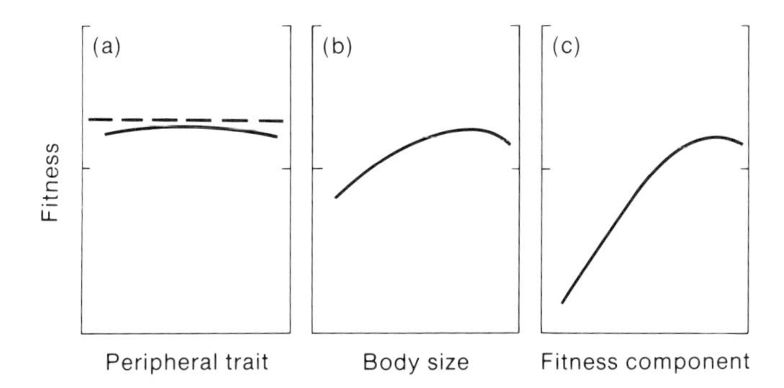

Figure 19.4. The hypothetical relationship between fitness and three types of traits (after Robertson, 1955).

increases until, at the upper limit, overall fitness declines as the fitness component increases. Assuming the fitness component is fecundity, this turnover may result from lowered viability associated with the physiological stress of high fecundity. When we consider life-history characteristics, the traits of major importance are components of fitness, such as viability and fecundity. However, other traits, such as body size, are important because they are simpler to measure and they may also indicate some relationship with fitness.

In a general sense, for a population to respond genetically (adapt) to a change in the environment, there must be genetic variation in the population. Obviously, if there is no genetic variation, there can be no response, and, as the variation increases, the rate of response should increase. This general principle was formulated algebraically by R. A. Fisher and is called *the fundamental theorem of natural selection.* For a simple genetic model, like that discussed below, this principle is that the rate of change in fitness is equal to the genetic variance in fitness.

To illustrate the fundamental theorem, let us consider a population that reproduces parthenogenetically (has no sexual reproduction) and that the ith genotype in the population has a relative fitness value and frequency of w_i and q_i, respectively. The mean fitness of the population is then

$$\bar{w} = \sum_{i=1}^{k} q_i w_i$$

where there are k different genotypes in the population (for a review of selection models, see Chapter 3). After one generation of selection, the frequency of genotype i is

$$q_i' = \frac{q_i w_i}{\bar{w}}.$$

Using the same definition of mean fitness as above and the new genotypic frequencies, the mean fitness in the next generation is (dropping the notation above and below Σ)

$$\begin{aligned} \bar{w}' &= \sum q_i' w_i \\ &= \sum \frac{q_i w_i}{\bar{w}} w_i \\ &= \frac{1}{\bar{w}} \sum q_i w_i^2. \end{aligned}$$

What is the change in relative fitness as the result of the selection? Let us standardize the relative fitness values so that $\bar{w} = 1$. Therefore, the change in fitness is

$$\begin{aligned} \Delta w &= \bar{w}' - \bar{w} \\ &= \sum q_i w_i^2 - 1. \end{aligned}$$

When the numbers in a particular class are given as frequencies, then the variance in fitness (using our notation here) is

$$\begin{aligned} V_w &= \sum q_i (w_i - \bar{w})^2 \\ &= \sum q_i w_i^2 - 2\bar{w} \sum q_i w_i + \bar{w}^2 \sum q_i. \\ &= \sum q_i w_i^2 - 2\bar{w}^2 + \bar{w}^2 \\ &= \sum q_i w_i^2 - 1. \end{aligned}$$

Therefore, the expected change in fitness due to genetic factors is equal to the genetic variance in fitness, or

$$\Delta w = V_w.$$

With several assumptions (no new genetic variation introduced into the population by migration, mutation, or recombination and a constant environment), then one might expect the population to become more adapted with time, asymptotically approaching a genetic constitution with a few highly adapted types. As a result, one would expect there to be little genetic variation in fitness or important components of fitness. Looking back at Table 8.2, it is obvious that those traits most closely related to fitness, such as viability in chickens and litter size in mice, do have the lowest heritabilities. One reason for a low heritability or low additive genetic variance for a trait in a natural population is that the trait has been strongly selected.

However, as we discussed before, there are limits to the extent of change that a trait may undergo. When we consider these limits from a genetic viewpoint, rather than an energetic viewpoint, the extent of response is influenced both by the amount of genetic variation in the trait and the genetic correlation between the trait and other traits undergoing selection—for

Table 19.6
Simple illustrations of pleiotropy and gametic disequilibrium, models that may explain negative correlations between traits. In (b) gametes B_1C_2 and B_2C_1 are very low in frequency.

(a) Pleiotropy

Genotype	A_1A_1	A_1A_2	A_2A_2
Fecundity	High	Intermediate	Low
Viability	Low	Intermediate	High

(b) Gametic disequilibrium

Genotype	B_1C_1	(B_1C_2)	(B_2C_1)	B_2C_2
Fecundity	High	(High)	(Low)	Low
Viability	Low	(High)	(Low)	High

example, other fitness components. If the genetic correlation between the traits is negative, then an increase in one, say fecundity, would result in a reduction in another, say viability, and no net increase would occur in overall fitness. In other words, these internal constraints would allow a selective change in a fitness component to the point at which negative genetic correlations would restrict any further changes.

What is the basis of such genetic correlations between traits? Two major factors are thought to be important in determining genetic correlations—*pleiotropy*, the situation in which alleles at a locus simultaneously affect two or more traits, and *gametic disequilibrium*, in which alleles at one locus are associated with alleles at another locus. The top half of Table 19.6 is a simple example of pleiotropy, in which genotype A_1A_1 results in both high fecundity and low viability, while the homozygote A_2A_2 has the opposite consequence.

The bottom half of Table 19.6 is an example of gametic disequilibrium resulting in a similar association of traits. In this two-locus example for a haploid organism, we will assume that the B_1 allele at the B locus confers high fecundity (B_2 low fecundity) and the C_2 allele at the C locus high viability (C_1 low viability). There are four possible genotypes (gametes, if the organism were diploid), but if we assume that only two of them, B_1C_1 and B_2C_2, are common, then individuals would generally either have a combination of high fecundity and low viability, B_1C_1, or low fecundity and high viability, B_2C_2.

Notice that in pleiotropy one locus has multiple effects, while in gametic disequilibrium each locus affects a different trait. Furthermore, the pleiotropic association is permanent, while the interlocus association may be broken up by recombination (or sexual reproduction). In both cases, as illustrated here, if there is selection for high fecundity, then there would be a correlated response resulting in a lowered viability. This antagonism between components of fitness (or other traits) may result in a reduced rate of response or in little response to directional selection (see last half of Example 21.3).

When considering life-history characteristics, we need to remember that these are quantitative traits and that, for example, environmental factors and genetic-environmental interaction may be important in determining these phenotypes (see Chapters 8 and 18). Furthermore, associations of traits measured in natural populations are phenotypic correlations and do not necessarily reflect genetic correlations. In other words, even though there appear to be general patterns of correlation of life-history characteristics, such as high fecundity and low viability in some species or low fecundity and high viability in other species, the extent to which these traits are genetically determined is not clear. Such correlation patterns may be the result of genetic factors, environmental factors, or both. Even if there is a negative genetic correlation between viability and fecundity, it may be obscured by environmental correlations. In fact, in some cases the environmental correlation between traits may be positive and the genetic correlation negative.

The measurement of genetic correlation between traits is based on the observed versus the expected resemblance of relatives for the traits of interest (see Falconer, 1981, for details). Using this genetic correlation and estimates of heritability based on resemblance between relatives, we can make a prediction for the expected response in a selection experiment. Although this is a common procedure in plant and animal breeding, in which many traits are important for market, the approach has not been used frequently in examining life-history evolution (however, see Example 19.7).

The extensive interest in life-history evolution in recent years stems from the predictions made first by R. H. MacArthur that the demographic or life-

---

Example 19.7. Recently, Rose and Charlesworth (1981a) estimated the genetic determination of life-history characteristics in a *D. melanogaster* population. By analyzing the variation among sibs, they estimated that the genetic variance of fecundity characteristics in their population was primarily additive and that of longevity variation was primarily composed of dominance variance. In addition, they found negative genetic correlations between several measures of fecundity and longevity. They then carried out several selection experiments to determine if the observed phenotypic change was consistent with the predicted changes, using these estimated genetic variances and correlations (Rose and Charlesworth, 1981b).

For example, three generations of selection for increased egg-laying in 1- to 5-day-old flies increased that trait significantly, as was predicted by the large amount of additive genetic variance and the heritability of 0.62. However, there was no correlated change in longevity in this experiment, even though the estimated genetic correlation between this fecundity trait and longevity was negative and quite large. In another selection experiment, however, there did appear to be antagonistic effects between early and late life-history characteristics, a finding consistent with the negative additive genetic correlations between fecundity and longevity.

---

history characteristics of a population may vary depending upon the type of environment it experienced. For example, in a constant environment limited by resources, the population may exist at or near the carrying capacity, K. In other environments, in which conditions change in an unpredictable manner, a population may seldom reach the carrying capacity and is generally in a high-population-growth phase. If we assume that these are two extremes of a continuum of environmental conditions found in nature, then we can say that populations in these two situations are subject to high-density conditions, or K-selection, and low density conditions, or r-selection, respectively (MacArthur, 1962).

In order to evaluate the evolutionary consequences of environmental patterns on life-history characteristics, Pianka (1970) categorized in a general way the population characteristics and the traits that selection would favor in the two types of environments (see Table 19.7). For example, he assumed that in an unpredictable environment the population density would be low and mortality type III (high early mortality), so that selection would favor those individuals with rapid development (early reproductive maturity) and high fecundity because these changes would enable a population to recover quickly after an episode of high mortality. On the other hand, in a stable environment, the population density would be near the carrying capacity and adult mortality concentrated in the older age classes, so selection would favor delayed reproduction and generally a more efficient use of resources to promote a longer life. Returning to the concept of energy allocation to reproduction versus growth or maintenance, for populations in unpredictable environments more energy is used for reproduction, increasing r, while for populations in stable environments more energy is used for growth and maintenance (survival), or increasing K (see Example 19.8).

Table 19.7
Factors that may be correlated with r- or K-selection (after Pianka, 1970).

	r-selection	*K-selection*
Environment	Variable and/or unpredictable	Constant and/or predictable
Population characteristics		
Mortality	Density independent	Density dependent
Survivorship	Type III	Type I or II
Population size	Variable, well below K	Constant, near K
Competition (intra- and interspecific)	Minimal	Strong
Selection favors	Short life	Long life
	Rapid development	Slower development
	Early reproduction	Delayed reproduction
	Semelparity	Iteroparity
	Small body size	Larger body size
		Higher competitive ability

In the effort to attain generalities concerning life-history evolution, these ideas have been overextended to classify species as *r*-selected or *K*-selected, depending upon the population characteristics of the species. In addition, it has been suggested that whole groups of organisms, such as insects or annual plants, can be categorized as *r*-selected and that other groups, such as terrestrial vertebrates or perennial plants, are *K*-selected. Although such classifications may generally agree with the life-history expectations based on *r*- and *K*-selection, one should be careful when inferring that the life-history patterns are a consequence of selection in response to an environmental pattern. As an example, Schaffer (1974) has shown that if a population has age-specific survival and that if environmental fluctuations influence primarily juvenile mortality, then selection should favor long life, slower development, and iteroparity. This conclusion is the opposite of that given in Table 19.7, in which unpredictable environmental variation (*r*-selection) favors a short life, rapid development, and semelparity. In other words, the one-dimensional *r*-*K* continuum is generally incapable of explaining the diversity of life-history patterns within a group (Tinkle et al., 1970).

Example 19.8. To demonstrate that environmental differences have led to a specific evolutionary response, it is appropriate to compare populations of the same species exposed for a number of generations to different environments. The actual comparison should take place in a common environment, so that any phenotypic differences observed would be of genetic origin. Law et al. (1977) compared populations of *Poa annua*, annual meadow grass, from "opportunist" populations, those appearing to experience primarily unpredictable environmental conditions and, consequently, density-independent regulation, and "pasture" populations, those experiencing primarily constant environments with density-dependent regulation. Samples from 14 populations of each type were grown together, initially in a greenhouse and later in an experimental field. Age-specific fecundity and age-specific viability measures and size characteristics were taken on these plants over a period of 10 months. For example, the diameter of the plants from the pasture populations was significantly larger than that of the plants from the opportunist populations, suggesting that these plants from a relatively constant environment were allocating more energy for growth. The fecundity and survival data were combined into an overall measure of population growth, λ, the finite rate of increase (see Chapter 11; the technique to estimate λ from life-history data is similar to that used to estimate r, see Chapter 12). Figure 19.5 gives the distribution of λ for the two sets of populations. The mean λ value for the opportunist population, 1.84, is significantly higher than the 1.65 value for the pasture population. The difference reflects a shorter prereproductive period and a higher seed output early in life for

Another approach to considering the evolution of life-history characteristics in unpredictable environments is the following. Given that there is unpredictable environmental variation, selection for increased fitness may be achieved by minimizing the between-generation variation in components of fitness. A number of life-history changes might reduce such variation over generations. For example, iteroparity would buffer an organism against environmental variations that are shorter than the generation length and would minimize variation in age-specific fecundity. Phenotypic plasticity that resulted in changes in life-history characteristics responsive to environmental patterns could reduce temporal variation in age-specific survival. Adaptations involving diapause, dormancy, and migration might also serve to reduce variation in age-specific survival or fecundity (see Example 19.9). In other words, many life-history characteristics may evolve because they reduce the variation in age-specific survival and fecundity in variable environments.

Besides fecundity and survival patterns, life-history patterns are composed of adaptations that enable an organism to survive or avoid unfavorable environments. Unfavorable environments may occur in a predictable pattern,

---

the density-independent populations, consistent with the predictions for an *r*-selected population.

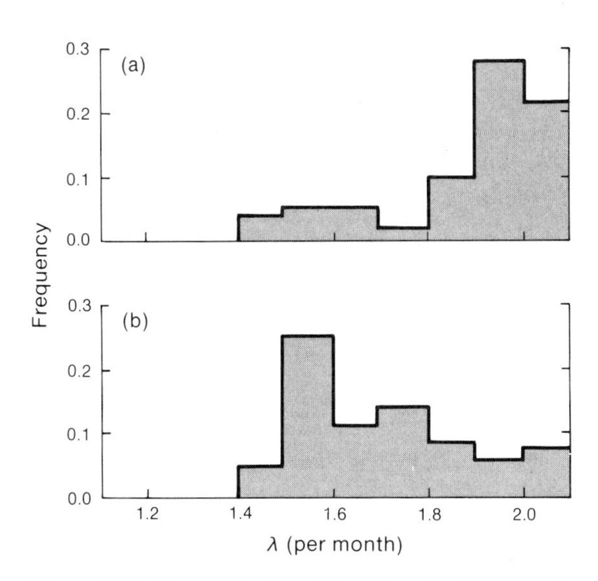

Figure 19.5. The distribution of the finite rate of increase in (a) opportunistic and (b) pasture populations of *Poa annuis* (Law et al., 1977).

such as seasonal variation with cold winters that affects many plants and insects, or in an unpredictable manner, such as the periodic droughts or great storms that occur in many environments. Organisms generally avoid seasonally adverse conditions either by dormancy, as in many plants, by diapause, as

Example 19.9. In many plants, seed dormancy is broken by a combination of cold temperatures followed by warm temperature or changing photoperiod. However, within some species variation occurs in the germination requirements, even among the seeds produced by one individual (Harper, 1977). For example, in some species, the seeds are produced in pairs of quite different sizes (seed dimorphism). The dormancy-breaking requirements are so different in the different sizes that a year generally separates germination in the two classes of seeds. Another example of seed polymorphism occurs when individual plants produce some seeds that require cold before germination and other seeds that are killed by cold. The cold-intolerant seeds that germinate in the fall may be selectively favored in some cases, while the seeds that endure the winter as seeds and germinate in the spring may be favored in other environments.

Many organisms that live in temporary ponds have adaptations to survive during dry periods. For example, many species of *Daphnia* produce resting (diapausing) eggs that can survive dry periods. In a temporary pond in Illinois, Lynch (1983) found that egg production varied seasonally among the common clonal groups as identified by electrophoresis (see Table 19.8). In three different years, clonal group *A* produced the most resting eggs early in the season, while the other clones generally produced more resting eggs in late summer just prior to the dry season. Resting eggs may be produced sexually or asexually. Clonal group *A* appears to produce resting eggs asexually, while both *B* and *C* produce resting eggs sexually. In other words, the temporal production of resting eggs is genetically determined, and, in addition, the means of producing them, sexual or asexual, may have genetic consequences for the population.

Table 19.8
Production of resting eggs in *Daphnia pulex* in three different years (after Lynch, 1983).

Clonal group	*1978*		*1979*		*1980*	
	Early	*Late*	*Early*	*Late*	*Early*	*Late*
A	0.96	0.0	0.88	0.14	1.00	0.08
B	0.04	0.0	0.00	0.79	0.00	0.01
C	0.00	1.0	0.12	0.07	0.00	0.91

in many insects, or by migration to other areas, as in birds. A life stage that is particularly well adapted to survival under the stringent conditions, such as seeds in plants or eggs or adults in insects, becomes dormant or diapauses.

Many organisms that can move or migrate in response to environmental adversity avoid unfavorable conditions that occur in an unpredictable pattern. For example, in response to environmental cues, aphids will develop winged forms in order to migrate. Plants and other sessile organisms cannot move in response to such a change but may develop mechanisms that ensure dispersal of seeds to favorable conditions. For example, most species of weeds that live in unpredictable environments have very efficient dispersal mechanisms, such as sticktight seeds that attach to animals or plumose seeds that may be dispersed great distances by the wind.

SUMMARY

Selection can occur at any or all life stages or ages of an organism. Overall selection may be divided into zygotic, fecundity, sexual, and gametic selection. Zygotic and fecundity selection have resulted in a large variety of life-history patterns in natural populations. We can understand these patterns on a phenotypic level by considering allocation of energy to reproduction or to growth and maintenance and on a genetic level by using the fundamental theorem of natural selection and genetic bases of correlated responses, pleiotropy and gametic disequilibrium.

PROBLEMS

1. A census of genotypic frequencies in the progeny of a cross between two heterozygotes gave the following results at different ages:

Age	A_1A_1	A_1A_2	A_2A_2
0	0.25	0.5	0.25
1	0.2	0.55	0.25
2	0.15	0.6	0.25
3	0.1	0.6	0.3

   What are the estimated overall relative viabilities and the relative viabilities from ages 1 to 2?

2. The frequency of melanic moths is lower in some cases than can be explained by zygotic selection alone. Design experiments to determine whether either fecundity or sexual selection is occurring in these populations. What results would be consistent with the higher-than-expected frequency of melanics?

3. A population was surveyed for its heritability of one trait related to fecundity and another related to viability, and their values were 0.1 and 0.4, respectively. Using the fundamental theorem of natural selection in its simplest form, what interpretation could you make about past selection in this population?

4. Assuming that the negative genetic correlation found between fecundity and viability is due either to pleiotropy or gametic disequilibrium, how could you experimentally differentiate between these causes?

5. If selection is acting to maximize r, this can be achieved either by increasing fecundity in all age classes or by reducing the age to initiation of reproduction and keeping fecundity the same. For example, assume the following data:

		m_x		
Age	l_x	a	b	c
0	1.0	0	0	0
1	0.5	0	0	5
2	0.25	5	10	5
3	0.125	5	10	0
4	0.0	0	0	0

Calculate the r values for each fecundity schedule and discuss the results with respect to life-history evolution.

CHAPTER 20

Coevolution and the Evolution of Interspecific Interactions

I conclude that species diversity of plant communities is an evolutionary product, subject to self-augmentation through time. There is no ceiling or saturation level for the diversity which results; the chemical differentiation of the higher plants implies virtually unlimited potentialities for the addition of different species....

ROBERT WHITTAKER

When we considered the interactions between competitors in Chapter 13 and the interactions between predator and resource populations in Chapter 14, we did not discuss the evolutionary aspects of these species interactions. As will become apparent in this chapter, the interactions between species can be affected by the phenotypic characteristics of the two species, and, in a general sense, characteristics such as interspecific competitive ability and predator avoidance patterns may be considered phenotypic traits even though they themselves are a composite of morphological, behavioral, and physiological phenotypes. Such phenotypic traits may be considered in the same vein as other quantitative traits. For example, phenotypic variation in interspecific competitive ability may be affected by the environment, and selective changes may occur in it as a result of competitive interaction (or independent of interactions). In fact, the concept of *coevolution* suggests that interacting species such as competitors, predator and resource populations, or mutualists may evolve in response to each other. A useful definition of coevolution is that it is an evolutionary change in a trait of one interacting species that is a response to a trait of the individuals in the other species and that is followed by an evolutionary response in the second species to the change in the first (Janzen, 1980). In some cases, the term coevolution may be used in a more restrictive sense to refer only to changes in mutualistic species, such as those between yucca plants and yucca moths or fig trees and fig wasps. In these obligatory relationships, the plants need the insects for pollination and the insects use the plants as larval feeding sites. Of course, though two species appear to have adaptive characteristics that are consistent with those expected of interspecific relationships, the possibility remains that these traits were present in the separate species before they came into contact.

I. INTERSPECIFIC COMPETITIVE ABILITY AND CHARACTER DISPLACEMENT

The outcome of interspecific competition in an experimental situation between two species may depend on the genetic constitutions of the competing species. In other words, genetic factors within a species may vary from population to population, resulting in different phenotypic characteristics and interspecific competitive ability. However, the determinant of competitive outcome may be not merely the ability to compete interspecifically (interference competition) but rather the overall adaptedness of different popula-

Example 20.1. As discussed in Chapter 13, the outcome of interspecific competition between the two sibling species of the flour beetle, *Tribolium castaneum* and *T. confusum*, is dependent upon the environment. In addition, the outcome is highly dependent upon the genetic strain of each species used. To investigate the impact of strain differences, Park et al. (1964) used four strains of each species that differed in their demographic characteristics. The outcomes of a number of experimental trials among these strains are given in Table 20.1. The winner is primarily

Table 20.1
The outcome of interspecific competition among four strains of two sibling species of *Tribolium* (from Park et al., 1964).

T. castaneum	*T. confusum*	*Proportion of trials that T. castaneum won*
cI	*bI*	1.0
	bII	1.0
	bIII	1.0
	bIV	1.0
cII	*bI*	0.11
	bII	0.0
	bIII	0.0
	bIV	0.4
cIII	*bI*	0.0
	bII	0.0
	bIII	0.0
	bIV	0.0
cIV	*bI*	0.9
	bII	1.0
	bIII	0.9
	bIV	0.8

tions in particular environments (exploitative competition). For example, if two species compete in an exploitative manner and one species uses the limited resources more efficiently, then, all other things being equal, the latter will outcompete the other species. Even changes in competitive dominance may be due to increases or decreases of within-species fitness that in turn lead to differential success in interspecific competition. To place such changes in the context of the logistic model, evolutionary change may result in a change in r or K for one of the species, as discussed in Chapter 19, which may lead to dominance in interspecific competition (see Example 20.1).

On the other hand, evolutionary changes may occur as the result of the

---

dependent upon which strain of *T. castaneum* is used, with *cI* and *cIV* nearly always winning and *cII* and *cIII* nearly always losing. An analysis of the strain differences showed that *cI* and *cIV* had the highest intrinsic rates of growth and the highest rates of cannibalism (*Tribolium* adults and larvae eat eggs and pupae), factors that appeared to enable them to win in interspecific competition.

Drosophila melanogaster and *D. simulans* are two sibling species that coexist throughout much of the world. Under normal *Drosophila* culture conditions, *D. melanogaster* usually easily outcompetes *D. simulans*, but a mutant stock of *D. melanogaster* was outcompeted by *D. simulans* (Hedrick, 1972). When this lab stock was tested against *D. simulans* two years later, the *D. melanogaster* stock easily outcompeted *D. simulans*. In other words, a change in some factors that reversed the interspecific competitive outcome occurred independently of contact with the competitor. Comparisons of the two *D. melanogaster* samples (Table 20.2) showed that the winner in interspecific competition had greater productivity (number of emerging progeny), a shorter emergence time, and an earlier fecundity schedule. The greater productivity of the sample that won in competition suggests that it had a higher K, and the faster rate of development and earlier fecundity schedule gave the winning sample a higher r value than the losing sample, likely explanations for the observed reversal in competitive outcome.

Table 20.2
Demographic characteristics of a *D. melanogaster* sample that won in competition with *D. simulans* and the same stock that did not have the adaptive change and consequently lost in competition (from Hedrick, 1972).

D melanogaster sample	*Productivity*	*Mean emergence time (days)*	*Number of eggs per female on day 1*
Winner	299.2	12.10	3.56
Loser	125.0	12.74	0.34

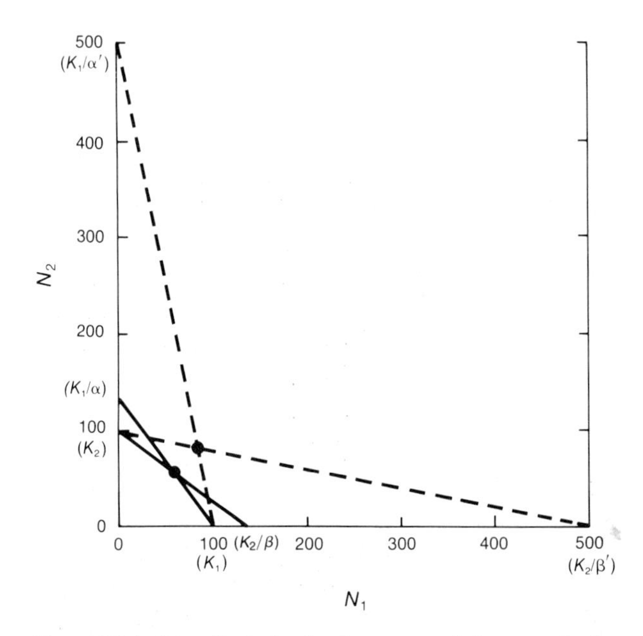

Figure 20.1. The effect of a simultaneous evolutionary reduction in α and β from 0.8 (solid lines) to 0.2 (broken lines) on the equilibrium numbers of two competitors.

presence of other competing species. For example, if we use the interspecific-competition equations given in Chapter 13, we may find that the evolutionary response results in a divergence of resource-utilization curves and thereby in a reduction of niche-overlap and competition coefficients. Figure 20.1 indicates the theoretical consequences of a reduction in α and β to 25 percent of their previous levels, from 0.8 to 0.2, so that two previously ecologically similar species evolve to use different resources. As a result of this change, if we assume that the carrying capacities of both species are 100, the equilibrium numbers of the two species increase from 55.6 to 83.3 (remember that their numbers would be equal to their respective carrying capacities if $\alpha = \beta = 0$). Of course, only one species might change in competitive ability, and that species might increase its competitive ability with the other species. For example, one species might acquire an allelopathic trait or exhibit agonistic behavior towards the other species resulting in an increase in α or β, depending upon whether species 2 or 1, respectively, were evolving.

Intuitively, we can conclude that the strongest competitive pressures generally occur intraspecifically, rather than interspecifically, because conspecific individuals are under more similar environmental constraints. However, when a species has small relative numbers, most of the competitors are members of other species, suggesting that overall interspecific selection might be stronger. On the other hand, for the species with the greatest relative numbers, intraspecific selection should be stronger. Such a frequency dependence of selection on interspecific competition may in theory lead to *genetic feedback*—that is, a genetic change in the species with smaller population numbers that increases its ability to compete interspecifically, thereby resulting in an increase in numbers and making it the predominant species. The other species, now the rarer species, may respond in a similar fashion, resulting in a periodic switch in numerical dominance (genetic feedback can also occur for other types of interspecific interactions, such as host–parasite interactions; see Example 20.4).

A change in interspecific competitive ability may result in a differential use of resources or in greater specialization on a specific resource. Such a response to interspecific competition may lead to an evolutionary change in a morphological or behavioral trait away from that of the competing species, a phenomenon called *character displacement* or *ecological character displacement*. Consistent with character displacement is the observation that mean phenotypic values differ for some ecologically meaningful measure in sympatric populations of the two species, while in areas in which the species are allopatric, the phenotypic values may be similar (see Example 20.2). However, it is often difficult to distinguish between a situation in which the phenotypic changes occur *in situ* and that in which there is constant colonization and local extinction and only those species that differ ecologically remain in sympatry. For example, one species may colonize an area and a second species, which differs from it ecologically, may subsequently colonize it successfully. In such a case, the second species would have had a preadapted ecology, which had evolved in another area, that allowed it to colonize. The observed phenotypic differentiation for these two sympatric species would not be the result of character displacement.

We have considered evolution in closely related species that are similar in morphology, behavior, and size. However, interspecific competition may occur between quite different species, as that among birds, ants, and rodents for seeds in some environments (see Example 13.1). In such instances, important competitive abilities, and any consequent evolutionary change and character displacement, may be quite different from that which might occur in competition with closely related species.

II. PREDATOR–RESOURCE EVOLUTION

Genetic factors may influence the ability of a prey to avoid predation or a host to avoid parasitism. On the other hand, such genetic factors may affect the ability of a predator to find and utilize a prey, or a parasite to find a host. Virtually every prey or predator has phenotypic traits that appear to be adaptations either to obtaining prey or to avoiding predators. Obvious antipredator devices are the cryptic behavior and morphology of many prey species that enable them to avoid predation (such as melanism in peppered moths; see Example 3.2). For example, many species have countershading that makes the upper body darker than the lower body. When light strikes from above, such a color pattern balances the effect of shadows, making the organism more difficult to perceive by predators. In some cases, the effect of selection by predators may be strongly counter to what is selectively advantageous within the species. For example, males of many frog species form choruses to attract female frogs. However, at least one predatory tropical bat uses frog vocalizations to locate prey, countering the innate mating advantage for calling frogs (Tuttle and Ryan, 1981).

Some resource species, particularly plants and some insects, have evolved chemical defenses against predator species (see Example 20.3). A number of plants have compounds that make their proteins and other nutrients less accessible to herbivores. For example, most legumes contain inhibitors of the

---

Example 20.2. In seed-eating birds, beak size is generally a good indicator of the size of seeds utilized, with larger beaked birds eating larger seeds. Some of Darwin's finches are seed eaters and exist without apparent competitors on certain of the Galápagos Islands and with competitors on other islands. Figure 20.2 gives the distribution of beak sizes for two finch species, *Geospiza fuliginosa* and *G. fortis*, which exist by themselves on the islands of Daphne Major and Los Hermanos, respectively, and which coexist on the islands of Isabela and Santa Cruz (Grant, 1981). When the species are allopatric, they have essentially the

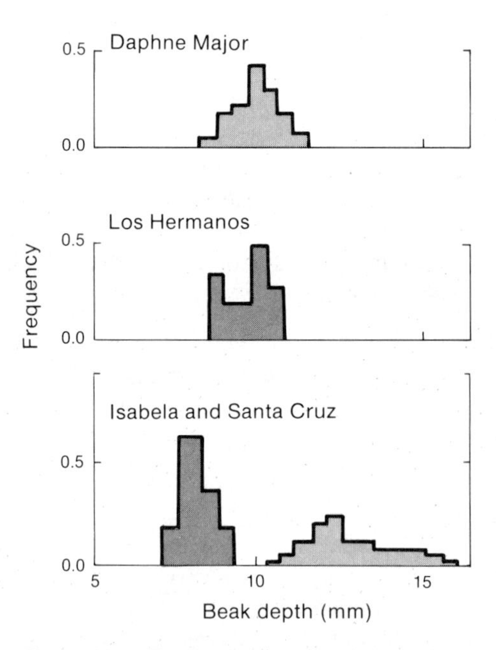

Figure 20.2. The distribution of beak sizes of two species of Galápagos finches when allopatric and when sympatric (after Grant, 1981).

digestive enzymes of herbivores, and potatoes, tomatoes, and other plants produce proteinase inhibitors in response to attack by chewing insects. Oaks deposit their proteins in such a way that most herbivores cannot obtain them separately from tannins, chemicals that combine with leaf proteins to make the proteins indigestible. In sugar maples, both the level of tannins and nutrients may vary from branch to branch and change on a regular basis or in response to insects. However, some beetle larvae avoid the tannins in oaks by "mining" the leaves and feeding on the inner tissues that contain low levels of tannins.

same mean and range of beak size. However, on the islands where they are sympatric, there is no overlap in the beak distributions, with *G. fortis* having a larger beak size and *G. fuliginosa* having a smaller beak size, observations generally consistent with character displacement (see the discussion in Grant, 1981).

Fenchel (1975) measured shell length in two congeneric species of mud snails, *Hydrobia ulvae* and *H. ventrosa*, that live in estuaries. When these species are allopatric, their shell sizes are quite similar (Figure 20.3). However, in sympatric populations, *H. ulvae* is always larger, and the size distributions of the two species generally do not overlap. Fenchel then measured the particle size of the food, mainly diatoms, ingested by the two species. In allopatric populations, the two species utilized food of nearly equal size, while in sympatry, *H. ulvae* utilized larger food than *H. ventrosa*. In other words, the difference in shell size was correlated with a change in resource utilization. In addition, none of these estuary populations was more than 100 to 150 years old, suggesting that the morphological change occurred very quickly in these populations. Presumably, the morphological shifts in both these finches and snails were primarily genetic, although environmental factors (including other species) could have affected phenotypic values.

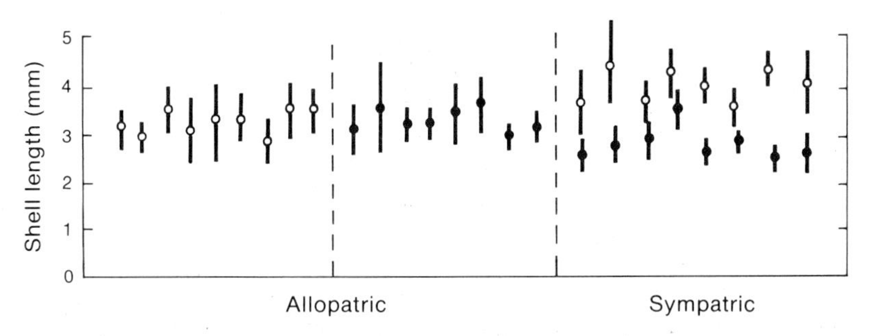

Figure 20.3. The shell length in two species of mud snails when allopatric (left) and sympatric (Fenchel, 1975). The circles indicate the mean shell length and the lines indicate the standard deviation for a given locality (open circles, *H. ulvae*, and closed circles, *H. ventrosa*).

Because such protective chemicals generally appear to have no fundamental role in plant growth and development in the absence of predators, they are called secondary chemical compounds. These chemicals may be costly to produce synthetically, and one would expect in the absence of predator pressure that their production would be reduced. Plant breeders interested in developing varieties resistant to insect pests may select for higher levels of these compounds, although this may lead to fruits or vegetables that are distasteful or toxic to humans.

As with interspecific competition, the numbers of a predator and its resource may be affected by changes that affect the demographic values of the species. For example, if we use the symbols of the Lotka-Volterra expressions given in Chapter 14, an increase in r, the intrinsic rate of increase for the resource species, results in an increase in the equilibrium numbers of

---

Example 20.3. Plants within a population or, as suggested, different parts of a plant may vary with respect to the level of various secondary chemical compounds they contain. Among plants, some of this variation is presumably determined genetically, although environmental factors, including the presence of pests, can greatly affect levels of such compounds. For example, within a ponderosa pine stand, during the winter tassel-eared squirrels feed frequently in some trees, while not in others (Farentinos et al., 1981). The squirrels generally prefer to feed on small twigs on fairly large cone-bearing trees, feeding on some trees preferentially over many years. Examination of squirrel preference in the laboratory determined that the animals could distinguish between twigs from those "feeding" trees and other "nonfeeding" trees.

Monoterpenes are secondary plant compounds present in many conifers that appear to have deterrent or toxic effects on many herbivores. A determination of the level of monoterpenes in feeding and nonfeeding trees showed that the level of monoterpenes in nonfeeding trees was approximately twice that in feeding trees. One monoterpene in particular, α-pinene, appears to be a very important cue for the squirrels. In a laboratory experiment, adding α-pinene to a mash resulted in captive squirrels preferentially feeding on mash without α-pinene (Farentinos et al., 1981).

It does not appear that all plants must be repellent to deter predation and achieve overall protection from a herbivore (this fact echoes the ideas of mimicry; see Chapter 16). Good support for this hypothesis comes from comparisons of predation in mixed versus pure stands of closely related species. For example, two species of clover, *Trifolium repens* and *T. fragiferum*, are commonly found intermixed in British grasslands. When these were grown in pure stands, hares ignored *T. repens* but competely defoliated the stands of *T. fragiferum* (Harper et al., 1961). However, in mixed stands, the *T. fragiferum* was also ignored and both species persisted.

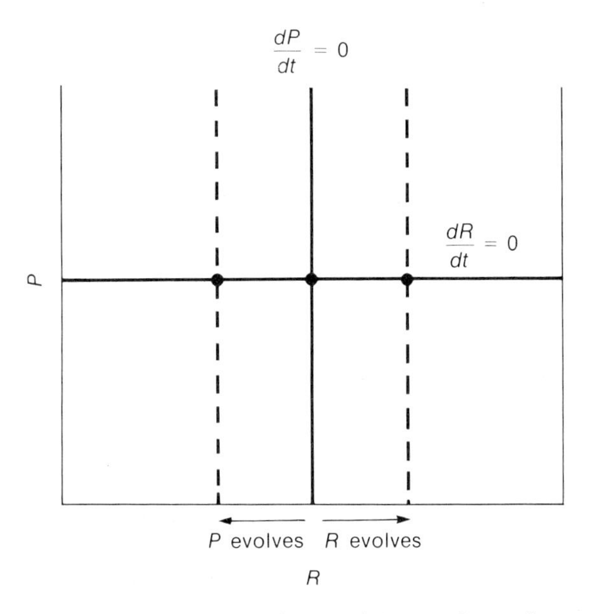

Figure 20.4. The effect of an evolutionary change in a predator or a resource species on the equilibrium numbers of the resource species.

the predator species. On the other hand, a decrease in d, the death rate of the predator species, results in a decrease in the equilibrium numbers of the resource species.

In a more general way, we may use the diagram of predator-versus-resource numbers to consider evolutionary responses that result because of an evolutionary change in the predator isocline. Figure 20.4 gives the simple graphical representation of the Lotka-Volterra predator–resource expressions; here the vertical line indicates the predator isocline—that is, where $dP/dt = 0$. If the predator evolves so that it can support itself at a lower resource density (let us say that it increases its ability to convert resource individuals into predators, b), then the isocline is moved to the left and results in a lowered equilibrium number in the resource population. On the other hand, if the resource population evolves so that it is less utilizable by the predator—say, a decrease in b because of a chemical defense change in the resource—then the predator isocline is moved to the right and the equilibrium number of the resource population is increased.

It appears that the result of evolutionary pressures would be either that the predator would become so efficient at catching the resource that it would eliminate the resource, or that the resource would become so efficient at avoiding the predator that the predator would switch to other resources or become extinct. (In fact, humans have become so efficient as predators, particularly after widespread firearm use, that they have made a number of resource species extinct; see Chapter 14.) However, in general there appears to be a balance in most situations between the predators' feeding ability and the

Example 20.4. Approximately two dozen rabbits were introduced into Australia from England in 1859. By 1950, the population of rabbits in Australia had become a plague, and the number of rabbits was estimated in the hundreds of millions. To control the rabbit numbers, a virus disease that occurs naturally in the Brazilian rabbit, myxomatosis, was introduced into Australia (Fenner, 1965). Initially, the virus was 100 percent lethal to the rabbit, making transmission quite ineffective. However, the virus shortly became less virulent, spread rapidly, and reduced the rabbit population to approximately 1 percent of its level in 1950 within several years. By the late 1950s, the virus had become much less virulent and more than half the myxoma viruses were only 70 to 90 percent lethal when tested in previously unexposed rabbits. The drop in lethality resulted from the fact that the viral strains causing lower mortality were at a selective advantage because they could be transmitted from live infected rabbits by mosquitoes or fleas. In addition, the rabbit host had developed some genetic resistance to the parasite. During the spread of myxoma virus, mortality of wild rabbits was continually monitored with a virus of known virulence. Figure 20.5 illustrates that the mortality of rabbits exposed to the virus repeatedly declined with increasing exposure so that rabbits from a population that had been exposed six times suffered only 25 percent mortality. In other words, both the myxoma virus and the rabbit

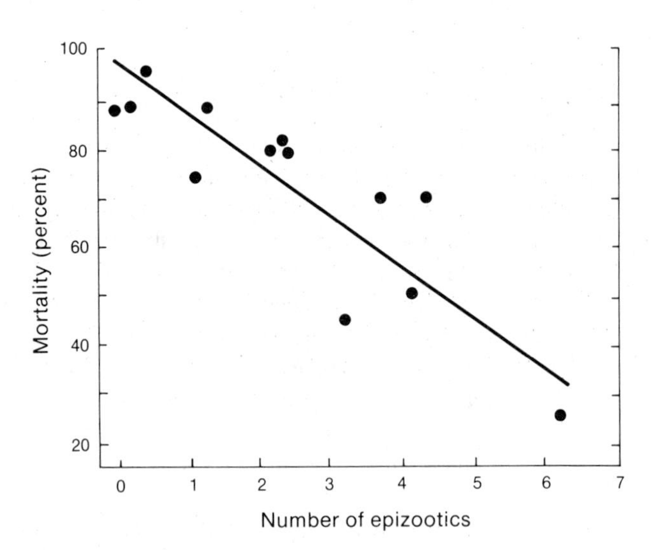

Figure 20.5. The relationship of mortality of wild rabbits to myxoma virus and the number of population exposures (after Fenner and Ratcliffe, 1965).

changed genetically in response to the presence of the other, and they now appear to have reached a kind of coevolutionary stability.

Pimentel et al. (1965) investigated the coevolution of a host species, the housefly (*Musca domestica*), and a parasitoid wasp (*Nasonia vitripennis*) in a series of laboratory experiments. To determine host evolutionary response, a control in which a constant number of houseflies, unexposed to wasps, were introduced each generation was compared to a condition in which houseflies were allowed to evolve in the presence of wasps. An evaluation of these strains after nearly three years revealed that the reproduction of wasps on the host decreased by approximately 70 percent. In addition, the average number of parasitoids in the evolved population was only 1,900, compared with 3,700 in the control experiment. Using Figure 20.4, we find that this result is comparable to a lowering of the horizontal resource isocline—that is, a lowered equilibrium parasitoid number is due to this evolutionary change. In a subsequent experiment, Pimentel et al. (1965) monitored the numbers of the two species when both were allowed to vary in populations initiated from the evolved sample and in populations from unexposed controls. The numbers in the control underwent large fluctuations, while the lines that had previously been exposed maintained quite stable numbers (see Figure 20.6.) These results are consistent with the hypothesis that coevolution in a predator–resource system may result in greater stability in population numbers over time.

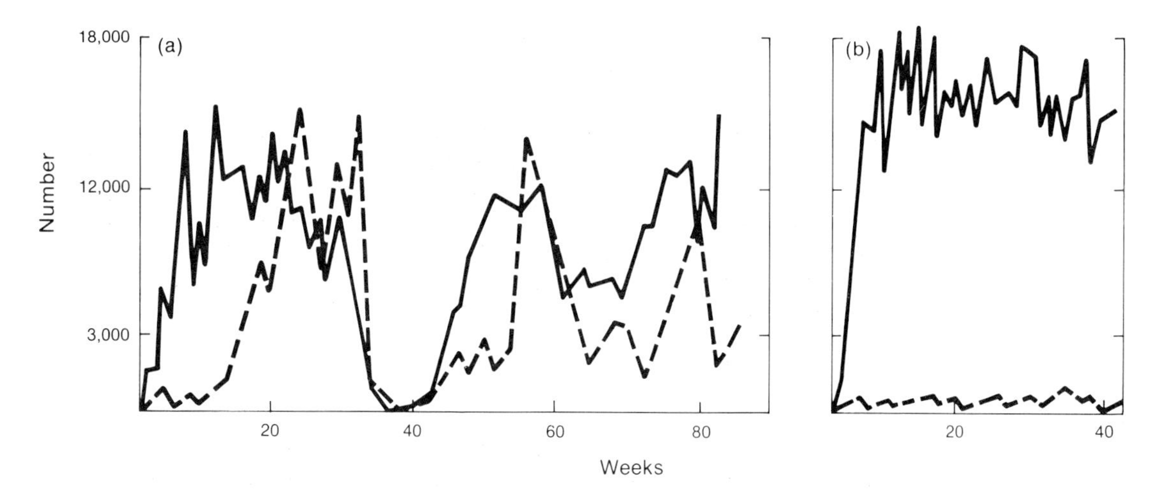

Figure 20.6. The numbers of a host (housefly) and a parasitic wasp (broken lines) in an laboratory system when (a) the populations had not been exposed to each other and (b) the populations had evolved with respect to each other in a previous experiment (after Pimentel et al., 1965).

resources' avoidance ability, although other factors are probably also important in determining the densities of predator and resource species, as we mentioned in Chapter 14.

Intuitively, we can conclude that, if the predator becomes rare, then predation as a selective force on the resource species should become less important. This development may lead to an increase in resource types more susceptible to predation (assuming that there is some cost in maintaining predator defense) and a consequent rise in the predator population number. The selective elimination of the predator-susceptible types may lead to a decline in the resource population, an increase in resistant types, and a subsequent decline in the predator population. Pimentel (1968) suggests that such genetic feedback cycles are important in the regulation of population numbers and that they will lead to stable proportions of resistant and susceptible resource individuals (see Example 20.4). However, the resistant-susceptible difference need not be the result of different alleles and may be an induced response, as in plants when attacked by herbivores (see Example 20.3 and the discussion of genotypes with flexible behavior in Chapter 18). Furthermore, the evolutionary response of a predator to a resource and vice versa may result in the damping of the oscillations predicted from the Lotka-Volterra model and could in part be responsible for the relative stability of predator and resource numbers in some environments (see Example 20.4).

It is possible that the evolution of predator–resource traits might occur more quickly than the evolution of interspecific competitive ability because the genes involved in predator–resource characteristics might differ sharply from the genes affecting intraspecific success. On the other hand, many of the genes involved in interspecific competitive ability might be the same as those involved in intraspecific competition—for example, those involved in fast growth. In addition, predator–resource traits often appear to be determined by only a few genes. For example, mimicry, a form of predator avoidance discussed in Chapter 16, may be the result of only a few genes, and host selection and host survival in parasities may be the result of a small number of genes. In fact, in many host–parasite systems there appears to be a widespread matching of host alleles conferring resistance to specific strains of parasites such as rusts, resulting in a "gene-for-gene" coevolutionary change in both species (see Example 20.5).

III. MUTUALISM

In some cases, the word coevolution may be used in a more restricted sense to refer to the evolution of mutualistic interactions between organisms. The interaction of plants and their pollinators, such as bees, birds, and bats, is a mutualistic association in which both species appear to benefit. In this case, the pollinator obtains nectar from the flowers it pollinates, and the plants are able to fertilize other flowers. An example of such a coevolutionary adaptation is in

Example 20.5. Many of the detailed studies of observed coevolutionary changes have focused on crop plants and their pests or diseases. For example, the Hessian fly, *Mayetiola destructor*, is a pest of wheat and has caused severe damage to susceptible varieties. Certain alleles in wheat will inhibit larvae development, but flies may have corresponding virulence alleles (Hatchett and Gallun, 1970). In fact, only when there is a virulence allele for each resistance allele can flies develop normally, an example of the "gene-for-gene" coevolutionary change.

The first definitive studies on the genetics of plant resistance to a rust were in flax (Flor, 1956). For example, two varieties of flax, Ottawa and Bombay, were tested for their susceptibility (S) or resistance (R) to two rust strains, 22 and 24. As shown in Table 20.3a, Ottawa is resistant to rust strain 24 and Bombay is resistant to rust strain 22. When the flax varieties are crossed, the F_1 is resistant to both rust strains, demonstrating that the alleles conferring resistance are dominant over the alleles allowing susceptibility. A cross of F_1 individuals results in segregation of four different types of individuals, those resistant to both rust strains, those resistant to 24 and not 22, those resistant to 22 and not 24, and those susceptible to both. Table 20.3b gives the numbers of these classes observed in an experimental cross and the expected number based on a 9:3:3:1 ratio of the segregation of two independent dominant genes. The expected values are very similar to the observed values, resulting in a small χ^2 value, suggesting that resistance to rust strain 22 is determined by a dominant allele at one locus and that resistance to rust strain 24 is determined by a dominant allele at another locus.

Table 20.3
The resistance (*R*) or susceptibility (*S*) of two flax varieties and their F_1 progeny (a) and their F_2 progeny (b) to two strains of rust (Flor, 1956).

(a)	*Flax*			
Rust strain	*Ottawa*	*Bombay*	F_1	
22	*S*	*R*	*R*	
24	*R*	*S*	*R*	
(b)	*F_2 progeny*			
22	*R*	*S*	*R*	*S*
24	*R*	*R*	*S*	*S*
Observed	109	36	36	12
Expected	108.6	36.2	36.2	12.1

Table 20.4
A hypothetical pattern of evolution of interspecific interaction leading to obligate mutualism (0 indicates no effect on the numbers of the other species and + indicates a positive effect of the other species on this species).

	Species 1	*Species 2*
No interaction	0	0
↓		
Commensalism	+	0
↓		
Nonobligate mutualism	+	+
↓		
Obligate mutualism	+	+

some orchids that have flowers closely resembling female bees, so that when male bees attempt to copulate, pollination occurs.

The evolutionary sequence of species interactions leading to mutualism may be envisioned in a general way (Table 20.4). First, assume that two species do not interact with each other and thus have no effect on the numbers of each other. This lack of effect may in part result from both species being extreme generalists and, therefore, having little effect on any other single species. Second, commensalism may evolve where one species benefits from the presence of the other while the second species is unaffected. Third, the second species may also begin to benefit from the presence of the first. However, at this stage the mutualisitic interaction is nonobligatory, or facultative, in that either species can exist without the other. Finally, the interaction between the two species may become obligatory, so that neither species can exist in the absence of the other. At this point, each species is a specialist and uses the other species as an essential resource in some sense (see Example 20.6).

The interaction of two potentially mutualistic species can be analyzed from a cost–benefit viewpoint. The benefit to one species from the other species may be some factor such as an added resource or mobility that may increase its numbers. On the other hand, there may be a cost to specializing on one mutualist, since it may be periodically rare and thereby simultaneously reduce the numbers of the other mutualist. Presumably, however, over time the costs of the interaction are minimized and the benefits increased, making the mutualistic interaction more beneficial than not.

As with interactions between competitors and between a predator and a resource species, the effect of two mutualistic species on each other's numbers can be predicted using mathematical models. For example, if we use an extension of the logistic model, the change in the numbers of two mutualists is

$$\frac{dN_1}{dt} = r_1 N_1 \left(\frac{K_1 - \alpha_{11} N_1 + \alpha_{12} N_2}{K_1} \right)$$

$$\frac{dN_2}{dt} = r_2 N_2 \left(\frac{K_2 - \alpha_{22} N_2 + \alpha_{21} N_1}{K_2} \right)$$

where r_i, N_i, and K_i are the intrinsic rates of increase, numbers, and carrying capacity of the ith species. The α_{11} and α_{22} coefficients are measures of intraspecific competition. For example, α_{11} expresses the extent of density dependence of species 1 on itself ($\alpha_{11} = 0$ indicates no density dependence,

---

Example 20.6. Many acacia species have facultative, mutualistic associations with ants. However, the bulls-horn acacia, *Acacia cornigera*, and the ant *Pseudomyrmex ferruginea* have a highly coevolved, obligatory mutualistic relationship (Janzen, 1966). In this case, the ant is dependent upon the acacia both for food and a domicile. These acacia have highly developed extrafloral nectaries that provide food for the ants and enlarged thorns with a tough cover that are used by the ants for reproduction. In addition, the acacia produces leaves year round, an unusual occurrence in tropical areas that have a dry season. In turn, these acacias are dependent upon the ants because they are very attractive to many herbivorous insects and the ants aggressively kill or remove most such intruders. In fact, *Pseudomyrmex* is active 24 hours a day, unusual among ants, thereby providing continuous protecton for the acacia.

To examine the beneficial effect of ants on acacia trees, Janzen (1966) artificially removed the ants from a number of trees. Table 20.5 gives some of the data comparing similar groups of trees with or without ants. In the absence of ants, the number of herbivorous insects increases more than twentyfold during the day and more than tenfold at night. The absence of the ants appears to have drastic effects on the trees, resulting in, for example, greatly reduced plant growth and an increase in tree mortality. Interestingly, several insect species mimic the ant so closely in behavior and appearance that the ants do not recognize them as intruders.

Table 20.5
The number of insects present, plant growth and mortality in acacias that have a mutualistic ant species either present or absent (Janzen, 1966).

	Ants present	*Ants absent*
Number of insects per shoot		
Day	0.039	0.881
Night	0.226	2.707
Total weight of growth (gm)	41,750	2,900
Tree mortality	0.28	0.43

---

and $\alpha_{11} = 1$ indicates logistic density dependence). The coefficients α_{12} and α_{21} indicate the positive per capita effect of species 2 on species 1 and that of species 1 on species 2, respectively. Note that the terms $a_{12}N_2$ and $\alpha_{21}N_1$ are positive, unlike similar terms that were used in the competition equations in Chapter 13.

Notice that if α_{11} or α_{22} is 0 (no density-dependent regulation of one or the other species), then this is a system with positive feedback—that is, the two species will continue to increase. For example, if $\alpha_{11} = 0$, then dN_1/dt is always positive, and the number of species 1 and consequently the numbers of species 2 will therefore continue to increase without bound.

If there is logistic density dependence in both species, $\alpha_{11} = \alpha_{22} = 1$, then the condition for stable coexistence of the two mutualistic species is $\alpha_{12}\alpha_{21} < 1$ (Vandermeer and Boucher, 1978). In this case, assuming that there is genetic variation affecting the level of mutualistic interactions, natural selection should act to increase $\alpha_{12}\alpha_{21}$ so that eventually this product would exceed unity and lead to unbounded growth of both populations. In other words, this model predicts that coevolutionary change for mutualists will not result in numerical stability as did genetic feedback for both competitor and predator—resource populations.

One way in which coevolutionary change between mutualists can continue to occur is if one of the mutualistic species is limited by predation from a third species (competition or resource limitation may also bound population growth). For example, in a simple extension of the model above a third species preys on species 2. In this case, a term $-\alpha_{23}N_3$ is added to the numerator of the equation for growth of species 2. If $\alpha_{23} = 1$, then for stable coexistence of the mutualists, $\alpha_{12}\alpha_{21}$ need only be smaller than 10 (Heithaus et al., 1980). An example of facultative mutualism, consistent with this model, is that between ants and spring herbs (the ants eat some seeds of these plants, but they also act as a major dispersal agent). In this case, the numbers of the mutualists appear to be stabilized by rodent seed predators (Heithaus et al., 1980).

In some cases, although evidence consistent with the coevolution of two mutualist species is substantial, excluding other possibilities is difficult. For facultative mutualists, the evolution of one species may have occurred independently of the other species and vice versa. For example, a mammal or bird may eat a particular fruit because it has dietary habits evolved in the absence of the fruit (see Example 20.7). The dispersal of the seeds not digested by the mammal or bird may benefit the plant, but the advantage would not confirm that the mammal and the plant had coevolved.

SUMMARY

The interaction between species may be affected by genetic variation. The outcome of interspecific competition may depend on the genetic composition of the species. Interspecific competition and predator–resource interactions may result in genetic changes in the species leading to character displacement

Example 20.7. An interesting example of a putative coevolutionary relationship is between the extinct dodo and the now nearly extinct tree *Calvaria major*. The dodo (Figure 20.7) was a large flightless bird endemic to Mauritius in the Indian Ocean that became extinct due to human predation in 1681. *C. major* is also endemic to Mauritius and now exists only as a few trees, all estimated to be more than 300 years old. *C. major* does produce sound seeds, but none has apparently germinated for several hundred years. The seeds even remain dormant in a variety of nursery conditions, probably because the seed is enclosed in a thick endocarp that prevents embryonic development. The dodo possessed a large beak and a gizzard with stones, suggesting that it could have used *C. major* seeds as a food source. The digestive system of the dodo and the coincidental extinction of the dodo and near extinction of *C. major* led to the hypothesis that *C. major* may have depended upon the dodo for scarifying the endocarp, thereby enabling the embryo to develop. To investigate this hypothesis, Temple (1977), force-fed fresh *C. major* seeds to turkeys, birds with digestive systems similar to that of the dodo. Of 17 seeds taken, 7 were crushed by the turkey's gizzard and 3 of the remaining 10 germinated under nursery conditions. Temple suggests that these may have been the first *C. major* seeds to germinate since the extinction of the dodo and that his results support the hypothesis that the fruits of *C. major* and the dodo were coevolved (however, see the comments of Owadally, 1979).

Figure 20.7. A drawing of the flightless dodo from Mauritius, now extinct (Ziswiler, 1967).

or stability of population numbers, respectively. Evolution of characters affecting mutualistic interactions in theory should lead to unbounded population growth, but a third predator species may limit growth. It is often difficult to differentiate between, on the one hand, traits that appear coevolved but that actually developed in the absence of the other species and, on the other hand, direct coevolutionary response to another species.

PROBLEMS

1. A change in interspecific competitive ability may be due to a number of factors. How would you determine the basis of such a change by examining pure and mixed cultures of two competitors?
2. Design an experiment using two competing species in which you might expect to observe ecological character displacement. If character displacement did not occur after a number of generations, what might be possible causes of the lack of response?
3. From Example 20.5, it appears that resistance in flax to two rust strains is dominant. Given your understanding of genetic change due to selection, why would you expect these alleles to be dominant? If the frequency of the two alleles for rust resistance were 0.1 and 0.2, what proportions of the population would be susceptible to both rust strains?
4. Some researchers suggest that certain secondary chemical compounds may be waste products that only incidentally confer protection against herbivores. Design an experiment or suggest observations that might determine whether these products are primarily waste products or protective compounds.
5. Using the equations given for two mutualists, and assuming $r_1 = r_2 = 0.1$, $N_1 = N_2 = 200$, $K_1 = K_2 = 100$, $\alpha_{11} = \alpha_{22} = 1$, $\alpha_{12} = \alpha_{21} = 0.5$, what are dN_1/dt and dN_2/dt? Assume that α_{12} and α_{21} change due to coevolution so that they are both equal to 2. Using the same values of the other parameters, what are the expected changes in the numbers of the two species? What conclusions can you draw from these results?

CHAPTER 21

Applied Population Biology

> We may reasonably expect to have eventually a complete theory of ecology that will not only provide a guide for the practical solution of land utilization, pest eradication, and exploitation problems but will also permit us to ... construct a model that will incorporate genetics and ecology in such a way as to explain the past and also predict the future of evolution on earth.
>
> LAWRENCE SLOBODKIN

The basic principles of population biology are applicable to a number of areas of human concern. We have already discussed how population genetics can help us to understand the incidence of genetic diseases and how population ecology is important in determining the factors leading to the extinction of particular species. In fact, insight from population biology appears necessary for a fundamental grasp of many problems in public health, agriculture, medicine, fisheries, and forestry.

Here we will concentrate on four aspects of applied population biology, primarily because they appear to be the ones that will have particular importance in the near future. These four aspects are the preservation of genetic variation, principally in crop plants; the development of crops and crop varieties; the control of public health and agricultural pests; and the harvesting of natural populations. Using a population biology perspective in our approach to these areas also gives some insight into the effects of contemporary practices in these areas.

I. THE CONSERVATION OF GENETIC VARIATION

In the past, a number of different varieties of each crop plant were grown, many of which were adapted to local environmental conditions. In recent years, fewer varieties of many crops are being used, resulting in a restriction of the gene pool of these crops. Similar problems are not as critical in domesticated animals, although endangered species of both plants and animals may become depleted of genetic variation (see Frankel and Soule, 1981). In fact, inbreeding depression in rare or endangered species kept in zoos is becoming an important problem in the maintenance of these species (see Chaper 7).

In many areas of the world, newly developed high-yield strains of crop plants have replaced the endemic varieties, or *land races*. This change is exemplified by the "green revolution" in areas where, for example, new strains of rice and wheat have replaced land races. The high-yield varieties often necessitate large-scale planting and the mechanized spreading of fertilizer and pesticides. As a result, many local varieties that were selected to yield well in a labor-intensive situation are no longer used. However, these local varieties often have alleles important to adaptation to local conditions, such as soil types, disease, and pests, that could make them valuable in the development of future varieties. Unfortunately, many land races may already have been lost, and the wild relatives of crop plants may be the only source for many advantageous alleles. Furthermore, limiting each crop species to only a few varieties, as in the Common Market, is drastically reducing the number of varieties that are widely used.

Two related phenomena are the possible results of such developments. First, a crop species can become genetically vulnerable to a disease because, in a large planting, the varieties used have the same alleles at disease- or insect-resistance loci (see Example 21.1). In 1970, such genetic uniformity in planted corn led to a loss of approximately 15 percent of the crop in the United States to the southern corn leaf blight. Recovery from this disaster only occurred when a blight-resistant strain of corn was developed (Ullstrop, 1972).

The second possible result of such developments is that genetic variation in the total species may be reduced. It is possible that variation may be lost in land races, germplasm collections, and wild progenitors alike. Several different

---

Example 21.1. Grapes are a very important economic crop, primarily owing to the production of wine. In the last half of the nineteenth century, much of the wine industry of Europe was decimated by plant lice of the phylloxera type from North America that attacked the roots of the grapevine (Wagner, 1974). Although a large variety of wine grapes had been developed in Europe, the grapes were all of one species, *Vitis unifera*, making the vineyards highly susceptible to infestation. In North America, there are more than 30 species of grapes, many of which are resistant to attack by these plant lice, but these American species were not selected for varieties that were capable of producing high-quality wine in a variety of climates. As a result, a program was initiated to make rootstocks resistant to plant lice compatible with grafts from the European species and adaptable to the European climate and soil. Using inter-specific crosses to combine characteristics of several American species, researchers finally developed resistant and compatible rootstocks, and these rootstocks are still used today in most of the vineyards of Europe. Of course, these rootstocks are now also used in other parts of the world, such as the United States, Australia, and Chile.

---

approaches have been advocated to ensure the maintenance of the alleles in crop species. The first approach is the conservation of genetic variation by growing the land races *in situ*. Although in practice this procedure may be difficult to carry out, where implemented the possibility would be greater of retaining particular combinations of genes at different loci and locally adapted alleles in the population than if the population were grown in other environments. Furthermore, only in this case would continued evolution in the natural environment be possible. For such a program to be effective, it would be desirable to maintain preserves composed of a number of separate areas characterized by ecological diversity, each with a sizable population and with exchange occurring between them.

The second approach to conserving genetic variation is the accumulation of world collections of many cultivated varieties and their wild relatives. Generally, seeds from such germplasm collections are stored at low temperatures, permitting preservation of these collections for extremely long periods. However, such stored seed may need to be grown periodically to ensure that the germination rate remains adequate. Obviously, seed storage can provide a reservoir for the genetic variation in collected varieties if care has been taken to collect a wide diversity in the species. For example, the Consultative Group on International Agricultural Research, a group responsible for the dissemination of the high-yielding wheat and rice varieties, maintains large numbers of accessions of different crop varieties at a number of international institutions (Table 21.1).

In either of these approaches, it is crucial that sampling of the species be adequate to ensure that the collection is sufficiently varied. The strategy of sampling depends upon the distribution of genetic variation in the population. For example, if all populations have nearly the same allelic frequencies, then thoroughly sampling one population will give a substantial portion of the genetic variation in the species. However, many species vary extensively among populations, making it necessary that many different populations be sampled to ensure adequate variation. Therefore, in general three major factors must be considered in such a sampling scheme: the number of seeds to

Table 21.1
The number of crop accessions in CGIAR germplasm centers and their location in 1981 (from Plucknett and Smith, 1982).

Crop	*Number of accessions*	*Location*
Rice	60,000	Philippines
Kidney beans	27,000	Columbia
Sorghum	20,000	India
Pearl millet	17,000	India
Cowpeas	15,000	Nigeria
Maize	13,000	Mexico
Chickpeas	12,000	India
Potatoes	8,000	Peru

be collected per local population, the number of populations sampled, and the distribution of these populations over the species range. Where there is inadequate knowledge of the species, it is usually advisable to sample as many populations as possible, basing the distribution of the samples upon both the structure of the species and the heterogeneity of the environment.

The adequacy of the sample size of a local population is a function of the frequency of the alleles that one wishes to include in the sample. Brown (1978) calculates that where a number of alleles of relatively low frequency exist, a random sample of 50 to 100 individuals would be adequate under most circumstances. That is, there is a high probability that such a sample would include a copy of each allele. As an example of this approach, assume that the frequency of allele A_2 is q and that the frequency of all other alleles is thus $1 - q$. Therefore, the probability of failing to include A_2 in a sample of n diploid individuals (getting only other alleles) is

$$\Pr(\text{no } A_2 \text{ alleles}) = (1 - q)^{2n}$$

assuming independence among alleles in the sample. The probability of including the A_2 allele in the sample is $1 - \Pr(\text{no } A_2 \text{ alleles})$. Therefore, if $q = 0.01$, then a sample of 10 individuals should include the A_2 allele 18 percent of the time.

In measurements of the genetic diversity in either land races or the wild relatives of crop plants, three major types of traits are considered: economic or other metric traits, disease- or pest-resistance genes, and Mendelian morphological markers. All these traits, plus the adaptation of different varieties to a range of environments, would be important to consider in a sampling regime. In addition, electrophoretic variants may be useful, since the distribution of alleles over populations may indicate the population structure of the species and help identify local populations. Electrophoretic variants may also be important in the identification of genotypes common in only one or a few populations.

II. THE DEVELOPMENT OF CROPS AND CROP VARIETIES

The domestication of plants is older than recorded history, and the time and even the area of domestication of many plants is unknown. (We will concentrate on crop plants here, but the same principles apply to agricultural animals.) Some species have been domesticated for more than 7,000 years, others have been domesticated much more recently, and still others are currently being evaluated or developed for domestication (Table 21.2). The process of domestication involves several steps. Initially, the plant must be recognized as a potential crop with food or fiber value, and techniques to cultivate it must be developed. Particular plants that have high levels of food or fiber or that are easy to cultivate are most likely to be selected. Later, the process becomes more systematic, with the focus on selection for certain attributes of agricultural value (see Example 21.2).

Table 21.2
The estimated era of domestication of a number of crop plants (from Simmonds, 1976).

Era	*Some examples*
Pre-7000 BP*	Barley, wheat, maize, beans
Pre-4500 BP	Potato, grapevine, rice, cotton
Pre-2000 BP	Soybeans, citrus, tea, flax
Pre-1000 BP	Sugarcane, coffee, tobacco
1000 BP–present	Rubber, sugarbeet, strawberry
Present	Winged bean, amaranthus, jojoba

* BP = years before present.

Once a species has been domesticated, then selection can take place to improve its yield. In the past, a combination of natural and artificial selection resulted in land races of many plants—that is, essentially in ecotypes for

---

Example 21.2. The tomato is a relatively new crop plant and was avoided even into the early twentieth century because of its alleged toxicity (it is a member of the nightshade family). Today the tomato yields more tons of produce and has a value greater than any other vegetable grown for human consumption in the United States. (Botanically, the tomato is actually a fruit, since it develops from an ovary. Popularly, though, it is considered a vegetable.) The tomato is presumed to have been domesticated from the wild forms present in the Andean region of South America (Rick, 1978).

A number of phenotypic qualities with agricultural importance have been incorporated into tomato cultivars in recent decades. These include an allele that results in uniform ripening of the fruit (no green area near the stem) and another that limits growth to prevent the plant from growing in an indeterminate fashion. Some tomato varieties now also have traits that make them easily harvestable by machine. For example, some cultivars are uniformly short and have tough-skinned fruit that can be harvested by machine and that ripen in a concentrated period.

The extensive phenotypic variation among the wild species of tomatoes has potential breeding importance. For example, one type of tomato lives only 5 m away from the high-tide line in the Galápagos Islands. These plants can survive in seawater, suggesting that the development of a tomato variety that uses water containing a high proportion of salt may be feasible. A second species exists in very dry habitats in Peru, where it obtains most of its water from fog or mist. The survival of this species appears to depend on extreme resistance to water loss, a trait of potential importance to irrigated tomatoes.

---

particular areas. The advent of modern plant-breeding practices has resulted in more efficient procedures designed to select for particular desirable attributes. On the simplest level, a systematic screening may be conducted of available lines for a particular trait. The lines having, for example, the highest yield or the greatest disease resistance may be used. If varieties with an even higher yield or greater disease resistance are desired, then recurrent selection schemes are utilized. As discussed in Chapter 8, such a scheme may involve selecting extreme individuals as parents generation after generation to improve the population mean for the desired trait. Other techniques, such as mutation breeding, induction of polyploids, and crosses between species, have sometimes been useful in the development of new characteristics.

When screening or selection is based on a single trait, then individuals

---

Example 21.3. For traits that are genetically determined and discrete, such as disease resistance often is, screening a large number of lines or varieties may give some individuals that combine all of a number of the desired traits. For example, Williams (1977) screened 5,000 lines of cowpeas for resistance to five different diseases. Of these lines, almost half were susceptible to all five diseases, while only 28 lines were resistant to all five (Table 21.3). Obviously, lines that have all the desired disease-resistance characteristics are uncommon—0.6 percent of this sample—suggesting that such screening programs must be large and include a wide diversity of lines.

However, most traits of agricultural importance are quantitative traits that are only partially genetically determined. When a particular combination of a number of such traits is desired, then selection response may be quite slow. An example of such a long-term selection experiment for multiple traits is that conducted for 11 characteristics, which together constituted "performance," in Leghorn chickens (Dickerson, 1955). This flock was selected for 20 years for viability, egg production, egg weight, freedom from blood spots in the eggs, and seven other traits. Heritability estimates for many of these traits were above 0.2, suggesting that

Table 21.3
The number and proportion of cowpeas lines from 5,000 tested that are resistant to zero to five diseases (after Williams, 1977).

	Number of diseases resistant to					
	0	*1*	*2*	*3*	*4*	*5*
Number of lines	2392	1083	788	581	128	28
Proportion	0.478	0.217	0.158	0.116	0.026	0.006

---

with the desired level of the trait are often obtained fairly easily (see Example 8.5). However, when a combination of traits is desired, the probability of finding an individual or line with that particular set of traits is generally lower, and response to simultaneous selection for several traits may be very slow (see Example 21.3).

It is possible that advances in molecular biology may revolutionize plant and animal breeding and the development of new varieties or even species. For example, screening for disease resistance might utilize single cells that then could be made to regenerate whole plants, or single genes of importance might be cloned from one species and incorporated into others. In fact, one can envision the development of new crops with heretofore unachieved combinations of traits. There will still be physiological and genetic limitations on

---

response to selection would be substantial over a period of time. However, the response to selection was much less than expected and, for example, both viability and egg production actually declined over the experiment. The egg weight and blood-spot traits showed response to selection (Figure 21.1), but it was much smaller than expected. These difficulties were probably due to strong negative genetic correlations among the traits and with fitness (see Chapter 19).

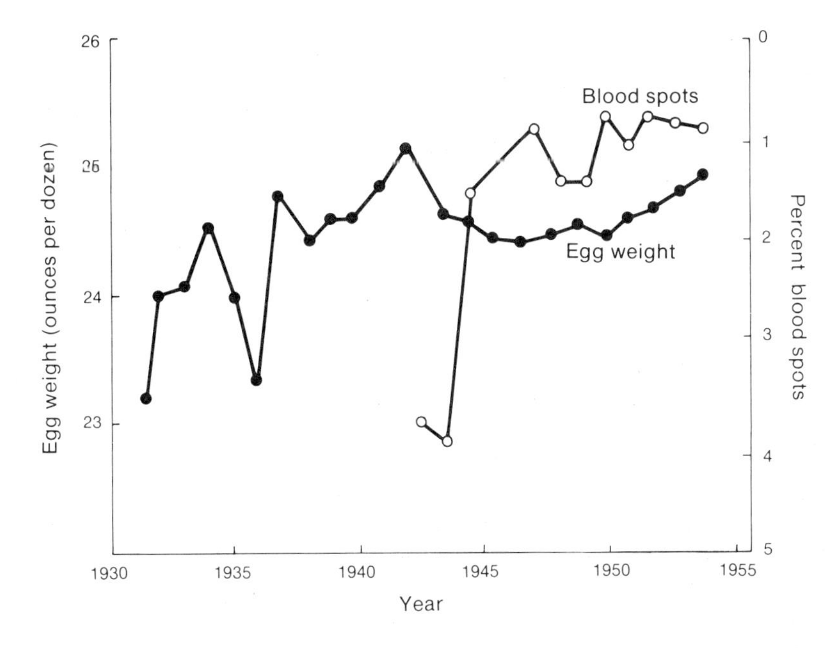

Figure 21.1. The egg weight and percentage of blood spots over time in a selected Leghorn population (Dickerson, 1955).

what phenotypic combinations are possible. For example, nitrogen fixation is a high-energy-requiring process, and it seems likely that the introduction of nitrogen-fixating genes into a crop such as corn may, at least initially, have a detrimental effect on yield.

III. THE CONTROL OF AGRICULTURAL AND PUBLIC–HEALTH PESTS

Agricultural crops have always been beset by insect and other pests, but modern agricultural techniques have changed the agricultural environment in a number of ways that have further necessitated diligent pest control. For example, the planting of large expanses of land with the same crop, a monoculture, allows a pest easier dispersal from plant to plant or from field to field. In addition, using only a few varieties of a particular plant reduces the natural genetic variation that confers resistance to pest infestation of various types. In a similar vein, human disease, passed by insect and animal vectors, has also become more of a problem with greater concentrations of people into cities.

The introduction of synthetic organic pesticides to control both agricultural and public-health pests after World War II greatly increased crop production and reduced the incidence of many diseases. The pesticides were inexpensive and very effective and quickly replaced other suppressive measures. However, the use of pesticides to control pest outbreaks has greatly altered the biological environment and in a number of cases has reduced or eliminated the predators or parasites that were originally important in the control of the population number of pest species.

The documentation of the evolutionary response to the presence of pesticides suggests that many species of pest organisms are capable of quickly evolving resistance to pesticides. Many of these pests, such as the mosquitoes that transmit malaria, are vectors of human diseases, but the problems of control are similar to those for agricultural pests. For example, in 1969, 15 species of anopheline mosquitoes, the malarial vector, were resistant to DDT, and in 1976, 43 species were resistant. A rather discouraging observation is that malaria is rapidly increasing in many areas where it had nearly been eradicated (from a half million cases in 1969 to 27 million cases in 1977 in India; Chapin and Wasserstrom, 1981), apparently because of the increased use of pesticides associated with new agricultural practices and the consequent increased pesticide resistance. The housefly seems to be the organism most broadly resistant geographically to pesticides, but *Tribolium*, a genus of flour beetle, and seven species of rodents including two rats (see Example 21.4) are widely resistant to various pesticides. Only among plant weeds is there apparently no substantial resistance to chemicals, although in a number of plant species, local types appear resistant to the effect of herbicide applications. Because weeds are often controlled by growth hormones rather than exotic chemicals, adaptive response may be somewhat slower.

A number of factors are potentially important in the evolution of pesticide resistance. A basic consideration is the presence of a physiological, anatomical, or other mechanism for the resistance to a pesticide. In numerous examples, resistance to pesticides takes different forms in different strains of the same species (see Example 8.1). Given that the potential for resistance is present, a number of factors are involved in a population becoming resistant to a pesticide. First, the new genetically advantageous variant or types must either be already present in the population, be generated by mutation, or migrate in from other populations. Once the resistant form is established in the population, its increase is dependent upon the factors discussed in Chapter 3, such as the level of dominance, selective advantage, and initial frequency. Of course, in many pest species generation length is short, there are multiple generations per year, and fecundity is extremely large, all of which are

---

Example 21.4. The Norway or brown rat, *Rattus norvegicus*, a widespread agricultural and domestic pest, was controlled for about a decade by the anticoagulant warfarin. However, resistance to this chemical evolved quickly and now has been found both in the United States and many countries in Europe. The mechanism of resistance has been examined in some detail and appears to be the result of a dominant resistant allele, Rw^2, at an autosomal locus. Most of the resistant animals are heterozygotes for this allele and the wild-type allele, Rw^1, because homozygotes for the resistant allele have quite low viability (see Table 21.4). The lowered viability in Rw^2Rw^2 individuals results from a twentyfold increase in vitamin K requirement (Hermodson et al., 1969). As a result, there appears to be a stable polymorphism when warfarin is applied, owing to the overall net survival advantage of the heterozygote. Further support for this conclusion comes from a survey in a population where the frequency of the resistant allele appeared to be stable over a number of years and showed a large excess of heterozygotes compared to Hardy-Weinberg expectations (Greaves et al., 1977). In this population, where warfarin is regularly applied, the overall relative fitnesses of the three genotypes were estimated to be 0.68, 1.0, and 0.37, respectively, a substantial advantage to the heterozygote.

Table 21.4
The factors affecting relative fitness and an estimate of relative fitness at the *Rw* locus in the Norway rat (Greaves et al., 1977).

	Rw^1Rw^1	Rw^1Rw^2	Rw^2Rw^2
Wafarin	Susceptible	Resistant	Resistant
Vitamin K dependence	No	Intermediate	Yes
Relative fitness	0.68	1.00	0.37

---

advantageous for the evolution of resistance in the population. In addition, it is possible that chemical-resistance genes may sometimes reside in plasmids (small nonnuclear pieces of DNA). Bacterial resistance to antibiotics is often in plasmids, and, because transmission of plasmids may occur horizontally (at times other than reproduction), both between members of the same species and of different species, antibiotic-resistant bacteria have become a major medical problem.

In recent decades, increased interest has been shown in biological, rather than chemical, control of pest species, particularly as many pests have become resistant to common pesticides. Integrating biological control programs into overall control programs results in a composite of different approaches that constitute *integrated pest management* (*IPM*). The basic approaches used in IPM, along with their specific uses, are given in Table 21.5. We will discuss some aspects of classical biological control in the prickly pear cactus (see also Example 13.7), competitive control in sterile male release in the screwworm, and cultural control in cotton pests (Example 21.5); we have already mentioned chemical control. It should be noted that in China and some other countries aspects of conservative biological control and cultural control are quite widely and effectively used.

In classical biological control, long-term success requires the grower to be aware of the important aspects of the new environment and the evolutionary status of the species involved. Where species are introduced in biological control, two major problems arise: initial survival of the colonizers and their continued existence and eventual adaptation to the new environment. The problem of the initial survival is basically an ecological one—that is, the introduced population must be obtained from an environment similar enough

Table 21.5
The concepts of integrated pest management and their use (after Batra, 1981).

Concept	*Use*
Classical biological control	Introduction of host-specific, self-reproducing, natural enemy of pest
Augmentative biological control	Periodic release of a natural enemy that cannot maintain itself
Conservative biological control	Management of agroecosystem or target area to conserve natural enemies, such as polyculture, hedgerows, or birdhouses
Competitive control	Pest-resistant crops, sterile-male introduction
Biorational control	Use of behavior-modifying compounds, such as pheromones
Chemical control	Use of natural or synthetic compounds that change metabolism
Cultural control	Management by physical techniques, such as quarantine, tilling, timing of operations, weeding, cleaning up garbage, and removing standing water

to that in which it is introduced to ensure its immediate survival and reproduction (see Example 21.6). The second aspect is more an evolutionary problem, and the potential for successful spread may be more dependent upon

Example 21.5. Cotton is heavily treated with insecticides, and almost 50 percent of all insecticides applied in the United States are used to control its pests (Adkisson et al., 1982). However, most of the major pests of cotton have developed resistance to insecticides. During the 1930s, the key pests on cotton were controlled by repeated applications of chemical dusts. After World War II, the use of chlorinated hydrocarbons afforded most complete control of the key pest insects and greatly increased cotton production. However, two secondary pests, the bollworm and the tobacco budworm, developed insecticide resistance, resulting in large crop losses. The natural enemies of these pests were killed by the insecticides.

In the early 1970s, to combat the effects of the bollworm, the tobacco budworm, and two other pests, the boll weevil and the cotton fleahopper, agriculturists in southern Texas initiated a new coordinated program. This program had several aspects: a control program at harvesting to kill most diapausing weevils by shredding and plowing under stalks, control of overwintering adults of weevils and fleahoppers by early use of insecticides before the natural enemies of the bollworm and the tobacco budworm were affected, and the use of a short-season variety of cotton. This newly developed variety flowers before the weevils can cause significant damage (Figure 21.2). The weevil density does not reach high levels until August, enabling a large percentage of the short-season yield to be already set. As a result of this new program, the profit per acre has increased severalfold and the use of insecticides reduced approximately 90 percent.

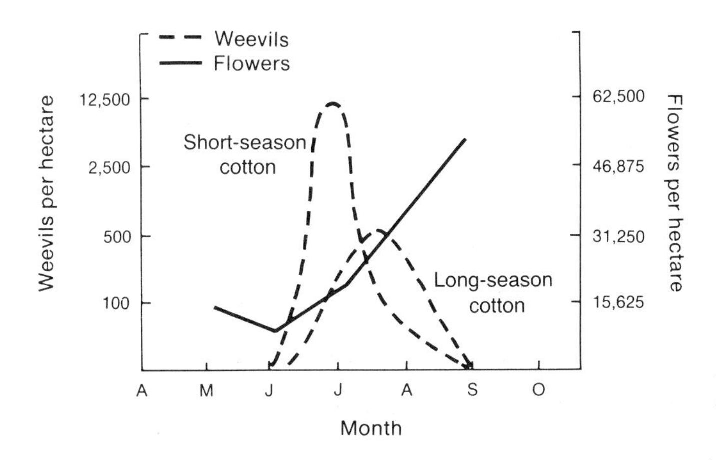

Figure 21.2. The weevil density and the flowering times of two varieties of cotton in Texas (after Adkisson et al., 1982).

the amount of genetic variation in the founders or the phenotypic plasticity of the founding genotypes. Perhaps the best populations to use for introductions would be large samples from ecological central areas, since this combination would result in a high probability of early survival and the fast evolution of adapted types.

One widely used approach to controlling insect pests is competitive control in the form of a release of large numbers of sterile individuals into the infested area, the *sterile-insect-release method* (*SIRM*). For example, sterile males may be released in large numbers in an effort to fertilize a large proportion of females, resulting in low progeny production. In a number of cases, SIRM has been very effective in reducing the levels of both agricultural and public-health pests. Obviously, sterile males must be effective in fertilizing the target females or the program will have little effect. The effectiveness of the sterile males is a function of their survival and mating success, and factors that would enhance either component would increase the effectiveness of the program. On the other hand, if the sterile males are unsuccessful in mating, because of premating mortality, lack of female receptivity, or poor male mating ability, then the program itself may have little success (see Example 21.7). The recent outbreak of the Mediterranean fruit fly in California illustrates another problem with SIRM—that is, when the numbers in the pest population are low, the mistaken release of unsterilized individuals may actually contribute to the pest problem.

Another possible approach to biological control is to use a genetic mechanism that increases the frequency of an allele that is susceptible to some environmental factor, such as temperature or an insecticide. Such alleles are

---

Example 21.6. The prickly pear is a cactus native to the Western Hemisphere. A number of species of this cactus were introduced to Australia in the nineteenth century as garden plants. The major pest species, *Opuntia stricta*, became a pest by 1900 and by 1925 covered 60,000,000 acres of Australia, often with a dense growth 1–2 m high. As a result, large expanses of grazing land and many homesteads had to be abandoned. In an effort to control prickly pear, a number of species of potential parasites were collected and more than a dozen were released. The most important species in controlling the prickly pear was a moth native to northern Argentina called *Cactoblastis cactorum*. *C. cactorum* females lay eggs on the cactus and the larvae feed on the pods, introducing bacterial and fungal infections that kill the cacti. Overall, 3 of the 12 potentially parasitic species became established. All the established species were common in their country of origin and could be considered to have a relatively broad niche. The founder numbers of two species, one of which was *C. cactorum*, were low (100 or less) and that of the third species was large (10,000), making it appear that founder size may not always be crucial in the establishment of a control species.

---

called *conditional lethals*, since their effect is expressed only under particular environmental conditions. In theory, one mechanism to increase such alleles is meiotic drive, a situation in which unequal numbers of the two types of gametes are produced in heterozygotes (see Chapter 19). If a conditional lethal could be increased in frequency because it is associated with a meiotic-drive element, then a major part of the population could be eliminated when the environment changes (or is artificially changed) and makes the lethal effective. However, most meiotic-drive elements themselves are associated with lethality or sterility of the homozygote, making it difficult to increase them in frequency.

Another approach possible for transporting conditional lethal alleles into a population is the use of chromosomal variants for which heterozygotes have lowered fertility. An unstable equilibrium is a result of such a relative fitness array (see Chapter 3). If large numbers of the chromosomal mutant type are released, then its frequency in the population may be greater than the unstable equilibrium, and the force of selection will further increase it (a laboratory test of this scheme is given in Figure 3.8). If the chromosomal mutant is associated with a conditional lethal, then virtually the whole population is susceptible to the environmental factor that causes lethality.

IV. THE HARVESTING OF NATURAL POPULATIONS

A number of nondomesticated species are harvested as human sources of food, fur, wood, or other commodities. To continue such utilization, the species must be allowed to reproduce at a level sufficiently high to maintain a population density or biomass that makes continued harvesting practical. In a number of instances, overutilization has resulted in a decline of a species to the point where harvesting is no longer economically feasible, and in some cases, such pressure has led to extinction or near extinction of species. Given that such commercial exploitation of natural populations will continue, what level of harvesting will be optimal, that is, will result in the *maximum sustainable yield*?

A general theoretical approach to this question is to assume a particular population-growth model and to calculate the density at which the greatest increase occurs in population numbers per unit time, assuming that this increase indicates the size of potential yield. Of course, if the population is not density regulated, as in an exponential model, then the maximum yield occurs at the highest density. However, one would expect the recruitment rate to decline and the loss to increase at higher densities, resulting in a lower yield. This general trend can be seen with the logistic growth model, which has the maximum population growth rate when the population size is half the carrying capacity (see discussion in Chapter 11). In other words, when $N = K/2$, then dN/dt has its maximum value for this model. Figure 21.5 illustrates the population numbers and growth for the logistic model at three different initial population densities when there is regular harvesting that brings the population back to its original density. At both the low and high densities, the amount of growth is much smaller than for a density of $K/2$. In other words, the logistic model predicts that the maximum sustainable yield is at a density half that of the carrying capacity.

Let us consider a general applied approach used to measure the yield in a population that is conceptually similar to the logistic model. If we call S_t the total stock or biomass of a species available at time t, then the change in stock from one time period to the next is

$$S_{t+1} - S_t = R + G - D - Y.$$

Here R indicates recruitment both by population growth and immigration, G the growth of individuals in size in the population, D the loss in the population

---

Example 21.7. One of the largest programs using SIRM is the screwworm eradication program in Texas. Screwworms are the larvae of blowflies, which infest wounds of cattle and sheep. The eradication program was begun in the 1950s, and involved the release of large numbers of sterile flies from airplanes. The program was quite effective until outbreaks occurred in several recent years, particularly in 1972 and 1976 (Figure 21.3; Richardson et al., 1982). Examination of screwworms from the 1978 outbreak suggested that they were different from released sterile flies, and subsequent chromosomal and electrophoretic analysis indicated that there may be several genetically isolated types of blowflies (see LaChance et al., 1982, for an alternative view). Furthermore, these

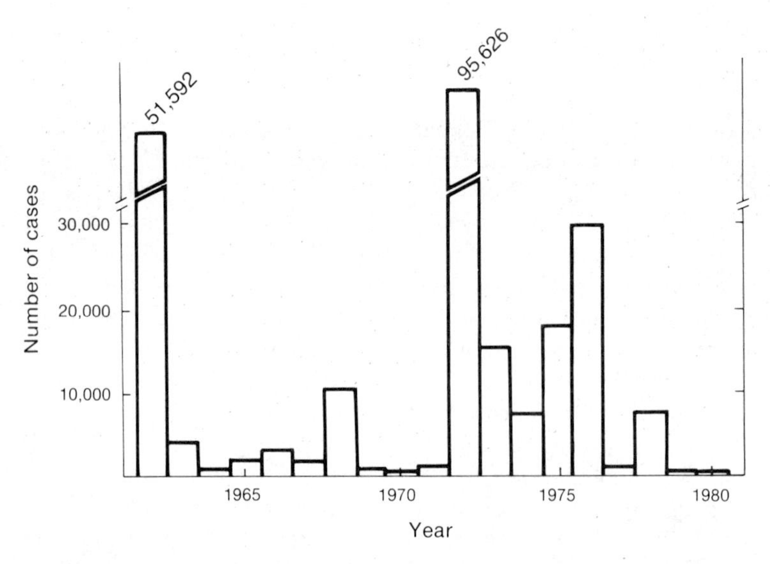

Figure 21.3. The reported cases of screwworms in the United States (Richardson et al., 1982).

---

due to natural mortality and emigration, and Y the yield or stock caught by humans. For the stock to remain at any given size, then

$$R + G = D + Y.$$

In other words, for an equilibrium stock size, the factors increasing stock size, R and G, are balanced by the factors reducing stock size, D and Y. Notice that a number of equilibrium values are determined by the size of Y but that we are interested in the equilibrium for which Y is a maximum.

Given such a general approach to stock changes, if we examine population density at low levels, the yield should also increase with increasing

---

types have a reservoir in Mexico from which they invade the southwestern United States (Figure 21.4). Obviously, when sterile flies are of one type, their mating effectiveness with individuals of another type may be low. Another suggestion for the cause of the outbreaks was that the blowflies propagated in captivity generation after generation may have changed in response to their laboratory conditions. In fact, Bush (1978) demonstrated that the captive screwworm population had changed over time at least at one electrophoretic locus, indicating that it may have changed at other loci as well. Currently, an attempt to eradicate the screwworm from northern Mexico is planned. If the diversity of blowflies is as great as suggested by Richardson et al., eradication may be slow. However, ensuring that the released sterile flies are of the same type as the infesting flies should help the success of the program.

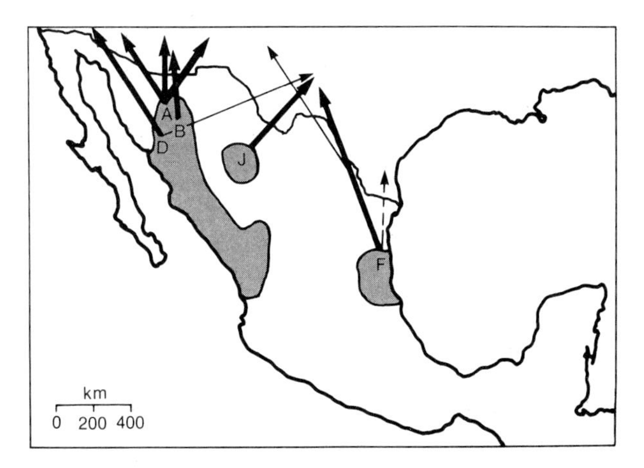

Figure 21.4. The possible origin and invasion routes of the five putative types of screwworms from Mexico (after Richardson et al., 1982).

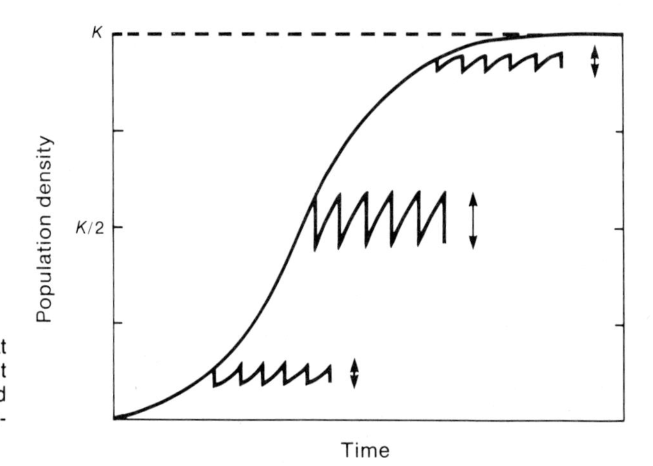

Figure 21.5. The effect of harvesting at regular time intervals at three different population densities. The size of the yield is indicated by the length of the two-headed arrows.

effort. However, at high densities, the yield may no longer increase with more effort because of density-dependent effects either decreasing R or G or increasing D. If the yield is plotted against the effort expended to obtain the catch (a measure assumed to be related to stock size), then a curve (sometimes called a *harvest parabola*) similar to that in Figure 21.6a is expected (see

---

Example 21.8. The harvest parabola approach is often applied to marine animals, particularly fish, to estimate the maximum sustainable yield. For example, in yellowfin tuna, the catch recorded for different amounts of fishing effort in different years is given in Figure 21.6b (after Watt, 1968). The catch of yellowfin increased from less than 100 million pounds in the 1930s to nearly 200 million pounds in the late 1940s and 1950s without ruining the fishery. In other words, the prediction from the fitted harvest parabola (broken line) that a higher catch could be achieved was supported by an increased catch with a higher fishing effort. In this case, the maximum sustainable yield appears to be approximately 180 million pounds as estimated from the fitted harvest parabola.

One would also predict from this model that if we knew the proportion of harvested individuals, then some proportion would provide the maximum sustainable yield in a density-regulated population. This prediction has been supported by experimental studies carried out by Silliman and Gutsell (1958) on guppies in laboratory aquaria. They maintained two aquaria as experimental controls that reached a carrying capacity (measured in fish weight) of approximately 30 g (see Figure 21.7a and 21.7b). Two other aquaria (Figure 21.7c and 21.7d) were subjected to different percentages of harvesting after 40 weeks. With 50 or 75 percent harvest every three weeks, the total fish weight was reduced to very low levels. Where there was a 10 or 25 percent harvest, the population was much below the carrying capacity, with a 25 percent harvest giving a

Example 21.8). To indicate the similarity to the prediction of the logistic model, the axes are also labeled as ΔN and N (see also Figure 11.9).

Remember, however, that we have been focusing on the situation in which the harvest exactly equals the excess production of the population for a given density. For a number of species, such as most whales, the recruitment has

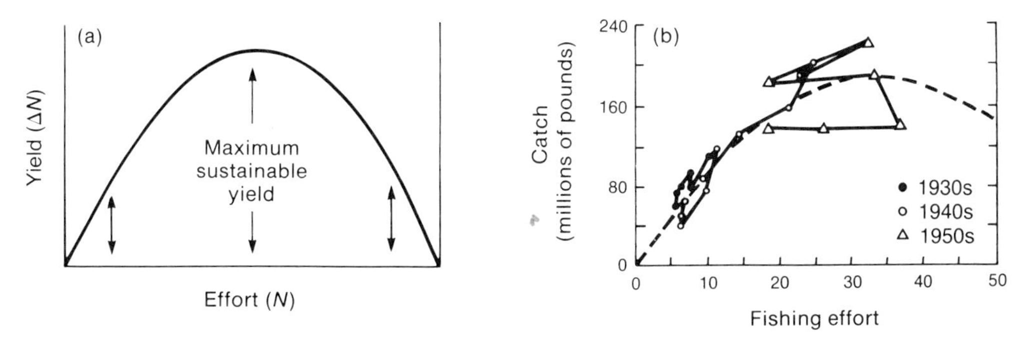

Figure 21.6. (a) A theoretical harvest parabola that plots effort versus yield and (b) the observed harvest parabola for yellowfin tuna (after Watt, 1968).

somewhat higher yield. A detailed examination of these data suggests that the maximum sustainable yield is at harvesting rates between 30 and 40 percent (Gulland, 1962).

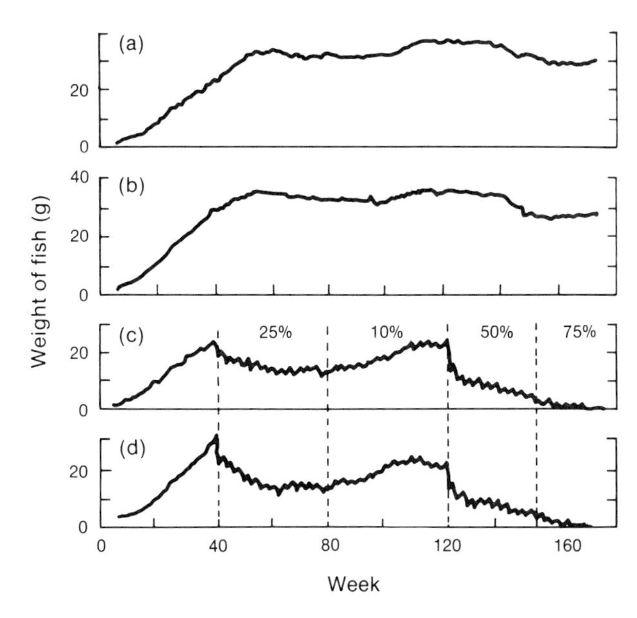

Figure 21.7. The total weight of guppies in two control aquaria (a and b) and at different levels of harvesting (b and c) (after Silliman and Gutsell, 1958).

been much smaller than the harvest, leading to a reduction in density over time (see Beddington and May, 1982). Similarly, as discussed in Example 14.8, the relatively low fecundity and late age to maturity in the common skate results in a low recruitment and, along with high fishing pressure, is the apparent cause of its extinction from the Irish Sea.

Obviously, other factors, such as interspecific competition, prey availability, or weather, may be important in determining the yield of a given species. For example, interspecific competition in conjunction with temperature appears important between herring and mackerel in the North Atlantic. In the last part of the nineteenth century, mackerel was the dominant species, while for most of the twentieth century, herring has been the dominant species. This reversal appears to have been initiated by environmental factors, such as temperature, and also influenced by the effect of the relative abundances of the two species on recruitment (Skud, 1982).

We have not considered any demographic aspects of the population being harvested. When the species has a complex life-history pattern, then age-specific survival and fecundity patterns need to be examined. For example, if a species has a high prereproductive mortality, then harvesting should focus on the younger age classes so that they will not be lost in terms of potential yield. On the other hand, if a species is semelparous, a wise harvesting strategy would be to harvest immediately after reproduction. One approach to evaluate the optimal age to harvest is to calculate the reproductive value for different ages for the species (see Chapter 12) and harvest at the age(s) when the reproductive value is lowest.

SUMMARY

A number of the principles of population biology are applicable to problems in medicine, agriculture, and fisheries. Understanding the genetic structure of a population and aspects of genetic variation is useful in genetic conservation. The development of crops has benefitted from an understanding of the genetic basis of phenotypic variation. The control of animal pests is best accomplished by understanding their population biology and devising a scheme to minimize their effect using integrated pest management. Finally, population ecology is applicable to predicting the effects of harvesting in natural populations and determining the maximum sustainable yield.

PROBLEMS

1. Assume that the frequency of an allele is 0.1, 0.06, and 0.02, respectively, in three populations and that a total of 30 individuals (seeds) can be stored. What is the probability of including the allele in a sample of 30, all taken from the population with a frequency of 0.06? What is the probability of including the allele in a sample of 30, 10 from each population? Which is the best sampling strategy?

2. Improvement of crops is generally based on a selection index that ideally includes information on the heritability of the important traits, genetic correlations among them, and the commercial value of each trait. Make an analogy of this approach with selection in natural populations and life-history evolution.
3. Calculate the expected equilibrium frequency of the Rw^2 allele in the rat, using the relative fitnesses in Table 21.4. Some populations do not have this allele. Give two possible reasons for this.
4. Discuss three aspects of integrated pest management, giving examples to illustrate different usage. Why do you think many agriculturists are reluctant to use IPM?
5. What information is needed to estimate the maximum sustainable yield in a population? What effect might harvesting of a species at this level have on predators, competitors, or other species interacting with the species?

CONCLUSIONS TO PART 3

This part covers various topics in population biology that combine concepts from both population genetics and population ecology. In Chapters 15 and 16, we discussed how ecological factors may result in genetic polymorphism, adaptation, and possibly speciation. In Chapters 17 and 18, we discussed the various factors that appear important in molecular and behavioral evolution, respectively. Chapters 19 and 20 focused on the genetic basis of adaptive changes in life-history characteristics and interspecific interactions. It is possible to gain a deeper insight into the dynamics of all these evolutionary topics by considering jointly the principles of population genetics and population ecology than by remaining within one or the other perspective exclusively.

An approach combining genetics and ecology is also of great potential significance for applied biological problems in agriculture and public health. For example, as discussed in the last chapter, to be successful, control programs for agricultural or public health pests should incorporate genetic information about an organism's pesticide resistance and adaptive potential as well as ecological information on its life-history characteristics and interactions with predators or competitors. A perspective that includes all these factors is not only likely to be more successful than traditional approaches to pest control, but is less likely to have detrimental effects on other organisms, including humans. As a result, I feel that an increased understanding of the principles of population biology and its application will result in a more harmonious relationship between humans and the natural world we live in.

GLOSSARY

This list gives general definitions for terms important in population biology and, in parentheses, the chapters in which the terms are introduced.

Abiotic factors. Nonliving factors, such as temperature and moisture, that affect the distribution and abundance of organisms (9).

Adaptation. The cumulative genetic response in a population resulting from selective pressures posed by the environment (16).

Allele. A particular form of a gene, presumably reflecting a certain DNA sequence, such as A_1 or A_2 (2).

Assortative mating. Mating that is nonrandom with respect to the phenotypes in a population—that is, where the probability of individuals of two given phenotypes mating is not equal to the product of their frequencies in the population (7).

Balancing selection. A general term for selection that by itself results in maintenance of different genetic morphs or alleles, such as heterozygote advantage and frequency-dependent selection (17).

Biotic factors. Intrinsic and extrinsic factors, such as dispersal ability and interspecific competition, that affect the distribution and abundance of organisms (9).

Bottleneck. A temporary reduction in population size from which future generations are derived (6).

Carrying capacity. The number of individuals that can be maintained by the resources available in a particular environment, symbolized by K (11).

Character displacement. An evolutionary change in interspecific competitive ability resulting from a differential use of resources (20).

Chi-square test. A statistical test to determine if the observed numbers deviate from those expected under a particular hypothesis, often expressed as χ^2 test (2).

Cline. Change in the frequency of an allele (or genotype) over space that is often attributed to the joint effects of gene flow and selection (5).

Coefficient of inbreeding. The probability that the two homologous alleles in an individual are identical by descent, symbolized as f (7).

Coevolution. The coordinated evolutionary change in two or more species in response to the presence of each other (20).

Competition. An interaction between species in which the organisms behaviorally exclude one another from resources (*interference competition*) or in which one species utilizes a resource important for the other species (*exploitative competition*) (13).

Competition coefficient. Relative per capita effect of one species on another species. The relative per capita effect of species 2 on 1 is symbolized by α, and the relative per capita effect of species 1 on 2 is symbolized by β (13).

Competitive exclusion. The principle that states that one competing species will eventually eliminate the other (13).

Competitive release. Change in resource utilization, population numbers, or geographic distribution of one species as the result of the absence of a competitor (13).

Conservation of genetic variation. An effort to maintain genetic variation in a species by managing natural or seminatural populations or by collecting and storing diverse genetic types of a species (21).

Demographic genetics. The study of genetic change that is caused by age-dependent selection on specific genotypes (19).

Demographic transition. The change in demographic characteristics of a human population from a population with high birth and death rates to one with low birth and death rates (12).

Demography. The study of birth and death rates in a population and their impact on population growth and age structure (12).

Density. Numbers of individuals of a species per unit area. *Absolute density* is based on a complete count of an area or sample area, while *relative density* is based on an indirect measure indicating the presence of the organism (10).

Density-dependent factors. Factors that affect the per capita mortality and fecundity of a population as a function of the population density (12).

Density-dependent selection. Selection in which the relative fitness values are a function of the population density (15).

Density-independent factors. Factors that affect the per capita mortality and fecundity independent of population density (12).

Dispersal. The movement of individuals from one place to another (9, 12).

Dispersion. The distribution of a population over an area (10).

Distribution. The total geographic area occupied by a species (9).

Effective population size. An ideal population of a given size in which all parents have an equal expectation of being the parents of any progeny individual, symbolized by N_e (6).

Electrophoresis. A biochemical technique that allows separation and identification of enzymes resulting from the presence of different alleles (1).

Equilibrium. *Ecological equilibrium*: a state at which there is no change in the number of individuals in a population (11); *genetic equilibrium*: a state at which there is no change in the allelic frequencies of the population (3). An equilibrium may be stable or unstable, and a perturbation will result in a return to or movement away from the equilibrium, respectively.

Exponential growth. Population growth that assumes continuous reproduction and no limit to population size (11).

Extinction. The disappearance of a species, in recent centuries often due to human predation or human-related changes in the environment (14).

Fecundity schedule. Values giving the average number of births per individual for all possible ages, the m_x values (12).

Founder effect. The chance changes in allelic frequency resulting from the initiation of a population by a small number of individuals (6).

Frequency-dependent selection. Selection in which the relative fitness values are a function of the frequency of the genotypes in the population (15).

Functional response. An increase in the predation rate as a response to an increase in the numbers in the resource population (14).

Fundamental theorem of natural selection. The theorem stating that the rate of change in fitness is equal to the genetic variation in fitness (19).

Gene flow. Movement of individuals from one population to another that results in the introduction of migrant alleles through mating and subsequent reproduction (5).

Gene pool. The total variety and number of alleles within a population or species (1).

Generalists. Species that have broad niche widths or tolerances, as contrasted with specialists (9, 14).

Genetic counseling. Counseling based on the risks for genetically defective infants, usually given where some familial history exists of a genetic disease (2).

Genetic drift. The chance changes in allelic frequency that result from the sampling of gametes from generation to generation (6).

Genetic feedback. A genetic change in one interacting species followed by a subsequent genetic change in the other interacting species (20).

Genetic liability. The underlying genetic disposition to a particular phenotype, usually when the phenotype is expressed in two or a few classes (8).

Genetic variance. The variation in a phenotype that results from genetic causes, symbolized by V_G. The extent of the *additive genetic variance*, symbolized by V_A, is important in determining the rate and amount of response to directional selection.

Genotype-environment interaction. The effect on a phenotypic value that results from a specific genotype and a specific environment that is not predictable from either separately, symbolized by GE_{ij} (8).

Geometric growth. Population growth that assumes a constant per capita growth rate for a given time period and no limitation to population size (11).

Group selection. Selection that depends upon differential survival or reproduction (splitting) of groups of individuals (18).

Hardy-Weinberg principle. The principle that states that after one generation of random mating, single-locus genotypic frequencies can be represented as a function of the allelic frequencies (2).

Heritability. The proportion of phenotypic variance that is genetic (*heritability in the broad sense*, h_B^2) or the additive genetic component (*heritability in the narrow sense*, h_N^2). *Realized heritability* is an estimate of h_N^2 from a selection experiment (8).

Heterozygosity. The proportion of heterozygotes in a population (2).

Heterozygote advantage. The situation in which the heterozygote has a higher fitness than either homozygote and leads to a stable equilibrium (3).

Heterozygote disadvantage. The situation in which the heterozygote has a lower fitness than either homozygote and leads to an unstable equilibrium (3).

Inbreeding. Nonrandom mating in which the mating individuals are more closely related than individuals drawn by chance from the population (7).

Inbreeding depression. The reduction of population fitness due to inbreeding (7).

Integrated pest management. An overall program to control pests that includes biological control as well as traditional means of pest control (21).

Intrinsic rate of increase. The rate of increase in a small time interval of a population uninhibited by density effects; also called *instantaneous rate of increase*, symbolized by r (11).

Kin selection. Selection in which individuals affect the fitness of relatives differently from that of unrelated individuals (18).

***K*-selection.** The type of selection thought to occur in constant and/or predictable environments, consequently favoring such characteristics as delayed reproduction and greater competitive ability (19).

Life expectancy. The expected duration of life from a given age x, symbolized as e_x (12).

Life-history evolution. Selection that results in coordinated changes in demographic values such as fecundity and survival schedules (19).

Logistic growth. Population growth that assumes that reproduction is continuous and that the total population size is limited (11).

Maximum substainable yield. An approach based on the logistic model to maximize the harvest of a natural population (21).

Mean fitness. The sum of the product of the relative fitness and the frequency of the genotypes in the population, symbolized as $\bar{w}$ (2).

Mimicry. An adaptation in which a group of repellent species resemble each other (*Müllerian mimicry*) or a nonrepellent species is similar to a repellent species (*Batesian mimicry*) (16).

Model. A verbal, mathematical, or graphical representation that allows for an accurate description, basic understanding, or prediction of future events of some process (Introduction).

Molecular clock. A hypothesis suggesting that there is a regular turnover or replacement of molecular variants (amino acid or nucleotide substitutions) over time (17).

Mutation. The original source of genetic variation caused, for example, by a change in a DNA base or a chromosome. *Spontaneous mutations* are those that appear without explanation, while *induced mutations* are those attributed to a particular mutagenic agent (4).

Mutation rate. The rate of mutational change from one allelic form to another, symbolized by u or v (4).

Mutation-selection balance. The equilibrium resulting from mutation introducing detrimental alleles into a population and selection eliminating them (4).

Mutualism. A relationship between species in which each benefits from the presence of the other (13, 20).

Net replacement rate. The number of offspring produced per individual per generation, symbolized as R_0 (11, 12).

Neutrality theory. The theory used to explain the existence of molecular variation in which there is no differential selection (alleles are neutral with respect to each other) and new variants are introduced into the population by mutation and eliminated by genetic drift (6, 17).

Niche. The total set of environmental conditions necessary for the persistence of a species. The *fundamental niche* is all the potential combinations of environmental conditions in which a species may exist. The *realized niche* is the conditions in which it really does exist (9).

Niche width. The range of tolerance or niche requirements for a particular species; also known as *niche breadth* (9, 13).

Numerical response. An increase in the number in a predator population as a response to an increase in the number in the resource population (14).

Phenotype. The morphological, biochemical, behavioral or other properties of an organism. Often only a particular trait of interest, such as weight, is considered (8).

Phylogenetic tree. A diagram that organizes the relationship between species or populations, putatively indicating ancestral relationships (17).

Polymorphism. The existence of two or more genetically determined forms (alleles) in a population in substantial frequency. In practice, a polymorphic gene is one at which the frequency of the most common allele is less than 0.95 or 0.99 (2).

Population. *Ecological population*: a group of individuals of a given species in a particular area (10); *genetic population*: individuals that exist together as a group in time and space and can interbreed (1).

Predator-prey interaction. A broadly defined interaction between two species in which one species is the predator and reduces the number of the other species, the prey or *resource*, by feeding upon it (14).

Probability. The proportion of the time that a particular event occurs. For example, the proportion of the days that it rains in the summer is the probability that it will rain on any given summer day (Introduction).

Quantitative trait. A trait that generally has a continuous distribution and is affected by many genes and many environmental factors.

Random-mating population. A group of individuals in which the probability of members mating with individuals of particular types is equal to their frequency in the population.

Rate of allelic substitution. The rate of replacement of alleles in a population. If amino acid or nucleotide sequences are known, the rate of amino acid or nucleotide substitution can be calculated (17).

Recurrence risk. The probability that a given genetic disease will recur in a family given that some individual in the family already had the disease (2).

Relative fitness. The relative ability of different genotypes to pass on their alleles to future generations; symbolized by w_{11}, w_{12}, and w_{22} for genotypes A_1A_1, A_1A_2, and A_2A_2, respectively (2).

Reproductive-isolation mechanism. An inherent biological factor that separates two related species. The mechanisms can be divided between prezygotic and postzygotic kinds (16).

Reproductive value. The expected number of future offspring for an individual of a given age x, symbolized as v_x (12).

Resource. A substance or object required by an organism for normal growth, maintenance, and reproduction, such as food or a nesting site (13); a species used by a predator for food (14).

Resource partitioning. The extent of the division of resources by two competing species; the greater the resource partitioning, the smaller the *niche overlap* (13).

***r*-selection.** The type of selection thought to occur in variable and/or unpredictable environments in which there is catastrophic mortality, consequently favoring such characteristics as early and high fecundity (17).

Selection. *Progressive selection*: an increase in alleles that are advantageous in a given environment; *purifying selection*: a decrease in the frequency of detrimental alleles in a population (3).

Selection components. Zygotic, sexual, gametic, and fecundity selection that altogether constitute the relative fitness of a genotype (19).

Selective disadvantage. The reduction in relative fitness of genotype A_2A_2 as compared with that of A_1A_1; also known as the *selection coefficient*, symbolized by s (3).

Sociobiology. The study of social behavior in an evolutionary context (18).

Specialists. Species that have narrow niche width or tolerance, often feeding on a very restricted resource, as contrasted with generalists (9, 14).

Speciation. The process of species formation. Speciation may occur when the incipient species are in the same area (*sympatric*), in different areas (*allopatric*), or in adjacent areas (*parapatric*) (16).

Species. A group of actually or potentially interbreeding populations; a group of populations that are reproductively isolated from other such groups (16).

Statistic. An estimate of the value of a particular measure based on a sample. For example, the estimate of the allelic frequency based on a sample from the population of interest is a statistic (Introduction).

Survival schedule. Values giving the probability of surviving from birth to a particular age x, symbolized as l_x (12).

BIBLIOGRAPHY

GENERAL REFERENCES

Connell, J. H., D. B. Mertz, and W. W. Murdock (eds.). 1970. *Readings in Ecology and Ecological Genetics*. Harper & Row, New York.

Dawson, P. S., and C. E. King (eds.). 1971. *Readings in Population Biology*. Prentice-Hall, Englewood Cliffs, N. J.

Ellseth, G. D., and K. D. Baumgardner. 1981. *Population Biology*. Van Nostrand, New York.

Harper, J. L. 1977. *Population Biology of Plants*. Academic Press, New York.

King, C. E., and P. S. Dawson (eds.). 1983. *Population Biology: Retrospect and Prospect*. Columbia University Press, New York.

Roughgarden, J. 1979. *Theory of Population Genetics and Evolutionary Ecology: An Introduction*. Macmillan, New York.

Solbrig, O. T., and D. J. Solbrig. 1979. *An Introduction to Population Biology and Evolution*. Addison-Wesley, Reading, Mass.

Wilson, E. O., and W. H. Bossert. 1971. *A Primer of Population Biology*. Sinauer, Sunderland, Mass.

Part 1: Population Genetics

Cavalli-Sforza, L. L., and W. F. Bodmer. 1971. *The Genetics of Human Populations*. W. H. Freeman, San Francisco.

Falconer, D. S. 1981. *Introduction to Quantitative Genetics*. Ronald Press, New York.

Hartl, D. L. 1980. *Principles of Population Genetics*. Sinauer, Sunderland, Mass.

Hedrick, P. W. 1983. *Genetics of Populations*. Jones and Bartlett, Boston, Mass.

Lewontin, R. C. 1974. *The Genetic Basis of Evolutionary Change*. Columbia University Press, New York.

Spiess, E. B. 1977. *Genes in Populations*. Wiley, New York.

Part 2: Population Ecology

Begon M., and M. Mortimer. 1981. *Population Ecology*. Sinauer, Sunderland, Mass.

Hutchinson, G. E. 1978. *An Introduction to Population Ecology*. Yale University Press, New Haven, Conn.

Krebs, C. J. 1978. *Ecology: The Experimental Analysis of Distribution and Abundance*. Harper & Row, New York.

Pielou, E. C. 1974. *Population and Community Ecology*. Gordon and Breach, New York.

Ricklefs, R. E. 1975. *Ecology*. Chiron Press, Portland, Ore.

Part 3: Ecological Genetics, Evolution, and Evolutionary Ecology

Ayala, F. J. (ed.). 1976. *Molecular Evolution*. Sinauer, Sunderland, Mass.

Dobzhansky, T., F. J. Ayala, G. L. Stebbins, and J. W. Valentine. 1977. *Evolution.* W. H. Freeman, San Francisco.
Ehrman, L., and P. A. Parsons. 1981. *Behavior Genetics and Evolution.* McGraw-Hill, New York.
Emlen, J. M. 1973. *Ecology: An Evolutionary Approach.* Addison-Wesley, Menlo Park, Calif.
Ford, E. B. 1971. *Ecological Genetics.* Chapman and Hall, London.
Futuyma, D. J. 1979. *Evolutionary Biology.* Sinauer, Sunderland, Mass.
Futuyma, D. J., and M. Slatkin. 1983. *Coevolution.* Sinauer, Sunderland, Mass.
Merrell, D. J. 1981. *Ecological Genetics.* University of Minnesota Press, Minneapolis, Minn.
Milkman, R. (ed.). 1982. *Perspectives on Evolution.* Sinauer, Sunderland, Mass.
Pianka, E. R. 1981. *Evolutionary Ecology.* Harper & Row, New York.
Wilson, E. O. 1975. *Sociobiology.* Belknap Press, Cambridge, Mass.
Wittenberger, J. F. 1981. *Animal Social Behavior.* Duxbury Press, Boston.

REFERENCES CITED IN TEXT

Adkisson, P. L., G. A. Niles, J. K. Walker, L. S. Bird, and H. B. Scott. 1982. Controlling cotton's insect pests: A new system. *Science* 216: 19–22.
Allard, R. W., and J. Adams. 1969. The role of inter-genotypic interaction in plant breeding. *Proc. XII Intern. Congr. Genet.* 3:349–370.
Anderson, W. W., T. Dobzhansky, O. Pavlovsky, J. Powell, and D. Yardley. 1975. Genetics of natural populations. III. Three decades of genetic change in *Drosophila pseudoobscura. Evolution* 29:24–136.
Avise, J. 1976. Genetic differentiation during speciation. Pp. 106–122. In F. J. Ayala (ed.), *Molecular Evolution.* Sinauer, Sunderland, Mass.
Ayala, F. J. 1972. Frequency-dependent mating advantage in *Drosophila. Behav. Genet.* 2:85–91.
Ayala, F. J., J. R. Powell, M. L. Tracey, C. A. Mouraõ, and S. Perez-Salas. 1972. Enzyme variability in the *Drosophila willistoni* group. IV. Genetic variation in natural populations of *Drosophila willistoni. Genetics* 70:113–139.

Baker, M. C., L. C. Mewaldt, and R. M. Stewart. 1981. Demography of white-crowned sparrows (*Zonotrichia leucophrys muttalli*). *Ecology* 62:636–644.
Baldwin, N. S. 1964. Sea lampreys in the Great Lakes. *Can. Audubon Mag.* Nov.–Dec.: 2–7.
Batra, S. W. T. 1981. Biological control in agrosystems. *Science* 215:134–139.
Battaglia, B. 1958. Balanced polymorphism in *Tisbe reticulata*, a marine copepod. *Evolution* 12:358–364.
Beddington, J. R., and R. M. May. 1982. The harvesting of interacting species in a natural ecosystem. *Sci. Amer.* 247:62–69.
Birch, L. C. 1953. Experimental background to the study of distribution and abundance of insects. I. The influence of temperature, moisture, and food on the innate capacity for increase of three grain beetles. *Ecology* 34:698–711.
Birch, L. C. 1960. The genetic factor in population ecology. *Amer. Natur.* 94:5–24.
Blauestein, A. R., and R. K. O'Hara. 1981. Genetic control for sibling recognition. *Nature* 290:246–248.
Bloom, W. L. 1984. Selection for flower color dominance during introgression in natural populations of *Clarkia. Amer. J. Bot.* (in press).
Blower, J. G., L. M. Cook, and J. A. Bishop. 1981. *Estimating the Size of Animal Populations.* George Allen and Unwin, London.

Boag, P. T., and P. R. Grant. 1981. Intense natural selection in a population of Darwin's finches (*Geospizinae*) in the Galapagos. *Science* 214:82–85.

Bodmer, W. F., and L. L. Cavalli-Sforza. 1976. *Genetics, Evolution, and Man.* W. H. Freeman, San Francisco.

Bonaiti-Pellie, C., and C. Smith. 1974. Risk tables for genetic counselling in some common congenital malformations. *J. Med. Genet.* 11:374–377.

Bonnell, M. L., and R. K. Selander. 1974. Elephant seals: Genetic variation and near extinction. *Science* 134:908–909.

Bosch, C. A. 1971. Redwoods: A population model. *Science* 172:345–349.

Bradshaw, A. D., T. S. McNeilly, and R. P. G. Gregory. 1965. Industrialization, evolution, and the development of heavy metal tolerance in plants. *Ecology and the Industrial Society. Brit. Ecol. Soc. Symp.* 5:327–343.

Brander, K. 1981. Disappearance of the common skate *Raia batis* from Irish Sea. *Nature* 290:48–49.

Brower, L. P. 1969. Ecological chemistry. *Sci. Amer.* 220:22–29.

Brower, L. P., L. M. Cook, and H. J. Croze. 1967. Predator responses to artificial Batesian mimics released in a neotropical environment. *Evolution* 21:11–23.

Brown, A. H. D. 1978. Isozymes, plant population structure and genetic conservation. *Theoret. Appl. Genet.* 52:145–157.

Brown, J. H. 1971. The desert pupfish. *Sci. Amer.* 225:104–110.

Brown, J. H., and D. W. Davidson. 1977. Competition between seed-eating rodents and ants in desert ecosystems. *Science* 196:880–882.

Buri, P. 1956. Gene frequency in small populations of mutant *Drosophila. Evolution* 10:367–402.

Bush, G. L. 1975. Modes of animal speciation. *Ann. Rev. Ecol. Syst.* 6:339–364.

Bush, G. L. 1978. Planning a rational quality control program for the screwworm fly. Pp. 37–84. In R. H. Richardson (ed.), *The Screwworm Problem.* University of Texas Press, Austin.

Cameron, A. W. 1964. Competitive exclusion between the rodent genera *Microtus* and *Clethrionomys. Evolution* 18:630–634.

Carlquist, S. 1974. *Island Biology.* Columbia University Press, New York.

Caughley, G. 1966. Mortality patterns in mammals. *Ecology* 47:906–918.

Chapin, G., and R. Wasserstrom. 1981. Agricultural production and malaria resurgence in Central America and India. *Nature* 293:181–185.

Charlesworth, B., R. Lande, and M. Slatkin. 1982. A neo-Darwinian commentary in macroevolution. *Evolution* 36:474–498.

Christiansen, F. B., O. Frydenberg, and V. Simonsen. 1973. Genetics of *Zoarces* populations. IV. Selection component analysis of an esterase polymorphism using population samples including mother-offspring combinations. *Hereditas* 73:291–304.

Clarke, C. A., and P. M. Sheppard. 1966. A local survey of the distribution of industrial melanic forms in the moth *Biston betularia* and estimates of the selective values of these in an industrial environment. *Proc. Roy. Soc. B.* 165:424–439.

Clausen, J., D. D. Keck, and W. M. Hiesey. 1941. Regional differences in plant species. *Amer. Natur.* 75:231–250.

Clegg, M. T., A. L. Kahler, and R. W. Allard. 1978. Estimation of life cycle components of selection in an experimental plant population. *Genetics* 89:765–792.

Clegg, M. T., J. F. Kidwell, M. G. Kidwell, and N. J. Daniel. 1976. Dynamics of correlated genetic systems. I. Selection in the region of the glued locus of *Drosophila melanogaster. Genetics* 83:793–810.

Cleghorn, T. E. 1960. MNSs gene frequencies in English blood donors. *Nature* 187:701.

Coale, A. J. 1983. Population trends in China and India (a review). *Proc. Nat. Acad. Sci.* 80:1757–1763.

Connell, J. H. 1961. Effects of competition, predation by *Thais lapillus*, and other factors on natural populations of the barnacle *Balanus balanoides*. *Ecol. Monogr.* 31:61–104.

Connell, J. H. 1970. A predator-pregy system in the marine intertidal region. I. *Balanus glandular* and several predatory species of *Thais*. *Ecol. Monogr.* 40:49–78.

Conover, D. O., and B. E. Kynard. 1981. Environmental sex determination: Interaction of temperature and genotype in a fish. *Science* 213:577–579.

Cooke, F., and F. G. Cooch. 1968. The genetics of polymorphism in the goose *Anser caerulescens*. *Evolution* 22:289–300.

Cooke, F., P. J. Minsky, and M. B. Seyer. 1972. Color preferences in the lesser snow goose and their possible role in mate selection. *Can. J. Zool.* 50:529–536.

Cooper, R. M., and J. P. Zubek. 1958. Effects of enriched and restricted early environments on the learning ability of bright and dull rats. *Can. J. Psychol.* 12:159–164.

Crombie, A. C. 1945. On competition between different species of granivorous insects. *Proc. Roy. Soc. B.* 132:427–448.

Crow, J. F. 1957. Genetics of insect resistance to chemicals. *Ann. Rev. Entomol.* 2:227–246.

Croze, H. 1970. Searching image in carrion crows. *Zeitschrift für Tierpsychologie*, Supplement 5.

DeBach, P., and R. A. Sundby. 1963. Competitive displacement between ecological homologues. *Hilgardia* 34:105–166.

Diamond, J. M. 1975. Assembly of species communities. Pp 342–444. In M. L. Cody and J. M. Diamond (eds.), *Ecology and Evolution of Communities*. Harvard University Press, Cambridge, Mass.

Diamond, J. M. 1982. Evolution of bowerbirds' bowers: Animal origins of the aesthetic sense. *Nature* 279:99–102.

Dickerson, G. E. 1955. Genetic slippage in response to selection for multiple objectives. *Cold Spring Harbor Symp. Quant. Biol.* 20:213–244.

Dickerson, R. C. 1971. The structure of cytochrome C and the rates of molecular evolution. *J. Molec. Evol.* 1:26–45.

Dobzhansky, T., and B. Spassky. 1967. Effects of selection and migration on geotactic and phototactic behavior of *Drosophila*. I. *Proc. Roy. Soc. B.* 168:27–47.

Dudley, J. W. 1977. 76 generations of selection for oil and protein percentage in maize. Pp. 451–473. In E. Pollak, O. Kempthorne, and T. Bailey (eds.), *International Congress Quantitative Genetics*. Iowa State University Press, Ames.

Duggins, D. O. 1980. Kelp beds and sea otters: An experimental approach. *Ecology* 61:447–453.

Eaton, G. G. 1976. The social order of Japanese macaques. *Sci. Amer.* 235:96–106.

Ehrlich, P. R., and L. C. Birch. 1967. The "balance of nature" and "population control." *Amer. Natur.* 101:97–107.

Ehrlich, P. R., and A. H. Ehrlich. 1972. *Population, Resources, Environment*. W. H. Freeman, San Francisco.

Ehrman, L. 1967. Further studies on genotype frequency and mating success in *Drosophila*. *Amer. Natur.* 101:415–424.

Ellseth, G. D., and K. D. Baumgardner. 1981. *Population Biology*. Van Nostrand, New York.

Embree, D. G. 1979. The ecology of colonizing species, with special emphasis on animal invaders. Pp. 51–65. In D. J. Horn, G. R. Stairs, and R. D. Mitchell (eds.), *Analysis of Ecological Systems*. Ohio State University Press, Columbus.

Farentinos, R. C., P. J. Capretta, R. E. Kepver, and V. M. Littlefield. 1981. Selective herbivory in tassel-eared squirrels: Role of monoterpenes in ponderosa pine chosen as feeding trees. *Science* 213:1273–1275.

Fenchel, T. 1975. Character displacement and coexistence in mud snails (Hydrobiidae). *Oecologia* 20:19–32.

Fenner, F. 1965. Myxoma virus and *Oryctolagus cuniculus*: Two colonizing species. Pp. 485–510. In H. G. Bales and G. L. Stebbins (eds.), *The Genetics of Colonizing Species*. Academic Press, New York.

Fenner, F., and F. N. Ratcliffe. 1965. *Myxomatosis*. Cambridge University Press, London.

Ferguson, M. W. J., and T. Joanen. 1982. Temperature of egg incubation determines sex in *Alligator mississippiensis*. *Nature* 296:850–853.

Ferris S. D., R. D. Sage, C. Huang, J. T. Nielsen, V. Ritte, and A. C. Wilson. 1983. Flow of mitochondrial DNA across a species boundary. *Proc. Nat. Acad. Sci.* 80:2290–2294.

Fisher, J., and R. A. Hinde. 1949. The opening of milk bottles by birds. *Brit. Birds* 42:347–357.

Flor, H. 1956. The complementary genic systems in flax and flax rust. *Adv. Genet.* 8:29–54.

Foster, G. G., M. J. Whitten, T. Prout, and R. Gill. 1972. Chromosome rearrangement for the control of insect pests. *Science* 176:875–880.

Frankel, D. H., and M. E. Soule. 1981. *Conservation and Evolution*. Cambridge University Press, London.

Frejka, T. 1973. The prospects for a stationary world population. *Sci Amer.* 228:15–23.

Fries, J. F. 1980. Aging, natural death, and the compression of morbidity. *N. Eng. J. Med.* 303:130–135.

Gaines, M. S., L. R. McClenaghan, and R. K. Rose. 1978. Temporal patterns of allozymic variation in fluctuating populations of *Microtus ochrogaster*. *Evolution* 32:723–739.

Gause, G. F. 1934. *The Struggle for Existence*. Williams and Wilkins, Baltimore.

Goldberg, D. E. 1982. The distribution of evergreen and deciduous trees relative to soil type: An example from the Sierra Madre, Mexico, and a general model. *Ecology* 63:942–951.

Goldsby, R. A. 1971. *Race and Races*. Macmillan, New York.

Grant, P. R. 1972. Interspecific competition among rodents. *Ann. Rev. Ecol. Syst.* 3:79–106.

Grant, P. R. 1981. Speciation and the adaptive radiation of Darwin's finches. *Amer. Sci.* 69:653–663.

Graves, J. E., and G. N. Somero. 1982. Electrophoretic and functional enzymic evolution in four species of eastern Pacific barracudas from different thermal environments. *Evolution* 36:97–106.

Greaves, J. H., R. Redfern, P. B. Ayres, and J. E. Gill. 1977. Warfarin resistance: A balanced polymorphism in the Norway rat. *Genet. Res.* 30:257–263.

Greenberg, L. 1979. Genetic component of bee odor in kin recognition. *Science* 206:1095–1097.

Gulhati, K. 1977. Compulsory sterilization: The change in India's population policy. *Science* 195:1300–1305.

Gulland, J. A. 1962. The application of mathematical models to fish populations. Pp. 204–217. In E. D. LeGrenand and M. W. Holdgate (eds.), *The Exploitation of Natural Animal Populations*. Blackwell, Oxford.

Hamrick, J. L. 1979. Genetic variation and longevity. Pp. 84–114. In O. Solbrig, S. Jain, G. Johnson, and P. Raven (eds.), *Topics in Plant Population Biology*. Columbia University Press, New York.

Hamrick, J. L., and R. W. Allard. 1972. Microgeographic variation in allozyme frequencies in *Avena barbata*. *Proc. Nat. Acad. Sci.* 69:2100–2104.

Harper, J. L., J. N. Clatworthy, I. H. McNaughton, and G. R. Sage. 1961. *Evolution* 15:209–227.

Harris, H., and D. A. Hopkinson. 1972. Average heterozygosity in man. *Ann. Hum. Genet.* 36:9–20.

Harrison, G. A., and J. J. T. Owen. 1964. Studies on the inheritance of human skin color. *Ann. Hum. Genet.* 28:27–37.

Hatchett, J. H., and R. L. Gallun. 1970. Genetics of the ability of the Hessian fly, *Mayetiola destructor*, to survive in wheats having different genes for resistance. *Ann. Entomol. Soc. Amer.* 63:1400–1407.

Hebert, P. D. N. 1974. Enzyme variability in natural populations of *Daphnia magna*. III. Genotypic frequencies in intermittent populations. *Genetics* 77:335–341.

Hedrick, P. W. 1973. Factors responsible for a change in interspecific competitive ability in *Drosophila*. *Evolution* 26:513–522.

Hedrick, P. W. 1982. Genetic hitchhiking: A new factor in evolution? *Bioscience* 32:845–853.

Heithaus, E. R., D. C. Culver, and A. J. Beattie. 1980. Models of some ant-plant mutualisms. *Amer. Natur.* 116:347–361.

Hermodson, M. A., J. W. Suttie, and K. P. Link. 1969. Warfarin metabolism and vitamin K requirement in the warfarin-resistant rat. *Amer. J. Physiol.* 217:1316–1319.

Hines, A. H., and J. S. Pearse. 1982. Abalones, shells, and sea otters: Dynamics of prey populations in central California. *Ecology* 63:1547–1560.

Holling, C. S. 1959. The component of predation as revealed by a study of small-mammal predation of the European pine sawfly. *Can. Entomol.* 91:293–320.

Holt, S. 1961. 1. Quantitative genetics of fingerprint patterns. *Brit. Med. Bull.* 16:247–250.

Hubbell, S. P. 1979. Tree dispersion, abundance, and diversity in a tropical dry forest. *Science* 203:1299–1309.

Huffaker, C. B. 1958. Experimental studies on predation: Dispersion factors and predator-prey oscillations. *Hilgardia* 27:343–383.

Jain, S. K., and D. R. Marshall. 1967. Population studies in predominantly self-pollinating species. X. Variation in natural populations of *Avena fatua* and *A. barbata*. *Amer. Natur.* 101:19–33.

Janzen, D. H. 1966. Coevolution of mutualism between ants and acacias in Central America. *Evolution* 20:249–275.

Janzen, D. H. 1980. When is it coevolution? *Evolution* 34:611–612.

Jarvis, J. V. M. 1981. Eusociality in a mammal; cooperative breeding in naked mole-rat colonies. *Science* 212:571–573.

Jensen, A. R. 1969. How much can we boost IQ and scholastic achievement? *Harvard Educ. Rev.* 39:1–123.

Johnston, R. F., and R. F. Sclander. 1971. Evolution in the house sparrow. II. Adaptive differentiation in North American populations. *Evolution* 25:1–28.

Jones, J. S., B. H. Leith, and P. Rawlings. 1977. Polymorphism in *Cepaea:* A problem with too many solutions. *Ann. Rev. Syst.* 8:109–143.

Kaneshiro, K. Y., and F. C. Val. 1977. Natural hybridization between a sympatric pair of Hawaiian *Drosophila. Amer. Natur.* 111:897–902.

Karban, R. 1982. Increased reproductive success at high densities and predator satiation for periodical cicadas. *Ecology* 63:321–328.

Kawai, M. 1965. Newly acquired pre-cultural behavior of the natural troop of Japanese monkeys on Koshiwa Islet. *Primate* 6:1–30.

Kephart, D. G., and S. J. Arnold. 1982. Garter snake diets in a fluctuating environment: A seven-year study. *Ecology* 63:1232–1236.

Kessel, B. 1953. Distribution and migration of the European starling in North America. *Condor* 55:49–67.

Kettlewell, B. 1973. *The Evolution of Melanism.* Oxford University Press, London.

Kidwell, M. G. 1983. Evolution of hybrid dysgenesis determinants in *Drosophila melanogaster. Proc. Nat. Acad. Sci.* 80:1655–1659.

Knight, G. R., A. Robertson, and C. H. Waddington. 1956. Selection for sexual isolation within a species. *Evolution* 10:14–22.

Koehn, R. K. 1969. Esterase heterogeneity: Dynamics of a polymorphism. *Science* 163:943–944.

Kornfield, I., D. C. Smith, and P. S. Gagnon. 1982. The cichlid fish of Cuatro Ciénegas, Mexico: Direct evidence of conspecificity among distinct trophic moths. *Evolution* 36:658–664.

Krebs, C. J., M. S. Gaines, B. J. Keller, J. H. Myers, and R. H. Tamarin. 1973. Population cycles in small rodents. *Science* 179:35–41.

LaChance, L. E., A. C. Bartlett, R. A. Bram, R. J. Gagné, O. H. Graham, D. O. McInnis, C. J. Whitten, and J. A. Seawright. 1982. Mating types in screwworm population? *Science* 218:1142–1145.

Lack, D. 1966. *Population Studies of Birds.* Oxford University Press, Oxford.

Langer, W. L. 1972. Checks on population growth: 1750–1859. *Sci. Amer.* 226:92–99.

Law, R., A. D. Bradshaw, and P. D. Putwain. 1977. Life-history variation in *Poa annua. Evolution* 31:233–246.

Le Boeuf, B. J., and S. Kaza (eds.). 1981. *The Natural History of Año Nuevo.* Boxwood Press, Pacific Grove, Calif.

Levings, S. C., and N. R. Franks. 1982. Patterns of nest dispersion in a tropical ground ant community. *Ecology* 63:338–344.

Levy, M., and D. A. Levin. 1975. Genic heterozygosity and variation in permanent translocation heterozygotes of the *Oenothera biennis* complex. *Genetics* 79:493–512.

Lewontin, R. C. 1970. Race and intelligence. *Bull. Atom. Scientists* 26:2–8.

Lewontin, R. C. 1978. Adaptation. *Sci. Amer.* 239:3–13.

Lewontin, R. C., and L. C. Dunn. 1960. The evolutionary dynamics of a polymorphism in the house mouse. *Genetics* 45:705–722.

Lewontin, R. C., and J. L. Hubby. 1966. A molecular approach to the study of genic heterozygosity in natural populations. II. Amount of variation and degree of heterozygosity in natural populations of *Drosophila pseudoobscura*. *Genetics* 54:595–609.

Ligon, J. D., and S. H. Ligon. 1982. The cooperative breeding behavior of the green wood hoopee. *Sci. Amer.* 247:126–134.

Little, E. L. 1971. *Atlas of United States Trees. I. Conifers and Important Hardwoods.* U. S. Dept. Agri. Misc. Publ. 1146.

Livingstone, F. B. 1969. Gene frequency clines of the locus in various human populations and their simulation by models involving differential selection. *Hum. Biol.* 41:223–236.

Loehlin, J. C., G. Lindzey, and J. N. Spuhler. 1975. *Race Differences in Intelligence.* W. H. Freeman, San Francisco.

Lowenstein, J. M., V. M. Sarich, B. J. Richardson. 1981. Albumin systematics of the extinct mammoth and Tasmanian wolf. *Nature* 291:409–411.

Lubchenco, J. 1978. Plant species in a marine intertidal community: Importance of herbivore food preference and algal competitive abilities. *Amer. Natur.* 112:23–39.

Lynch, M. 1983. Ecological genetics of *Daphnia pulex*. *Evolution* 37:358–374.

MacArthur, R. H. 1960. On the relative abundance of species. *Amer. Natur.* 94:25–34.

MacArthur, R. H. 1962. Some generalized theorems of natural selection. *Proc. Nat. Acad. Sci.* 48:1893–1897.

MacArthur, R. H., and E. O. Wilson. 1967. *The Theory of Island Biogeography.* Princeton University Press, Princeton, N. J.

Mares, M. A., T. E. Lacher, M. R. Willig, N. A. Bitar, R. Adams, A. Klinger, and D. Tazik. 1982. An experimental analysis of social spacing in *Tamias striatus*. *Ecology* 63:267–273.

Marshall, D. R., and S. K. Jain. 1969. Interference in pure and mixed populations of *Avena fatua* and *A. barbata*. *J. Ecol.* 57:251–270.

May, R. M. 1975. *Stability and Complexity in Model Ecosystems.* Princeton University Press, Princeton, N. J.

McGregor, R. L., and T. M. Barkley (eds.). 1977. *Atlas of the Flora of the Great Plains.* Iowa State University Press, Ames.

McKusick, V. A. 1978. *Medical Genetics Studies of the Amish.* Johns Hopkins University Press, Baltimore, Md.

McKusick, V. A. 1983. *Mendelian Inheritance in Man.* Johns Hopkins University Press, Baltimore, Md.

Merrell, D. J., and C. F. Rodell. 1968. Seasonal selection in the leopard frog, *Rana pipiens*. *Evolution* 22:284–288.

Miller, R. S. 1964. Ecology and distribution of pocket gophers (Geomyidae) in Colorado. *Ecology* 45:256–272.

Miller, R. S., and W. J. D. Stephen. 1966. Spatial relationships in flocks of sandhill cranes (*Grus canadensis*). *Ecology* 47:323–327.

Montagu, A. (ed.). 1980. *Sociobiology Examined.* Oxford University Press, Oxford.

Mooreale, S. J., G. J. Ruiz, J. R. Spotila, and E. A. Standora. 1982. Temperature-dependent sex-determination: Current practices threaten conservation of sea turtles. *Science* 216:1245–1247.

Moran, G. F., and A. H. D. Brown. 1979. Temporal heterogeneity of outcrossing rates in alpine ash (*Eucalyptus delegatensis* R. T. Bak.). *Theoret. Appl. Genet.* 57:101–105.

Morin, P. J. 1981. Predatory salamanders reverse the outcome of competition among three species of anuran tadpoles. *Science* 212:1284–1286.

Mukai, T., S. I. Chigusa, L. E. Mettler, and J. F. Crow. 1972. Mutation rate and dominance of genes affecting viability in *Drosophila melanogaster*. *Genetics* 72:335–355.

Murdoch, W. W. 1966. "Community structure, population control, and competition"—a critique. *Amer. Natur.* 100:219–226.

Nei, M., and A. K. Roychoudhury. 1980. Genetic relationship and evolution of human races. *Evol. Biol.* 14:1–59.

Neilson, M. M., and R. F. Morris. 1964. The regulation of European spruce sawfly numbers in the maritime provinces of Canada from 1937 to 1963. *Can. Entomol.* 96:773–784.

Nevo, E. 1978. Genetic variation in natural populations: Patterns and theory. *Theoret. Pop. Biol.* 13:121–177.

Nevo, E., G. Naftali, and R. Guthran. 1975. Aggression patterns and speciation. *Proc. Nat. Acad. Sci.* 72:3250–3254.

Nevo, E., and C. R. Shaw. 1972. Genetic variation in a subterranean mammal, *Spalax ehrenbergi*. *Biochem. Genet.* 7:235–241.

Nichols, P. L., and V. E. Anderson. 1973. Intellectual performance, race, and socioeconomic status. *Soc. Biol.* 20:367–374.

Orians, G. H., and M. H. Willson. 1964. Interspecific territories of birds. *Ecology* 45:736–745.

Overland, L. 1966. The role of allelopathic substances in the "smother crop" barley. *Amer. J. Bot.* 53:423–432.

Owadally, A. W. 1979. The dodo and the tambalacoque tree. *Science* 203:1363–1364.

Owen, D. F. 1966. Polymorphism in pleistocene land snails. *Science* 152:71–72.

Paine, R. T. 1974. Intertidal community structure. Experimental studies on the relationship between a dominant competitor and its principal predator. *Oecologia* 15:93–120.

Park, T. 1954. Experimental studies of interspecies competition. II. Temperature, humidity, and competition in two species of *Tribolium*. *Physiol. Zool.* 27:177–238.

Park, T., D. B. Mertz, and K. Petrusewicz. 1964. Genetic strains and competition in populations of *Tribolium*. *Physiol. Zool.* 37:97–162.

Paterniani, E. 1969. Selection for reproductive isolation between two species of maize, *Zea mays* L. *Evolution* 23:534–547.

Pearl, R. 1927. The growth of populations. *Quart. Rev. Biol.* 2:532–548.

Perrins, C. M. 1965. Population fluctuations and clutch size in the great tit, *Parus major* L. *J. Anim. Ecol.* 34:601–647.

Pianka, E. R. 1970. On r- and K-selection. *Amer. Natur.* 104:592–597.

Pimentel, D. 1968. Population regulation and genetic feedback. *Science* 159:1432–1437.

Pimentel, D. E. H. Feinburg, D. W. Wood, and J. T. Hayes. 1965. Selection, spatial distribution and the co-existence of competing fly species. *Amer. Natur.* 95:65–79.

Platt, J. R. 1964. Strong inference. *Science* 146:347–353.

Plucknett, D. L., and N. J. H. Smith. 1982. Agricultural research and third world food production. *Science* 217:215–220.

Prokopy, R. J., A. L. Averill, S. S. Cooley, and C. A. Roitberg. 1982. Associative learning in egg-laying site selection in maggot flies. *Science* 218:76–77.

Prout, T. 1971. The relation between fitness components and population prediction in *Drosophila*. II. Population prediction. *Genetics* 68:151–167.

Ralls, K., K. Baugh, and J. Ballou. 1979. Inbreeding and juvenile mortality in small populations of ungulates. *Science* 206:1101–1103.
Reed, T. E. 1969. Caucasian genes in American Negroes. *Science* 165:762–768.
Richardson, R. H., J. R. Ellison, and W. W. Averhoff. 1982. Autocidal control of screwworms in North America. *Science* 215:361–370.
Rick, C. M. 1978. The tomato. *Sci. Amer.* 239:76–87.
Ricklefs, R. E. 1975. *Ecology*. Chiron Press, Portland, Ore.
Ricklefs, R. E. 1976. *The Economy of Nature*. Chiron Press, Portland, Ore.
Roberts, D. F., and R. W. Hiorns. 1962. The dynamics of social intermixture. *Amer. J. Hum. Genet.* 14:261–277.
Rose, M., and B. Charlesworth. 1981a. Genetics of life history in *Drosophila melanogaster*. I. Sib analysis of adult females. *Genetics* 97:173–186.
Rose, M., and B. Charlesworth. 1981b. Genetics of life history in *Drosophila melanogaster*. II. Exploratory selection experiments. *Genetics* 97:187–196.
Rosensweig, M. L. 1971. *And Replenish the Earth*. Harper & Row, New York.

Sage, R. D., and R. K. Selander. 1975. Trophic radiation through polymorphism in cichlid fishes. *Proc. Nat. Acad. Sci.* 72:4669–4673.
Salt, G., and F. S. J. Hollick. 1944. Studies of wireworm populations. I. A census of wireworms in pasture. *Ann. Appl. Biol.* 31:52–64.
Saura, A., D. Halkka, and J. Lokki. 1973. Enzyme gene heterozygosity in small island populations of *Philaenus opumarius* (L.) (Homoptera). *Genetica* 44:459–475.
Schaal, B. A., and D. A. Levin. 1976. The demographic genetics of *Liatris cylindracea* Michx. (compositae). *Amer. Natur.* 110:191–206.
Schaffer, W. M. 1974. Optimal reproductive effort in fluctuating environments. *Amer. Natur.* 108:783–790.
Schlager, G., and M. M. Dickie. 1971. Natural mutation rates in the house mouse. Estimates for five specific loci and dominant mutations. *Mutation Res.* 11:89–96.
Schoener, T. W. 1982. The controversy over inter-specific competition. *Amer. Sci.* 70:586–595.
Schull, W. J., and J. V. Neel. 1965. *The Effects of Inbreeding on Japanese Children*. Harper & Row, New York.
Schwartz, D., and M. J. Mayaux. 1982. Female fecundity as a function of age. *N. Eng. J. Med.* 306:404–406.
Schwartz, O. A., and K. B. Armitage. 1980. Genetic variation in social mammals: The marmot model. *Science* 207:665–667.
Selander, R. K., and D. W. Kaufman. 1975. Genetic structure of populations of the brown snail (*Helix aspera*). I. Microgeographic variation. *Evolution* 29:385–401.
Silliman, R. P., and J. S. Gutsell. 1958. Experimental exploitation of fish population. *Fish Bull.* 58:215–252.
Simberloff, D. S. 1978. Using island biogeographic distributions to determine if colonization is stochastic. *Amer. Natur.* 112:713–726.
Simmonds, N. W. 1976. *Evolution of Crop Plants*. Longmans, London.
Sinclair, A. R. E. 1974. The natural regulation of buffalo populations in East Africa: III. Population trends and mortality. *East Afr. Wildl. J.* 12:185–200.
Sittman, F., H. A. Abplanalp, and R. A. Fraser. 1966. Inbreeding depression in Japanese quail. *Genetics* 54:371–379.

Skud, B. E. 1982. Dominance in fishes: The relation between environment and abundance. *Science* 216:144–149.

Slade, N. A., and D. F. Balph. 1974. Population ecology of Uinta ground squirrels. *Ecology* 55:989–1003.

Smith, F. E. 1963. Population dynamics in *Daphnia magna* and a new model for population growth. *Ecology* 44:651–663.

Snell, T. W., and C. E. King. 1977. Lifespan and fecundity patterns in rotifers: The cost of reproduction. *Evolution* 31:882–890.

Spiess, E. B. 1968. Low frequency advantage in mating of *Drosophila pseudoobscura* karyotypes. *Amer. Natur.* 102:363–379.

Stern, C. 1973. *Human Genetics.* W. H. Freeman, San Francisco.

Tamarin, R. H. 1978. *Population Regulation.* Dowden, Hutchinson, and Ross. Stroudsburg, Pa.

Tauber, C. A., and M. J. Tauber. 1978. Sympatric speciation based on allelic changes at three loci: Evidence from natural populations in two habitats. *Science* 197:1298–1299.

Temple, S. A. 1977. Plant-animal mutualism: Coevolution with dodo leads to near extinction of plant. *Science* 197:885–886.

Thatcher, D. R. 1980. The complete amino acid sequences of three alcohol dehydrogenase alleloenzymes (Adh^{N-11}, Adh^{S}, and Adh^{UF}) from the fruitfly *Drosophila melanogaster*. *Biochem. J.* 187:875–886.

Thompson, S. D. 1982. Microhabitat utilization and foraging behavior of bipedal and quadrupedal heteromyid rodents. *Ecology* 63:1303–1312.

Tizard, B. 1974. I.Q. and race. *Nature* 247:316.

Toft, C. A. 1980. Feeding ecology of thirteen syntopic species of anurans in a seasonal tropical environment. *Oecologia* 45:131–141.

Trimble, B. K., and J. A. Doughty. 1974. The amount of hereditary disease in human population. *Ann. Hum. Genet.* 38:199–223.

Tuttle, M. D., and M. J. Ryan. 1981. Bat predation and the evolution of frog vocalizations in the neotropics. *Science* 214:677–678.

Ullstrop, A. J. 1972. The impact of the southern corn leaf blight of 1970 and 1971. *Ann. Rev. Phytopath.* 10:37–50.

Usher, M. B. 1974. *Biological Management and Conservation.* Chapman and Hall, London.

van Delden, W., A. Keming, and H. van Dijk. 1975. Selection at the alcohol dehydrogenase locus in *Drosophila melanogaster*. *Experimentia* 31:418–420.

Vandermeer, J. H. 1972. Niche theory. *Ann. Rev. Ecol. Syst.* 3:107–132.

Vandermeer, J. H., and D. H. Boucher. 1978. Varieties of mutualistic interaction in population models. *J. Theoret. Biol.* 74:549–558.

Vogel, F., and A. G. Motulsky. 1979. *Human Genetics.* Springer-Verlag, New York.

Wagner, P. 1974. Wines, grape vines, and climate. *Sci. Amer.* 230:106–115.

Watt, K. E. F. 1968. *Ecology and Resource Management.* McGraw-Hill, New York.

Wells, P. V. 1983. Paleobiologeography of montane islands in the Great Basin since the last glaciopluvial. *Ecol. Monogr.* 53:341–382.

Werner, E. E., and D. J. Hall. 1976. Niche shifts in sunfishes: Experimental evidence and significance. *Science* 191:404–406.

Werner, P. A. 1978. On the determination of age in *Liatris aspera* using cross-sections of corms: Implications for past demographic studies. *Amer. Natur.* 112:1113–1120.

Westoff, C. F. 1978. Marriage and fertility in the developed countries. *Sci. Amer.* 239:51–57.

Whittaker, R. H. 1956. Vegetation of the Great Smoky Mountains. *Ecol. Monog.* 26:1–80.

Wickler, W. 1970. *Mimicry in Plants and Animals.* McGraw-Hill, New York.

Williams, R. J. 1977. Identification of multiple disease resistance in cowpea. *Trop. Agric. Trinidad* 54:53–59.

Wynne-Edwards, V. C. 1962. *Animal Dispersion in Relation to Social Behavior.* Oliver and Boyd, Edinburgh.

Zimmerman, D. R. 1977. A technique called cross-fostering may help save the whooping crane. *Audubon* 52–63.

Ziswiler, V. 1967. *Extinct and Vanishing Animals.* Springer-Verlag, New York.

ANSWERS TO NUMERICAL QUESTIONS

Chapter 2

1.

	p	q	p^2	$2pq$	q^2	p^2N	$2pqN$	q^2N	χ^2
a	0.4	0.6	0.16	0.48	0.36	16	48	36	3.04
b	0.68	0.32	0.462	0.435	0.102	23.1	21.7	5.1	20.6
c	0.467	0.533	0.218	0.498	0.284	62.1	141.9	80.9	5.63

1 degree of freedom; *a*—not significant, *b*—significant at 0.01 level, *c*—significant at 0.05 level.

3. A_1A_1 with a frequency of 0.49.
4. 0.1, 0.18
5.

	p_1	p_2	p_3	$E(H)$	$O(H)$	χ^2
a	0.35	0.30	0.35	0.665	0.7	2.23
b	0.2	0.2	0.6	0.56	0.0	4.00

3 degrees of freedom; *a*—not significant, *b*—significant at 0.01 level.

Chapter 3

1. 50, 150
2. 0.0, −0.019, −0.056, −0.073, 0.0; 0.198, 0.196
3. 0.75, 1.0; 0.57
4. 0.2
5. 0.25

Chapter 4

1. 0.000001, 0.000000725, 0.00000045, 0.000000175, −0.0000001; 0.909
2. 0.9999 p_0
3. 0.0000167, 0.00408; 0.0000333, 0.00577
4. 3×10^{-5}
5. 0.0001

Chapter 5

1. 0.575, 0.551; 0.1
2. 0.2, 0.8, 0.98
3. 0.707; 0.0736, −0.0056

Chapter 6

1. 0.3; 0.7
2. 0.0012, 0.0346; 0.478, 0.475; 0.456, 0.433
3. 6
4. 0.0909

Chapter 7

1. 0.12, 0.16, 0.72; 0.01
2. 0.408, 0.384, 0.208; 0.792
3. 0.125; 0.125
4. 0.00124, 0.0312

Chapter 8

2. 0.333
3. 0.25

Chapter 10

1. 3,000
2. 1,000
3. 0.267; 16.4, 22.3, 9.1
4. 0.56

Chapter 11

1. 810; 14.9
2. 0.075, 0.05, 0.025; 7.5, 10.0, 7.5
3. 230; 900; 3
4. 400, 400, 0

Chapter 12

1. 1.0, 0.2, 0.04, 0.008, 0.0016
2. 1.25; 2.2; 0.101, 0.102; 0.85, 1.25, 1.0, 0.5; 1, 5, 10, 4
3. 87, 75, 2, 5, 5, 0; 38, 20, 15, 1, 2, 0; 92, 80, 4, 8, 0, 0

Chapter 13

3. 62.5, 62.5; −12, −2
4. 0.111, 1, 9; 0.25, 0.667, 2.33
5. 0.636, 0.519

Chapter 14

2. −30, 3

Chapter 15

3. 0.5
4. 0.0

Chapter 17

3. 0.588, 3.0×10^{-7}; 0.288, 1.4×10^{-7}
4. $D_{AC} = 0.15$, $D_{(AC)B} = 0.25$, $D_{(AC)D} = 0.6$, $D_{(ACB)D} = 0.65$

Chapter 18

4. 0.5, No

Chapter 19

1. 0.333, 1.0, 1.0; 0.688, 1.0, 0.917
5. 0.27, 0.57, 0.99

Chapter 20

3. 0.49
5. −40, −40; 60, 60

Chapter 21

1. 0.976, 0.976
3. 0.323

AUTHOR INDEX

Abplanalp, H. A., 109
Adams, J., 287
Adams, R., 339
Adkissom, P. L., 403
Allard, R. W., 282, 287
Anderson, W. W., 18, 19
Anderson, V. E., 132
Armitage, K. B., 82
Arnold, S. J., 250
Averhoff, W. W., 406, 407
Averill, A. L., 340, 341
Avise, J., 312
Ayala, F. J., 287, 320, 421, 422
Ayres, P. B., 401

Baker, M. C., 210
Baldwin, N. S., 260
Ballou, J., 111
Balph, D. F., 205, 211
Barkley, T. M., 146, 147
Bartlett, A. C., 406
Batra, S. W. T., 402
Battaglia, B., 290
Baugh, K., 111
Baumgardner, K. D., 151, 152, 155, 421
Beattie, A. J., 390
Beddington, B., 410
Begon, M., 220, 264, 421
Birch, L. C., 1, 142, 143, 259
Bird, L. S., 403
Bishop, J. A., 168
Bitar, N. A., 339
Blavestein, A. R., 346, 347
Bloom, W. L., 310
Blower, J. G., 168
Boag, P. T., 298
Bodmer, W. F., 421
Bonaiti-Pellie, C., 130
Bonnell, M. L., 97
Bossert, W. H., 421
Boucher, D. H., 390
Bradshaw, A. D., 300, 370, 371
Bram, R. A., 406
Brander, K., 269
Brower, L. P., 302, 303
Brown, A. H. D., 358
Brown, J. H., 154, 227
Buri, P., 94, 95
Bush, G. L., 314, 407

Cameron, A. W., 242
Capretta, P. J., 382
Carlquist, S., 147
Caughley, G., 213
Cavalli-Sforza, L. L., 421
Chapin, G., 400
Charlesworth, B., 307, 368
Chigusa, S. I., 64
Christiansen, F. B., 362
Clarke, C. A., 56
Clatworthy, J. N., 382
Clausen, J. D., 120
Clegg, M. T., 51, 356
Cleghorn, T. E., 33
Coale, A. J., 221
Connell, J. H., 184, 207, 246, 421
Conover, D. O., 297
Cooch, F. G., 360
Cook, L. M., 168, 303
Cooke, F., 360
Cooley, S. S., 340, 341
Cooper, R. M., 122
Crombie, A. C., 231
Crow, J. F., 64, 99, 117
Croze, H., 288, 303
Culver, D. C., 390
Crick, F. H. C., 326

Daniel, N. J., 51, 356
Davidson, D. W., 227
Dawson, P. S., 421
DeBach, P., 245
Diamond, J. M., 242, 344
Dickerson, G. E., 398, 399
Dickerson, R. C., 329
Dickie, M. M., 70
Dobzhansky, T., 18, 19, 343, 421
Doughty, J. A., 41
Dudley, J. W., 127
Duggins, D. O., 266
Dunn, L. C., 361

Eaton, G. G., 338
Ebling, F. J., 158

Ehrlich, A. R., 195
Ehrlich, P. R., 193, 259
Ehrman, L., 286, 422
Ellison, J. R., 406, 407
Ellseth, G. D., 151, 152, 155, 421
Embree, D. G., 262
Emlen, J. M., 422

Falconer, D. S., 125, 126, 368, 421
Farentinos, R. C., 382
Feinberg, E. H., 385
Ferchel, T., 381
Fenner, F., 384
Ferguson, M. W., 297
Ferris, S. D., 325
Flor, H., 387
Ford, E. B., 422
Foster, G. G., 61
Frankel, D. H., 393
Franks, N. R., 163
Fraser, R. A., 109
Frejka, T., 197, 198
Fries, J. F., 208, 209
Frydenberg, O., 362
Futuyma, D. J., 422

Gagne, R. J., 406
Gagnon, P. S., 284, 285
Gaines, M. S., 2, 22, 23
Gallun, R. L., 387
Gause, G. F., 228, 237
Gill, J. E., 401
Gill, R., 61
Goldberg, D. E., 156
Goldsby, R. A., 133
Graham, O. H., 406
Grant, P. R., 242, 298, 312, 380, 381
Graves, J. E., 323
Greaves, J. H., 401
Greenberg, L., 347
Gregory, R. P. G., 300
Gulhati, K., 197
Gulland, J. A., 409
Guthran, R., 316
Gutsell, J. S., 408, 409

Halkka, D., 320, 321
Hall, D. J., 244
Hamrick, J. L., 24, 282
Harper, J. L., 372, 382, 421
Harris, H., 24
Harrison, G. A., 118
Hartl, D. L., 421
Hatchett, J. H., 387
Hayes, J. T., 385

Hebert, P. D. N., 37
Hedrick, P. W., 296, 377, 421
Heithaus, E. R., 390
Hermodson, M. A., 401
Hiesey, W. M., 120
Hines, A. H., 258
Hiorns, R. W., 79
Hollick, F. S. J., 165
Holling, C. S., 262, 264
Holt, S., 128, 129
Hopkinson, D. A., 24
Huang, C., 325
Hubbell, S. P., 240
Hubby, J. L., 318
Huffaker, C. B., 256
Hutchinson, G. E., 180, 201, 259, 421

Jain, S. K., 104, 240
Janzen, D. H., 375, 384
Jarvis, J. V. M., 348
Jensen, A. R., 131
Joanen, T., 297
Johnston, R. F., 301
Jones, J. S., 16, 17

Kahler, A. L., 356
Kaza, S., 4, 5
Kaneshiro, K., 312, 313
Karban, R., 165, 264, 265
Kaufman, D. W., 22, 23
Kawaii, M., 341
Keck, D. D., 120
Keller, B. J., 2
Keming, A., 322
Kephart, D. G., 250
Kepver, R. E., 382
Kessel, B., 148
Kettlewell, B., 17
Kidwell, J. F., 51, 356
Kidwell, M. G., 51, 326, 356
Kimura, M., 99
King, C. E., 364, 421
Kitching, J. A., 158
Klinger, A., 339
Knight, G. R., 306
Koehn, R. K., 278, 279
Kornfield, I., 284, 285
Krebs, C. J., 2, 164, 165, 200, 240, 421
Kynard, B. E., 297

LaChance, L. E., 406
Lacher, T. E., 339
Lack, D., 162, 304
Lande, R., 307
Langer, W. L., 196

Law, R., 370, 371
LeBoeuf, B. J., 4, 5
Leith, B. H., 16, 17
Levin, D. A., 40, 41, 358, 359
Levings, B. J., 163
Levy, M., 40, 41
Lewontin, R. C., 128, 134, 296, 318, 361, 421
Ligon, J. D., 339
Ligon, S. H., 339
Link, K. P., 401
Little, E. L., 146, 147
Littlefield, V. M., 382
Livingstone, F. B., 84
Lokki, J., 320, 321
Lowenstein, J. M., 333, 335
Lubchenco, J., 266, 267
Lynch, M., 372

MacArthur, R. H., 149, 153, 185, 369
Mares, M. A., 339
Marshall, D. R., 104, 240
May, R. M., 190, 243, 410
Mayaux, M. J., 223
McClenaghan, L. R., 22, 23
McGregor, R. L., 146, 147
McInnis, D. O., 406
McKusick, V. A., 41, 90
McNaughton, I. H., 382
McNeilly, T. S., 300
Mertz, D. B., 376, 421
Mettler, L. E., 64
Mewaldt, L. C., 210
Milkman, R., 422
Miller, R. S., 156, 157, 174
Minsky, P. J., 360
Montagu, A. 345
Mooreale, S. J., 297
Moran, G. F., 358
Morin, P. J., 267
Morris, R. F., 219, 220
Mortimer, M., 220, 264, 421
Motulsky, A. G., 110
Mouraõ, C. A., 320
Mukai, T., 64
Murdoch, W. W., 259, 421
Myers, J. H., 2

Naftali, G., 316
Neel, J. V., 111
Nei, M., 333, 334
Neilson, M. M., 219, 220
Nevo, E., 24, 316
Nichols, P. L., 132
Nicholson, A. J., 190
Nielsen, J. T., 325
Niles, G. A., 403

Odum, E. P., 90
O'Hara, R. K., 346, 347
Orgel, L. E., 326
Orians, G. H., 226
Owadally, A. W., 391
Owen, D. F., 16, 17
Owen, J. J. T., 118
Overland, L., 157

Paine, R. T., 261
Park, T., 230, 376
Parsons, P. A., 422
Paterniani, E., 306, 307
Pavlovsky, O., 18, 19
Pearl, R., 188
Pearse, J. S., 258
Perez-Salas, S., 320
Perrins, C. M., 218
Petrvsewicz, K., 376
Pianka, E. R., 369, 422
Pielou, E. C., 167, 170, 421
Pimentel, D., 385, 386
Platt, J. R., 3
Plucknett, D. L., 395
Powell, J., 18, 19, 320
Prokopy, R. J., 340, 341
Prout, T., 59, 61
Putwain, P. D., 370, 371

Ralls, K., 111
Ratcliffe, F. N., 384
Rawlings, P., 16, 17
Redfern, R., 401
Reed, T. E., 80
Richardson, R. H., 406, 407
Rick, C. M., 397
Ricklefs, R. E., 144, 222, 421
Ritte, V., 325
Roberts, D. F., 79
Robertson, A., 306
Rodell, C. F., 34, 35
Roitberg, C. A., 340, 341
Rose, M., 368
Rose, R. K., 22, 23
Rosensweig, M. L., 161
Roughgarden, J., 421
Roychoudhury, A. K., 333, 334
Ruiz, G. J., 297
Ryan, M. J., 380

Sage, G. R., 382
Sage, R. D., 284, 325

Salt, G., 165
Sarich, V. M., 333, 335
Saurd, A., 320, 321
Schaal, B. A., 358, 359
Schaffer, W. M., 370
Schlager, G., 70
Schoener, T. W., 245, 246
Schull, W. J., 111
Schwartz, D., 223
Schwartz, O. A., 82
Scott, H. B., 403
Seawright, J. A., 406
Selander, R. K., 21–23, 97, 284, 301
Seyer, M. B., 360
Shaw, C. R., 316
Sheppard, P. M., 56
Silliman, R. P., 408, 409
Simberloff, D. S., 242
Simmonds, N. W., 397
Simonsen, V., 362
Sinclair, A. R. E., 218
Sittman, F. 109
Skud, B. E., 410
Slade, N. A., 205, 211
Slatkin, M., 307, 422
Smith, C., 130
Smith, D. C., 284, 285
Smith, F. E., 187, 189
Smith, N. J. H., 395
Snell, T. W., 364
Solbrig, D. J., 421
Solbrig, O. T., 421
Somero, G. N., 323
Soule, M. E., 393
Spassky, B., 343
Spotila, J. R., 297
Spiess, E. B., 19, 286, 421
Standora, E. A., 297
Stebbins, G. L., 422
Stephen, W. J. D., 174
Stern, C., 8, 9, 70
Stewart, R. M., 210
Sundby, R. A., 245
Suttie, J. W., 401

Tamarin, R. H., 2, 216
Tauber, C. A., 314
Tauber, M. J., 314
Tazik, D., 339
Temple, S. A., 391
Thatcher, D. R., 323
Thompson, S. D., 145, 146
Tilley, S. G., 370
Tinkle, D. W., 370
Tizard, B., 133
Toft, C. A., 246
Tracey, M. L., 320
Trimble, B. K., 41
Tuttle, M. D., 380

Ullstrop, A. J., 394

Val, F. C., 312, 313
Valentine, J. W., 422
van Delden, W., 322
van Dijk, H., 322
Vandermeer, J. H., 143, 390
Vogel, F. 110

Waddington, C. H., 306
Wagner, P., 394
Walker, J. K., 403
Wasserstrom, R., 400
Watt, K. E. F., 408, 409
Wells, P. V., 148, 149
Werner, E. E., 244
Werner, P. A., 358
Westoff, C. F., 221
Whittaker, R. H., 154, 155
Whitten, C. J., 406
Whitten, M. J., 61
Wickler, W., 299
Wilbur, H. M., 370
Williams, R. J., 398
Willig, M. R., 339
Willson, M. H., 226
Wilson, A. C., 325
Wilson, E. O., 149, 421, 422
Wittenberger, J. F., 422
Wood, D. W., 385
Wynne-Edwards, V. C., 219

Yardley, D., 18, 19

Zimmerman, D. R., 270
Ziswiler, V., 192, 269, 391
Zubeck, J. P., 122

SUBJECT INDEX

Abalone, 258
Acacia cornigera, 389
Adaptation, 293–305
Admixture, 80, 81
Age pyramid, 222
Agelaius phoeniceus, 226
Agriculture, 393–403
  See also various crop plants
Agrostis tenuis, 300
Algae, 157, 158, 266
Allelic frequency, 28, 29, 34, 35
Allelopathy, 157
Allen's rule, 299
Alligators, 297
Altruism, 347–350
Ants, 162, 163, 226, 227, 389
Aphytis, 245
Applied population biology, 393–409
Ash, 358
Avena, 104, 240, 282

Barberry, 161
Barley, 356
Barnacles, 184, 207, 208
Barracuda, 323
Beech, 165
Bees, 346, 347
Beetles, 143, 165, 230, 231
Bentgrass, 300
Bergman's rule, 299, 301
Birds, *See* various species
Biston betularia, 17, 18, 56, 57
Blackbirds, 226
Blarina, 264
Blood groups, 33, 39, 79, 80
Bottleneck, 89, 97
Bowerbirds, 305, 344
Buffalo, 218
Bufo, 267

Cactoblastis cactorum, 404
Cactus, 153, 404
Calvaria major, 391
Capture-recapture method, 166–168
Caribou, 262
Carrying capacity, 183–187, 216–220, 231, 238, 289–291, 369–371
Cepadea nemoralis, 16, 17
Character displacement, 379–381
Chemical defenses, 380–382
Chickens, 398, 399
Chromosomes, 18–20, 61, 117, 286, 287, 310, 316, 405
Chrysopa, 314, 315
Cicada, 264, 265
Cichlids, 284, 285
Clarkia, 310
Clethrionomys gapperi, 242
Climatograph, 144, 152
Cline, 86, 278, 282, 310
Clove, 382
Coevolution, 375–392
Cohort, 201, 207, 208, 355, 357
Commensalism, 225, 226, 388
Common-garden experiment, 119, 120
Competition, 156, 157, 225–247, 376–381
  coefficients, 231, 232, 243–245, 378
  evolution of, 376–381
  exploitative, 226–228, 377
  interference, 226, 228
Competitive exclusion, 228, 231, 245, 376–378
Competitive release, 241, 242
Copepods, 290
Corn, 127, 134, 306, 307, 394
Cotton, 403
Cowpeas, 398
Cranes, 174, 270
Crops, 393–400, *See also* various crop plants
Crows, 288

Daphnia, 37, 39, 183, 187, 189, 372
DDT, 116, 117, 400
Deer, 262
Degrees of freedom, 36, 38
Demographic genetics, 353–362
Demographic transition, 220–221
Demography, 199–223

Density, 141, 142, 161–169
absolute, 163–168
capture-recapture method, 166–168
quadrat method, 164–165
relative, 168, 169
Density-dependent factors, 216–220, 390, 407–409
Density-independent factors, 216–220
Diapause, 372, 373
Dipodomys deserti, 145, 146
Disease resistance, 384, 387, 398
Dispersal, 75, 150–152, 177, 178, 219
Dispersion, 162, 163, 169–175
index of, 169–172
Distribution, 141, 142
changes in, 146–149
factors affecting, 149–158
DNA, 63, 64, 321, 324–326
Dodo, 391
Dominance hierarchy, 337–339
Dormancy, 372, 373
Doves, 242
Drosophila
genetic drift in, 94, 95
heteroneura, 312, 313
inversions, 18–20, 286, 287
melanogaster, 51, 59, 61, 64, 94, 95, 117, 306, 322, 326, 368, 377
pseudoobscura, 18–20, 286, 287, 343
selection in, 51, 59, 61, 286, 287, 306, 368, 377
silvestris, 312, 313
simulans, 376, 377
willistoni, 320

Ecotypes, 298–300
Eelpout, 362
Effective population size, 96–98
Electrophoresis, 20, 21
Elephant, 334, 335
Elephant seals, 4, 5, 97
Endangered species, 91, 270
See also Extinction
Environmental variation, 119–124, 144–146, 277–283
Equilibrium
competition, 232–238
heterozygote advantage, 55–58
heterozygote disadvantage, 60, 61
logistic, 183–186
migration, 76, 77
migration-selection, 83–86
minimum critical density, 192–194
mutation, 65, 66
mutation-selection, 67–69
predator-resource, 253–257
punctuated, 306
Eucalyptus delegatensis, 358
Evolution
coevolution, 375–392
life-history, 353, 362–373
molecular, 317–335
social behavior and, 337–351
See also Selection
Extinction, 192, 263–265, 268–270, 391

Fecundity schedules, 199, 200, 202–205, 210–216, 362, 363
Finches, 298, 312, 380, 381
Finite rate of increase, 181, 182, 189–192, 370, 371
Fish, 167, 244, 278, 284, 285, 297, 323, 362, 408–410
Fitness, 46–48, 278–280, 283–291, 346, 365, 366
See also Selection
Flax, 387
Fixation index, 359
Founder effect, 90, 91, 404
Frogs, 34, 35, 346, 347
Functional response, 260–262, 264, 265

Gametic disequilibrium, 367
Geese, 360
Gene flow, 75–86
continent-island model, 76–78
estimation of, 80–82
general model, 78–80
selection and, 83–86, 281–283
Generalists, 145, 251
Generation length, 203, 205
Genetic conservation, 393–400
Genetic counseling, 40–43
Genetic distance, 330–335
Genetic drift, 89–100
mutation and, 98–100, 317–321, 328
See also Founder effect
Genetic feedback, 378, 379, 384–386
Genetic variation, 15–25, 117–130
chromosomal, 18–20
electrophoretic, 20–25
quantitative, 117–130
visible, 15–18
Genotype-environment interaction, 119–122
Genotypic frequency, 28–39
See also Heterozygosity
Geospiza, 380, 381
Germplasm collection, 395
Gophers, 156, 157

Grapes, 394
Guppies, 408, 409

Habitat selection, 150, 283
Haplo-diploids, 349, 350
Hardy-Weinberg principle, 29–36
Harvest parabola, 408, 409
Hawthorn, 314, 315, 340, 341
Heath hen, 192
Heritability, 123–125, 366, 368, 398, 399
  broad-sense, 123
  estimation of, 124–130
  narrow-sense, 123–126
  realized, 125
Hessian fly, 387
Heterocephalus glaber, 348
Heterozygosity, 22, 24, 25, 38–40
  genetic drift, 94–96
  genetic drift and mutation, 99, 100
  inbreeding, 102–105
Home range, 338–340
House fly, 385
Humans, 8, 9, 52
  blood groups, 33, 39, 79, 80
  demography, 200, 208, 209, 220–223
  disease, 32–34, 45, 70–73, 84, 85, 90, 108–111
  genetic counseling, 40–43
  genetic variation, 24, 25, 128, 129
  population growth, 194–198
  predation, 263, 268, 270, 391, 405–410
  race, 118, 130–135, 333, 334
Hybrid zone, 310, 311, 316
Hydrobia, 381
Hyla, 267

Inbreeding, 101–112, 358, 359
  coefficient of, 103–108, 348, 349
  disease, 108–111
  depression, 109–111
  selection and, 108–111
Input-output ratio diagrams, 238–240, 287
Instantaneous rate of increase, 181, 182, 203–205, 289–291
Integrated pest management, 402–404
IQ, 131–135
Island biogeography, 149

Kelp, 266
K-selection, 369–371, 377

Lacewings, 314, 315
Lake trout, 260
Laminaria, 266
Lampreys, 260
Land races, 394–396
Lasioglossum zephyrum, 346, 347
Learning, 340–342
Lepomis, 244
Lethals, 48–52, 404, 405
Liatris cylindracea, 358, 359
Life expectancy, 204–210
Life-history evolution, 353, 362–373
Littorina littorea, 266
Locusts, 162
Loss, 177, 406, 407
Lotka-Volterra equations
  competition, 231–238, 378
  predator-resource, 251–262, 382, 383, 386
Lucilia cuprina, 190
Lynx, 89, 90, 259

Macaques, 338, 340, 341
Macroevolution, 306
Magicicada, 264, 265
Mammoth, 334, 335
Manatee, 334, 335
Marmots, 82
Mating
  assortative, 101, 306, 307, 360
  random, 27, 30, 31, 284, 285
  See also Inbreeding
Maximum sustainable yield, 405–409
Mayetiola destructor, 387
Maze, 122, 343
Meadow grass, 370
Meiotic drive, 357, 361, 405
Metal tolerance, 300
Mice, 70, 262, 325, 361
Microtus, 2, 22, 23, 242
Migration, *See* Gene flow
Milk-pilfering, 341
Mimicry, 299, 302, 303
Minimum critical population density, 192–194, 217
Mites, 256
Models, 3–6
Moisture, 154, 155
Molecular clock, 327–330
Mole rats, 316, 348
Monarchs, 302, 303
Moose, 262
Moths, 17, 18, 56, 57, 303, 404
Mus, 70, 325, 361
Musca domestica, 385
Mussel, 261

Mutation, 63–73
  estimation, 69–71
  genetic drift and, 98–100, 317–321, 328
  selection and, 67–69
Mutualism, 227, 228, 386–391
Mytilus californianus, 261
Myxomatosis, 384

Nasonia vitripennis, 385
Nearest-neighbor distance, 172–175
Nematode, 262
Net replacement rate, 4–6, 178, 179, 202–205
Neutrality, 98–100, 317–321, 328
Newts, 267
Niche, 142, 294
  breadth, 144, 145
  empty, 143, 144
  fundamental, 143
  overlap, 229, 230, 243, 246
  preferred, 143
  realized, 143
  width, 144, 145, 241, 243
Notophthalmus, 267
Numerical response, 259, 264, 265

Oaks, 165
Oenothera, 40, 41
Optimal foraging strategy, 251
Opuntia stricta, 404
Oryzaephilus suranimensis, 231
Osage orange, 146, 147, 152
Ovenbirds, 185

Paramecium, 237
Parasites, 249, 262
Parasite-host interactions, 249
Parasitoids, 249, 385, 386
Partridges, 152
Passenger pigeon, 192
Pedigree, 43, 105–110
Perch, 167
Perdix perdix, 152
Perognathus longimembris, 145, 146
Pesticide resistance, 116, 117, 400–402
Pheasants, 180
Phenotype, 113–125, 340–345, 363–365
Phenotypic plasticity, 296, 340–342, 371, 386
Philaenus spumarius, 320, 321
Phoeniculus purpureus, 339
Phototaxis, 343
Phylogenetic trees, 330–335

Pines, 170, 171
Pisaster ochraceous, 261
Plant breeding, 397–400
Plant lice, 394
Pleiotropy, 367
Poa annua, 370, 371
Poisson distribution, 170–172
Polymorphism, 22, 38–40
Population
  ecological, 163
  genetic, 27, 96–98
  projection, 212–216
  regulation, 216–223
Population growth, 4–6, 177–198
  exponential, 181, 182
  geometric, 178–181
  human, 194–198, 220–223
  logistic, 182–187, 388–390, 405–409
  zero, 196–198
Potential evapotranspiration, 154, 155
Potentialla glandulosa, 120
Predator-prey interactions, 16–18, 157, 158, 249–271, 379–387, 390, 391, 405–410
  evolution of, 379–387
Probability, 10–12, 170–172
Pseudomyrmex ferruginea, 389
Pupfish, 153, 154

Quadrat method, 164–166
Quantitative traits, 113–135, 340–345, 363–365

Rabbits, 384
Rana, 34, 35, 346, 347
Rare-type male mating advantage, 285–287
Raspberry, 161
Rats, 122, 401
Rattus norvegicus, 401
Recruitment, 177, 406, 407
Recurrence risk, 41–43
Refuge, 257, 258
Reproductive isolating mechanisms, 307–309
Reproductive value, 210–212, 410
Resource, 143
  partitioning, 241
  population, 249–262, 382–386
  utilization, 240–244
Rhagoletis pomonella, 314, 315, 340, 341
Rodents, *See* various species
Rotifers, 364
r-selection, 369–371, 377
Rust, 387

Sawfly, 219, 220, 262, 264
Scaphiopus, 267
Screwworms, 406, 407
Sea otter, 258, 266
Search image, 262, 287, 288
Sea urchins, 157, 158, 266
Seiurus auroeapillus, 185
Selection, 45–61
  balancing, 277–291, 317, 318
  basic model, 43–48
  correlated response to, 366–368
  density-dependent, 289–291
  directional, 294, 295, 298, 398
  disruptive, 294, 295, 299, 302, 303
  estimation, 56, 57, 355, 356
  fecundity, 358–360, 362
  frequency-dependent, 283–288
  fundamental theorem of natural, 365, 366
  gametic, 357, 361
  gene flow and, 83–86, 281–283
  group, 350, 351
  heterozygous advantage, 55–58, 318
  heterozygous disadvantage, 58–61
  kin, 345–350
  lethals, 48–52
  mutation and, 67–69
  progressive, 54, 55
  purifying, 54, 55
  recessives, 52–54
  sexual, 304, 344, 357, 360
  stabilizing, 294, 295, 299, 302, 304
  zygotic, 354–359
Self-fertilization, 102, 103, 282, 356, 358, 359
Sex ratio, 222
Sex determination, 297
Sheep, 213
Sheep blowfly, 190
Shrew, 262, 264
Skate, 269, 410
Snails, 16, 17, 20, 24, 266, 381
Snowshoe hare, 89, 90, 259
Social behavior, 337–351
Sociobiology, 345
Soil, 155–157
Sorex, 264
Spalax ehrenbergi, 316
Sparrows, 210, 301
Specialists, 145, 150, 151
Speciation, 305–316
  allopatric, 309–313
  parapatric, 309, 313, 316
  sympatric, 309, 312–315
Species, 305
  endemic, 269
  sibling, 230, 231, 305, 376, 377
Sphyroena, 323
Spittlebug, 320, 321
Squirrels, 205, 211, 382
Stable age distribution, 215
Starfish, 261, 266
Statistics, 7–10
  χ^2 test, 36–39
  degrees of freedom, 36–38
Sterile-insect-release method, 404, 406
Storks, 162
Survival schedules, 199–216, 362
Swifts, 304

Tadpoles, 267
Temperature, 153, 154
Territoriality, 337–340
Tisbe reticulata, 290
Tits, 218
Threshold, 115, 116
Tomato, 397
Transplant experiment, 150–152, 299
Transposable elements, 324, 326
Trees, 154, 155, 165, 170, 171, 358, 389, 391
Tribolium, 230, 231, 376, 377
Trifolium, 382
Trophic level, 263
Tuna, 408, 409
Turtles, 297

Vegetation patterns, 150, 151
Viceroys, 302
Vitis unifera, 394
Voles, 2, 22, 23, 242, 259

Wasp, 245, 385
Weevils, 403
Whales, 410
Wheat, 387
Wild oats, 104, 240, 282
Woodhoopoe, 339

Xanthocephalus xanthocephalus, 226

Yeast, 188

Zero isoclines
  competition, 232–236, 378
  predator, 253, 257, 382, 383
  resource, 252–257, 382, 383
Zoarces viviparus, 362
Zoo animals, 111, 270